Last Water on the Devil's Highway

THE SOUTHWEST CENTER SERIES

Joseph C. Wilder, Editor

Last Water on the Devil's Highway

A Cultural and Natural History of Tinajas Altas

Bill Broyles
Gayle Harrison Hartmann
Thomas E. Sheridan
Gary Paul Nabhan
Mary Charlotte Thurtle

With Appendices by Richard S. Felger, David J. Griffin, Robert S. Henry, Jacquie Gage Kahn, Gary Paul Nabhan, Philip C. Rosen, and Thomas R. Van Devender

The University of Arizona Press
Tucson

The University of Arizona Press

First issued as a paperback edition 2013

www.uapress.arizona.edu

Library of Congress Cataloging-in-Publication Data
Last water on the Devil's Highway : a cultural and natural history of Tinajas Altas / Bill Broyles ... [et al.].
p. cm. – (The Southwest Center series)
Includes bibliographical references and index.
ISBN 978-0-8165-2964-3 (hard cover : alk. paper)
ISBN 978-0-8165-3083-0 (paperback : alk. paper)
1. Tinajas Altas Site (Ariz.) 2. El Camino del Diablo (Ariz. and Mexico)–History.
3. Water supply–Arizona–Tinajas Altas Site–History. 4. Natural history–Arizona–Tinajas Altas Site. I. Broyles, Bill.
F817.Y9L38 2012
979.1'71–dc23 2011034941

Manufactured in the United States of America on acid-free, archival-quality paper containing
a minimum of 30% post-consumer waste and processed chlorine free.

17 16 15 14 13 6 5 4 3 2

Tinajas Altas and El Camino del Diablo.
(Photo by Bill Broyles)

Salutation

Our ancestral sites are never abandoned but are instead imbued with life and with spiritual forces of enormous innermost significance to living people. Hia C'ed O'odham traditional places, called archaeological sites in some other cultures, include water sources, landmarks, camps, shrines, trails, burials, hunting grounds, gathering places, landscapes, and even particular trees and boulders where special events—births, deaths, meetings—once occurred. For example, we call saguaro cactus *ha:san* and think of them as family ancestors. All of these places and things hold great meaning in the history and survival of the living Hia C'ed O'odham and other native nations. Tinajas Altas has this great meaning for us, and perhaps for you. We request that the sacred ground of our ancestors and our living family be respected as would be places of worship in other cultures. Please assist us in preserving our heritage and land so our children can know and enjoy this sacred and beautiful land given us by our Creator.

—Lorraine Marquez Eiler, Hia C'ed O'odham Alliance

Contents

List of Maps xi

List of Tables xii

Preface xiii

Introduction: Crossroads 3
Thomas E. Sheridan

1 Desert Water 9
Bill Broyles, Gayle Harrison Hartmann, and Thomas E. Sheridan

2 Native Peoples of the Tinajas Altas Region: Prehistoric to Present 43
Gayle Harrison Hartmann, Mary Charlotte Thurtle, and Gary Paul Nabhan

3 First Europeans to Forty-Niners: 1540–1854 71
Thomas E. Sheridan and Bill Broyles

4 Surveyors to Campers: 1854 to the Present 115
Bill Broyles and Gayle Harrison Hartmann

5 Natural History 179
Bill Broyles

Conclusion: Reflections on an Exceptional Place 191
When You Go

Acknowledgments 195

Appendices

Appendix 1. Some Native American Names and Uses of Plants, Mammals, and Reptiles of the Tinajas Altas Region 197
Gary Paul Nabhan and Jacquie Gage Kahn

Appendix 2. Mammals at Tinajas Altas 205
Robert S. Henry

Appendix 3. Birds at Tinajas Altas 207
David J. Griffin

Appendix 4. Amphibians and Reptiles at Tinajas Altas 215
Philip C. Rosen

Appendix 5. Vascular Plants of the Tinajas Altas Region 217
Richard S. Felger and Thomas R. Van Devender

References 239

About the Authors 259

About the Appendix Contributors 261

Index 263

Maps

Tinajas Altas and El Camino del Diablo 4

Natural waterholes of the region 29

Portion of Kino's 1701 map, redrawn to feature selected landmarks 30

Kino's routes, 1699 and 1700 38

Distances between historical water sources 39

Common routes from Sonoyta, Sonora, to Yuma, Arizona 41

Major Cultures of the Ceramic Period in the Southwest, AD 800–1450 47

Key to eighteenth-century tribes of the Southwest 50

Padre Eusebio Kino's 1701 map 70

Possible routes taken by Melchior Díaz 72

Eusebio Kino's routes, 1701 and 1702 76

Jacob Sedelmayr's possible routes, 1750 79

Juan Bautista de Anza's route, 1774 84

Juan Bautista de Anza's route, 1776 88

Pedro Fages's routes, 1781 and 1782 94

Romero and Caballero route, 1823 98

The region, showing Caborca, Altar, Cieneguilla, Altar Valley, and the Sonoran Trail to Quitovac and Sonoyta 102

From Stieler's map of Arizona, 1875, including the routes used by Kippen 20 years earlier 118

Sketch map of Tinajas Altas 135

Government Land Office map, 1879 137

Detail from the A. B. Gray map, 1857 138

Detail from Lumholtz "Map of Southwestern Arizona and Northwestern Sonora comprising the region formerly called Papaguería" 139

Nominally protected Desert Lands around Tinajas Altas 172

The proposed Tinajas Altas addition to Cabeza Prieta National Wildlife Refuge 174

Tables

1.1 Depths and Estimated Volumes of Tinajas Altas Basins 24

1.2 Names Applied to Watering Places Used by Early Spanish Explorers in the Tinajas Altas Region 33

1.3 Walking Mileages Between Key Waterholes 40

2.1 Tribal Names 54

3.1 Population of Altar District, 1834 101

Preface

Here [at Tinajas Altas] water usually could be depended upon, being stored after showers in natural cavities of the rock, but between Sonoita River and Yuma there is no running water for one hundred and thirty-five miles.

—Carl Lumholtz, *New Trails in Mexico*, 1912, p. 197

This book tells the story of a waterhole, actually a series of waterholes, in the hottest and most arid part of North America. The waterholes, Tinajas Altas in Spanish and High Tanks in English, are located in the southwest corner of Arizona, a few miles north of the international border with Mexico. For thousands of years they have been the most reliable source of water in a virtually waterless region—and thus have had enormous significance both for animals and humans. These waterholes are not associated with an active stream, nor are they a result of springs; instead, they are large natural catchments hollowed out over millennia by occasional desert storms. The tanks collect rainwater from the mountains above them and hold this water for an extended period.

The research that led to this book began in the late 1990s as part of a historical and archaeological project sponsored by the US Air Force, then overall manager of the Barry M. Goldwater Range, in which the Tinajas Altas now lie. We were approached by archaeologists Adrianne Rankin and Bruce Masse, with the Goldwater Range Cultural Resources office, and asked if we would like to be part of a project that would research the history and significance of these life-giving waterholes. We each agreed and divided up the story of Tinajas Altas among us.

The research effort was divided into two basic parts. First, historical documentation was to be pursued primarily through parallel avenues: study of written documents and interviews with individuals who had personal knowledge of Tinajas Altas. Second, an on-the-ground archaeological survey was conducted from January through early April 1998. As the Acknowledgments note, the written result of this work was a two-volume report entitled *The Only Water for 100 Miles*, with the first volume devoted to the documented history of Tinajas Altas and the second volume devoted to what was learned from the archaeological survey.

We then wanted to turn this scientific report into a more accessible book that would explain why Tinajas Altas is a magical place worthy of respect and preservation. So, we worked together to blend the chapters into cohesive topics, and we added a new introductory chapter, "Desert Water," a new chapter on natural history, and a reflective conclusion, plus we invited other experts to provide appendices. Our varied backgrounds allowed us to focus on viewing Tinajas Altas from several experiential perspectives. For example, Gayle Hartmann and Mary Charlotte Thurtle spent three months living in tents near Tinajas Altas while conducting the archaeological survey; historian Tom Sheridan carefully searched the documents of the early Spanish priests and explorers and spent a night at Tinajas Altas sharing his campsite with a kit fox; Gary Paul Nabhan, who has considerable reputation as an ethnoecologist, consulted Native Americans who had personal recollections of the region; and "desert rat" Bill Broyles visited Tinajas Altas on scores of occasions, sometimes arriving by vehicle and sometimes by foot, and researched the recent history of the area as well as conducting many interviews with others who have visited the site or worked in the area.

We hope these accounts will provide you, the reader, with a variety of informative and interesting perspectives on this place that, to us, is not only a place of water, but also a place of perpetual wonder and stark majesty.

Last Water on the Devil's Highway

Thirsty travelers filling canteens and water barrels at a desert waterhole along El Camino del Diablo. Tule Tank, 1910. (Lumholtz 1912:240)

INTRODUCTION

CROSSROADS

Thomas E. Sheridan

I share my campfire with a kit fox, the white bones of Tinajas Altas Canyon wrapped around us in the moonlight. The desert is silent and seems to stretch on forever, and it is easy to pretend that we are the only two souls on earth. But the kit fox is an accomplished beggar. It has haunted many campfires before mine.

I hunger for places like Tinajas Altas. Like many of my generation, I want to believe in wilderness, places empty of human presence except mine. But people have been drinking from the staircase of pools carved into the rocks above me for thousands of years. People whose names and languages have been forgotten. The echoes of their presence are frozen in stone projectile points and bedrock metates. In a network of trails that snake through the spare creosote of desert valleys—trails that later came to be called El Camino del Diablo. How many thousand generations of kit foxes have hovered in the shadows of campfires here, begging for scraps?

Their oversize ears heard them all coming—Quechan, Pee-Posh, O'odham, Spaniard, and the Babel of other nationalities trudging to the California gold fields. Tinajas Altas was a node in that network, perhaps the most important one between the Río Sonoyta and the Colorado River. The trails leading to and fro speak of the overpowering need for water in a land that makes you understand the word "desert" at a cellular as well as metaphysical level.

Those trails also whisper of pilgrimage, hunters' visions, blood vengeance. Of greed for silver and gold. And fear. Until the last century, the trails you followed depended upon whether you were friend or foe of powerful peoples living along the Colorado River. On one side were the Hàmakhá'v (Mojaves) and Quechan (Yumas), who dominated the river from Mojave Valley near modern Bullhead City to the junction of the Gila River with the Colorado near modern Yuma. On the other were the Halchidhoma, Kavelchadom, Hahuan, and Halyikwamai—people who moved off the Colorado and became the Pee-Posh (Maricopas). Like their enemies, they spoke languages from the River branch of the Yuman language family. But they settled along the Gila and allied themselves with the Akimel O'odham (Pimas), who speak a Uto-Aztecan language. Those alliances were being forged before Spaniards penetrated the region. Warfare between them did not cease until 1857, when O'odham and Maricopas killed nearly 100 members of a Quechan and Mojave raiding party (Kroeber and Fontana 1986).

Tinajas Altas may not have been a battle site, but it was close enough to the Quechan homeland to make it dangerous for Pee-Posh (Maricopas) and many O'odham. One of the contentions of this book is that European sojourners like Eusebio Francisco Kino, the great Jesuit missionary to the Pimería Alta, did *not* visit Tinajas Altas because his O'odham guides may have feared encountering their Quechan enemies there. Instead, Kino and his companions followed a route to the east, drinking from Heart and Cabeza Prieta Tanks.

The Spaniards made one attempt to bring the River Yumans under their yoke, but the Quechan brought that to a bloody end one hot July morning in 1781. Seven years earlier, Juan Bautista de Anza had led a small party of soldiers along the Camino del Diablo to establish a land route between Sonora and Alta California. Key to that route was Yuma Crossing on the Colorado River, which lay in the heart of Quechan territory. In 1775, Anza returned with a much larger colonizing expedition, bypassing the Camino and Tinajas Altas in favor of the more well-watered trail along the Santa Cruz and Gila rivers. Those colonists settled at Monterey and San Francisco, where Anza founded a *presidio* (military garrison) to anchor Franciscan missions as they leapfrogged northward.

To secure the route, the Spanish government established two missions and a village of Hispanic settlers near the junction of the Gila and Colorado rivers in 1780. The Quechan were friendly at first but soon grew tired of Spanish livestock eating their crops and Franciscan missionaries telling them how to live. On July 17, 1781, they fell upon the newcomers during mass and killed more than 100 of them with their arrows and clubs. The land route was severed. For the next seven decades, Yuma Crossing remained under Quechan control.

Nonetheless, a global hunger for gold gnawed away at the region's isolation. In 1771, Spanish soldiers chasing Seri and O'odham guerrillas found traces in the dry washes of Cieneguilla southwest of the presidio of Altar, Sonora. Within a year, 5,000 *gambucinos* (prospectors) and the people who preyed on them had descended on the strike. One strike led to another as prospectors fanned out in all directions to dry wash for gold in other arroyos. Geographer Robert West (1993:79) claims that Sonora triggered the first "authentic gold rushes in North America." Some of those gambucinos may even have slaked their thirst at Tinajas Altas. Gradu-

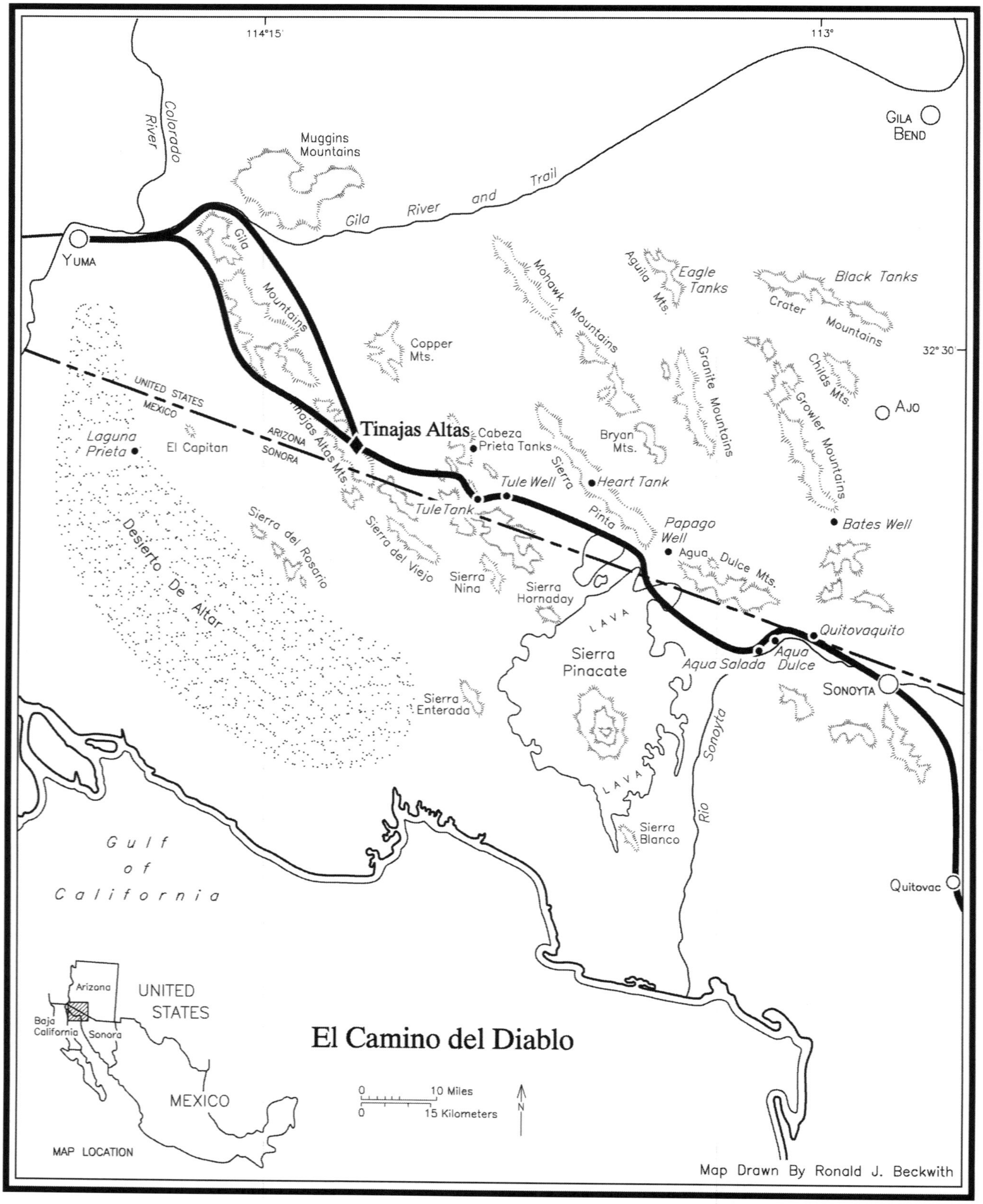

Tinajas Altas and El Camino del Diablo. (Map by Ronald J. Beckwith based on Sykes 1927:63 and Ives 1989:157).

ally, those Hispanic pioneers acquired their own knowledge of trails and waterholes between the Altar Valley and the Colorado River.

Meanwhile, the world's economy was changing, shrinking distances. The mill towns of Great Britain gave birth to industrial capitalism. Factories powered by coal sprouted like black mushrooms in Western Europe and the Eastern United States. But investment depended upon capital, capital that could flow from financial centers to frontiers. Banks and governments had no problem printing money as a medium of circulation. The real problem was guaranteeing its value. Britain decided to back its currency in gold in 1816, but nations like the United States hesitated to follow suit because the world did not produce enough of the precious metal to grease its growth in the first half of the nineteenth century. Adopting the gold standard would have paralyzed the US economy.

Then James Marshall saw flecks of gold while supervising the construction of John Sutter's mill in northern California. By 1849, the California Gold Rush was sucking "argonauts" from Canada to Chile into its maw. At least 20,000 followed the Southern Route, a network of trails converging on the Colorado River at Yuma Crossing (Etter 1998). One of those trails was the Camino del Diablo. The amount of gold mined in the United States skyrocketed from less than $900,000 in 1847 to $50 million in 1850. When another gold rush in Australia erupted in 1851, industrial capitalism finally had a standard of value to underwrite its voracious appetite.

Traffic along the Camino peaked between the Gold Rush and the Civil War. Hundreds died along the way, some of them buried at Tinajas Altas. This is when the trail acquired its hellish name. By then, Yuman hegemony had been overwhelmed by Manifest Destiny. Yuma Crossing belonged to the United States.

Once the Southern Pacific Railroad was built, however, global forces receded for a while. The Camino continued to be used and Tinajas Altas remained its most reliable source of water, but the trail no longer served as a vital corridor. Instead it became a curiosity, the stuff of legend, and the ultimate adventure for desert rats. Some of those adventurers, like W. J. McGee or Godfrey Sykes, also were scientists. Tinajas Altas wormed its way into two very different literatures—Western pulp fiction and scientific journals. Both genres emphasized the bone-dry and often deadly desolation of the landscape.

In an effort to preserve that desolate landscape, scientists and conservationists petitioned the federal government in the mid-1930s to create an enormous national monument or game range of 4,000,000 acres in southwestern Arizona. The protected area would have included Tinajas Altas. That plan stalled, but two areas within it—Organ Pipe Cactus National Monument (1937) and Cabeza Prieta Game Range (1939), now a national wildlife refuge—were protected.

Paradoxically, that very desolation attracted the US military in World War II. The army needed huge empty spaces with clear weather to train pilots and test bombs and artillery. In the early 1940s, the federal government withdrew nearly two million acres of southwestern Arizona from the public domain. Yuma Army Air Base took shape with astonishing speed characteristic of the war effort and graduated its first class of pilots in early 1943. The base was deactivated when World War II ended but reactivated in 1951 at the height of the Cold War. Travelers no longer encountered the desiccated carcasses of those who had died of thirst or heat stroke. But they occasionally found rocket fragments or cartridge casings on the desert floor.

Today, four forces converge on the desert. The military continues to conduct exercises on the Barry M. Goldwater Range. Tourists, campers, and desert rats flock to the area, leading conservationists to argue that the Camino del Diablo and Tinajas Altas should become part of an enormous Sonoran Desert National Park and Preserve—one that includes Organ Pipe Cactus National Monument, Cabeza Prieta National Wildlife Refuge, and the Goldwater Range itself. The military, campers, and conservationists value the desert for what it lacks—a pervasive human presence. Recently, a fourth, more insidious force has come into play: *narcotraficantes* (drug traffickers), undocumented migrants, and *la migra* (Border Patrol) trying to catch them. Tire tracks now scar the desert floor.

I did not hear any helicopters that night, so my illusion of wilderness still flickered. The fox with its far keener hearing probably knew better. Tinajas Altas and the Camino have been threads of larger cultural fabric ever since the first humans entered the region. This volume introduces you to the weavers of that enduring cloth.

The way to water. Former USGS junction sign at Tule Tank. (Photo by Juel and Madeleine Rodack, February 1967. Courtesy Gayle H. Hartmann)

El Camino del Diablo, near Tule Tank, 1981.
(Photo by Bill Broyles)

Bottom pool at Tinajas Altas, in a refreshing rain. (Photo by Juel and Madeleine Rodack, November 1973. Courtesy Gayle H. Hartmann)

1 Desert Water

Bill Broyles, Gayle Harrison Hartmann, and Thomas E. Sheridan

Tinajas Altas, a usually reliable waterhole, consists of a series of small rock-basins located in the hottest and driest part of the Sonoran Desert. The word *tinaja,* widely used for rock pools in the Southwest, is a Spanish word for earthenware jar, and the name *Tinajas Altas* is generally translated as "High Tanks." The name also applies to the mountain range containing this string of pools, a nearby mountain pass, and an adjacent archaeological site used extensively in both prehistoric and historic times. Although the name Tinajas Altas appears on most maps, the name High Tanks is often used by people of the region. In spite of being arid and remote, it is a place not only of intriguing and dramatic history, but also impressive biological character and geological charm.

Some of the tanks, or catchments, generally hold water year-round. For centuries they have been the most reliable source of water in the region known as the Western Papaguería: the swath of northwestern Mexico and southwestern Arizona bounded by the Sonoyta River on the east, the Colorado River on the west, the Gila River on the north, and the Gulf of California on the south. They are also the most reliable water between the Gila River and the freshwater *pozos,* or seeps, along the coast of the Gulf of California. For these reasons, the tinajas have been a major rest stop for diverse travelers during many centuries.

These tinajas, stacked one above another in a steep defile in the east face of the Tinajas Altas Mountains, consist of 15 natural water catchments, varying in size from a wash-bucket to a backyard pool. Five have been most useful to humans. The tinajas are estimated to hold at least 20,000 gallons of water when full (Broyles 1996b; Gaillard 1896; McGee 1906a, 1906b). Though usually reliable, they have gone dry several times in the past century, at least in 1905, 1946, 1957, and 1989 (Broyles 1996b).

The water in these tanks has been critically important to humans for millennia. For Native Americans, Tinajas Altas was a camp where pools quenched their thirst and trees provided shade. Within walking distance the people harvested plants and then processed them at the tinajas by pounding with heavy stone pestles and grinding with hand-sized rocks in depressions in the bedrock. These stone tools were critical components of an outdoor kitchen. Hunters killed big and small game nearby, and ambushed bighorn sheep as they came to the tanks for water. The Native Americans accentuated the importance of Tinajas Altas by painting and pecking designs onto the area's granite walls. The Hia C'ed O'odham, a western branch of the O'odham, camped on the terrace east of the tanks.

Tinajas Altas became an indispensable stop on a historical route that came to be called El Camino del Diablo, the Devil's Highway. This route was important for Spanish- and Mexican-era explorers as well as for later traders and wayfarers. It became infamous during the 1850s when travelers on their way to and from the gold fields of California stopped to replenish their canteens and memorialize their dead. In the twentieth century the tinajas served as a watering stop for explorers and scientists, a location for a Bureau of Animal Industry station set up to combat hoof-and-mouth disease, and a memorable destination for desert aficionados.

For 49ers and other early travelers crossing the forbidding Sonoran Desert from Sonoyta, Sonora, to the confluence of the Gila and Colorado rivers at Yuma, Arizona, the stop at Tinajas Altas marked the most reliable and best-known water on the trek. Many died seeking it and yet, ironically, other, smaller waterholes lie scattered in the ranges nearby. Some of these may have been unknown to travelers, some were very difficult to reach, and virtually all were less reliable than Tinajas Altas, making it *the* crucial stop for virtually everyone who traveled this route.

This book took shape from a project documenting the human presence in and around Tinajas Altas, from the earliest archaeological evidence to the present. We have pulled together an exceptionally wide net of what is known about this place. Many articles, newspaper stories, personal stories, and not a few tall tales record the Tinajas Altas story, and many of the books and journals that include pieces of this history are difficult to find. Our book is for the general readers and selective scholars alike. An archaeological survey in 1998 documented 30 distinct sites, including Native American rock circles, trails, bedrock mortars, and rock art, as well as more recent evidence of mining camps, historical structures, and shrines. Tinajas Altas is a signature waterhole on a heritage trail that covers millennia of Arizona and Southwest history. We have strived to put a face on this legendary place, for Tinajas Altas has witnessed a fascinating procession of people on the move. In these pages we hope that our readers can feel like pioneers, following in the footsteps of indigenous Native Americans, hardy Spanish explorers

The rugged upper ridges and peaks of Tinajas Altas Mountains. (Photo by Charles Sheldon, November 24, 1913. Courtesy Neil B. Carmony.)

and priests, adventurous 49ers, and inquisitive scientists walking into history.

THE PLACE

Tinajas Altas is located in extreme southwestern Arizona four miles north of the border with Mexico. It lies within the Barry M. Goldwater Range, 1,733,921 acres of arid desert valleys and steep-sided mountains managed by the US military to train its pilots. The Air Force trains on the eastern portion of the range, and the Marine Corps trains on the western part where Tinajas Altas is located. Goldwater-West comprises 691,758 acres; Goldwater-East, 1,042,163 acres.

Tinajas Altas owes its existence and much of its historical interest to its geology (Kresan 2007). The Tinajas Altas Mountains, a rugged granitic range, lie within the Basin-and-Range province, as defined by geologist Nevin Fenneman (1931). This region is characterized by linear mountain ranges, most of which are oriented northwest-southeast and separated by broad, flat valleys or basins. Some of the ranges are composed of light-colored granite and gneiss, appearing almost white from a distance. Others are composed of basalt and are reddish-brown to black in color, such as the conspicuous dark head of Cabeza Prieta Peak visible from Tinajas Altas on the eastern horizon. The mountain slopes are very steep and cliffy, bolting abruptly from the valley floor. Approaching them, one has the feeling of walking on a flat floor and suddenly having to climb a stepladder. The summit of the Tinajas Altas range is Tinajas Altas Peak at 2,764 feet, 2.7 miles (4.3 km) south of the tinajas. The elevation at the lowest tinaja is 1,180 feet.

Noted geologist Kirk Bryan (1925), who had studied water throughout the West, explained how these tinajas were born as plunge pools below intermittent waterfalls:

> Most of the mountains of southwestern Arizona are narrow uplifted strips of the earth's crust bounded by faults on one or both sides. After each uplift streams cut canyons and established

> smooth grades from the crest of the mountains to the adjacent valleys. . . . However, in . . . the Tinajas Altas Mountains . . . the new grade has not yet reached the crest of the range and is separated from the old by falls. . . . The falls occur in an indentation of the mountain front and . . . a considerable drainage area lies above the falls. This upland valley . . . has relatively gentle slopes, and the grade of the stream through it is in marked contrast to the grade of the falls section, which is approximately 1 to 1. The rock of this part of the mountain is a coarse pegmatitic granite. Some of the surfaces are stained brown from the deposition of limonite, especially in the cove of Tinajas Altas. There are many caves, reentrants, and pinnacles, which add much to the picturesqueness of the mountain slopes. (Bryan 1925:131, 133–134)

Few visitors have been lucky enough to be here when the skies open and those falls gush freely. One might need to wait several years or more to witness such a rare sight.

A Difficult Climate

The extreme climate plays an important role in determining which plants and animals can live in the area. Vegetation generally is sparse, but is dense along sandy washes. Two staple food plants of the wider region, saguaro cactus and mesquite, are present but not common. During wet springs and following summer and fall rains, annual plants are abundant (Felger, Rutman, et al. 2007). Large native mammals are limited to coyotes, desert bighorns and, rarely, javelinas and mountain lions. Small mammals include the desert cottontail, black-tailed jackrabbit, and numerous species of rodents (Hoffmeister 1986; Turner and Brown 1982:200). Geographically the region lies within the Lower Colorado River Valley subdivision of the Sonoran Desert; the valleys feature creosotebush and white bursage (Shreve 1951; Turner and Brown 1982).

The present climate can best be described as hot and dry, with marked seasonal variations. At the weather station at Yuma, Arizona, approximately 40 miles to the northwest, the average July high temperature is 106° F and the average December-January low temperature is 44°F. Temperatures exceeding 110° F are common and are measured in the shade, so objects as well as people in the direct sun can feel considerably warmer. Ground temperatures may exceed 180° F. Cooler temperatures occur from November to February, with average lows in the 40s and highs near 70° F. Freezes are rare and frost can damage many of the plant species here.

Rainfall, which is erratic and spotty, averages three to four inches (7.6 to 10 cm) per year, and the evaporation rate is 120 inches—10 feet (3.05 m)—per year (Sellers and Hill 1974). A single cloudburst may drop a deluge in excess of six inches; in other years, only a trace may fall. January is the only month in the last 30 years that has averaged close to three days of precipitation. Rainfall in this portion of the Southwest is bimodal, occurring during mid-winter and late summer. The summer rains are the result of localized thunderstorms associated with warm, moist air moving northwest from the Gulf of Mexico (Adams and Comrie 1997). The amount and timing of rains from these storms vary wildly and unpredictably, a fact of life that strongly affects the kinds and sizes of plants that can grow here (Comrie and Broyles 2002; Ezcurra and Rodrigues 1986; Felger 2000). Winter rains, on the other hand, come from storms moving eastward over California from the Pacific Ocean (Sellers and Hill 1974). Occasionally tropical storms and spent hurricanes roll up from the Gulf of California. However, some native plants, such as saguaros, rely on warm-season rainfall, not winter rains, for growth (Turner et al. 1995).

The unpredictability of the rainfall patterns played an important role in the way Native Americans used local plant resources. According to one Native American song, "I'itoi, the elusive Elder Brother of the Tohono O'odham, told the people that the world would burn without rain" (Sheridan 1996:116). Botanist Richard Felger (Felger, Broyles et al. 2007:5) noted that "the unreliability of rainfall is probably of greater significance to plant life than yearly averages." This unreliability requires that plants have the ability to capture rain, whenever it falls, and to produce leaves, stems, limbs, flowers, and seeds quickly. During the occasional wet winters and springs, such as the El Niño event that occurred in 1998 during the Tinajas Altas archaeological survey, annuals useful to humans and wildlife can be abundant (Bowers 2004), as they can be following summer-fall rains.

A Different Past

Using packrat midden data, paleoecologist Thomas R. Van Devender (1990, 2007) learned that 12,000 years ago, during the late Pleistocene, the Tinajas Altas Mountains, supported conifer woodland species of single-leaf piñon and California juniper. Single-leaf piñon disappeared from Tinajas Altas about 12,000 years ago, whereas juniper endured a few thousand years longer. Six to eight thousand years ago, catclaw acacia and blue paloverde were growing on slopes; today they are restricted to washes. Foothill paloverde does

A typical hit-and-miss summer rainstorm over the Lechuguilla Valley east of Tinajas Altas. (Photo by Bill Broyles)

not appear in the assemblages until a few thousand years ago. Van Devender (1990:148) notes that "in the middle Holocene nearly twice as many species were growing near the rock shelters [that yielded the Tinajas Altas samples] as occur there today." The modern Sonoran Desert climatic regime was in place 3,000 to 4,000 years ago. In the Tinajas Altas Mountains and elsewhere, "impoverished modern floras . . . suggest that the present climate is as hot and dry today as at any time in the Holocene" (Van Devender 1990:159; see Van Devender 2007). Botanist Richard S. Felger and Van Devender have reconstructed a unique, innovative flora—plant list—for Tinajas Altas that covers the past 40,000 years (Felger and Van Devender in press).

Geology and Landscape

Geology is a study of change and process. The Earth as we see it has been this way for only a short slice of time, and daily we read about earthquakes, floods, and volcanic eruptions that change the face of our planet. So it is with Tinajas Altas. What we see here today shows a stage of mountain building and erosion characteristic of Basin-and-Range geography. The North American plate rode westward over the smaller Farallon plate, pinching it against the Pacific plate. This produced a wave of volcanic mountain building from the Pacific coast eastward to central New Mexico. It was a slowly moving episode, called the Laramide Orogeny, occurring from 70 to 55 million years ago. The light-colored Gunnery Range granite batholith that we see exposed as the Tinajas Altas Mountains rose from the mantle about 53 million years ago. It is distinct from the darker and older pre-Cambrian metamorphic rocks found in the Sierra Viejo and Sierra Tuseral/Tule southeast of the Tinajas Altas Mountains, and in the Gila Mountains and Mohawk Mountains north and northeast of Tinajas Altas (Shafiqullah et al. 1980).

Then the movement of the plates slowed and the land was relatively quiet for about 20 million years. But, when the North American plate again sped up and slid over part of the Farallon and Pacific plates, a second fury of volcanism erupted, rolling from east to west this time. Explosive eruptions from several volcanoes blanketed this region with ash and sloshed molten rhyolite and basalt across the land. This was the Mid-Tertiary Orogeny, which rattled the Southwest from 36 until 14 million years ago and gave birth to the Basin-and-Range topography we see today (Kresan 2007). The basalt flows seen on Tordillo Mountain and Raven Butte, as well as the dark andesite and latite flows of Cabeza Prieta Peak, date from about 16 million years ago. As the continental plates moved, they stretched the land crust. Block faulting from 15 to 8 million years ago lifted the mountains and deepened the valleys (Shafiqullah et al. 1980).

About 12 million years ago, the Baja California Peninsula began inching northwestward away from mainland Mexico, and by 6.5 million years ago an inland sea formed in this subsidence rift. At about the same time, the Tinajas Altas region settled downward, and the sea filled a large basin that reached the area of present day Phoenix, Arizona, Las Vegas, Nevada, and California's Salton Sea trough (Ledesma-Vásquez and Carreño 2010; Lucchitta et al. 2001; Spencer and Pearthree 2001). This proto-sea was an ancestor of the Gulf of California, and it may have formed at least part of the Bouse Formation between 5.5 and 4.5 million years ago, remnants of which extend from Yuma to the southern tip of Nevada. At that time, the Colorado River had a much different course than today (Young and Spamer 2001).

Regional uplift and sediments, eroded from mountains and carried by the relatively young Colorado and lower Gila rivers, eventually pushed the sea coast southward to its present northern shoreline, but a number of times the Colorado River has burst its banks and re-flooded the Salton Trough. The Baja California Peninsula and southern California continue to move northwestward at about two inches (5 cm) per year.

The lavas of Sierra Pinacate in Sonora date from roughly 1.7 million to 12,000 years ago. Most likely they derive from a deep-mantle magma plume similar to the shield volcano forming the Hawaiian Islands and may or may not be directly related to the faults creating the Gulf of California rift (Gutmann 2007). Some of the northern lava flows cross what became El Camino del Diablo, and they altered how water flowed from the land, creating Las Playas and Pinta Playa in the United States. The Pinacate lava flows have even changed the course of the Sonoyta River. At one time it drained into Adair Bay, west of Sierra Pinacate, but now its usually dry riverbed runs southward on the east side of the mountain and meets the gulf at Estero Morúa near Puerto Peñasco (Gutmann 2002, 2007). The Sierra Pinacate currently is dormant, but its basalts host a number of important tinajas that served native and historical people. On clear days Pinacate Peak can be seen from the ridges at the mouth of Tinajas Altas canyon.

Regionally within the past 1.6 million years, vertical block-faulting has given way to horizontal strike-slip faulting. For example, at Cipriano Pass north of Tinajas Altas, the linear mountain mass has been offset by about 200 meters, making a gap now used by travelers (Tucker 1980). Briggs and Demsey (2000) suggest that Tinajas Altas Pass was created by similar strike-slip faulting.

Briggs and Demsey (2000:27, 29) point out that the modern landscape is marked by three primary elements: first, the mountain slopes intersect the desert floor at a sharp angle, indicating that erosion occurs during infrequent bursts of rainfall runoff. Second, the broad valley floors are characterized by large, ephemeral washes that flow roughly along the central axis of the valley. Tributary channels are dendritic, meaning that the sediments are of similar materials and that they are not underlain by bedrock. And third, the piedmonts, those broad plains from the mountain slopes to the center of the valleys, deepen away from the mountains to depths of 500 meters or more. As you would expect, sediment becomes shallower closer to the mountains, where occasional inselbergs (isolated islands of bedrock) rise above the surface. One inselberg can be found at the mouth of Borrego Canyon, south of Tinajas Altas. Granite bedrock has been the source for virtually all of the sediment in the Tinajas Altas area.

Visitors to the Tinajas Altas area are intrigued by the many side canyons, or embayments. These indicate that the mountains have been weathering since the late-Quaternary period without renewed tectonic activity. This quiet span of time has allowed erosion to create sinuous ridges and embayments (Bull 1973). One ridge in the neighboring Gila Mountains is so crooked and serrated that a geologist named it with the Tohono O'odham word for lightning, Vopoki (Wilson 1931). Even the relatively straighter ridges of the western side of the Tinajas Altas Mountains show little recent uplift or lateral movement (Briggs and Demsey 2000).

Visitors may occasionally feel earth tremors from quakes along the San Andreas fault system that runs through the Colorado River delta, but by looking at the shallow gullies and arroyos around Tinajas Altas, geologists can tell that there have been few large earthquakes here in the past few million years to lift the land or increase the gradient of rainfall runoff. The shallow bajada slopes, coupled with relatively arid climate, have left in place much of the sediment eroded from the mountains, though some of it has been moved by winds, creating areas such as the Mohawk Dunes. Wind-blown dust—loess—is a factor in the slow evolution of desert soils.

Among the striking geologic features at Tinajas Altas are the numerous cavities—little caves—in the granitic bedrock and boulders. Known as wind caves, niches (Bryan 1925) or tafoni (Goudie 2004a), they range in size from a few cubic inches to caves large enough to shelter people or bighorn sheep. The interior of a typical tafone is rough, flaky, and lacking desert varnish, a dark coating typical of exposed desert rocks.

Tafoni, an Italian word meaning "windows," "perforations," or "caverns," have been described in different environments and rock types around the globe. They appear in granite, sandstone, and limestone, and the term tafoni covers a suite of features including wind caves, fretwork, honeycombs, niches, "helmets," and stone lace. For example, at Papago Buttes in Phoenix, Arizona, red sandstone and granite cobbles are held together by calcium carbonate. Tafoni form in the buttes when moisture dissolves the calcium carbonate and then deposits it on the outer rock surface, hardening it; as the calcium carbonate leaches out of the sandstone breccia's lower surfaces and relatively softer core, the rock in those spots weathers first and at a faster rate, causing the rock to disintegrate (see Campbell 1999). In the Atacama Desert of Chile tafoni form due to differential wetting and drying of feldspar crystals that chemically change to softer clay minerals, which are removed by wind erosion (Segerstrom and Henriquez 1964). At some sites in other locales, crystals of various salts form within minute voids in freshly exposed granite; as they grow or swell with moisture, they expand and spall the rock (Bradley et al. 1978).

At Tinajas Altas the bedrock is granite, and it is likely that several mechanisms have worked in concert to create these tafoni. Granite, a hard, igneous rock composed of a variety of minerals, is grainy, and the word "granite" comes from the Latin *granum*, meaning a grain or seed. The basic ingredients of granite are varying proportions of quartz, potassium feldspar, and plagioclase, with scattered grains of mica, hornblende, and other minerals. The particular mix of minerals at Tinajas Altas is called the Gunnery Range Granite, and the mix may be uneven, so that sections may weather at different rates.

Granite is a symbol for being "rock solid," but it is also slightly porous and permeable, meaning that moisture can enter the rock and create mischief by combining with some of the minerals, dissolving others, and expanding during freezing or heating cycles. Chemical and physical weathering are relentless: in testament, the Lechuguilla Valley east of Tinajas Altas is filled thousands of feet deep with sediments eroded from the surrounding mountains.

Even in the relatively benign climate inside a home, granite countertops are vulnerable to chemical weathering from spilling acids like orange juice and applying caustic cleaners. Physical weathering also occurs when something is dropped (chipping) or a hot pan is put on the countertop (fracturing). Even water can alter and weaken the chemistry of the rock, as well as encouraging mold, bacteria, and fungus to grow on and within the rock. Homeowners fight these processes by sealing pores in the granite with acrylics, oils, or silicone, and by guarding against cracking from excessive heating, cooling, or stress.

In nature, alteration of rock may begin underground long before it is exposed. The granite at Tinajas Altas is heavily jointed in three planes. Water percolates along those large

Distinctive tafoni, overhangs and caves. For scale, note two people at center left. (Photo by Gayle H. Hartmann)

A "helmet" tafoni cave. (Photo by Bill Broyles)

joints as well as tiny cracks, attacking the rock by hydration, hydrolysis, oxidation-reduction, ion exchange, and solution. This chemical assault is called deep weathering (Thomas 2004) and tends to round the granite into boulders, towers, and curved ledges even before the surfaces reach daylight. Chemical constituents leached into the granite, such as iron, silicate, manganese, or sulfate, may harden some surfaces of the rock in a process called case hardening, or induration; the shell becomes harder than the core, and the softer core or underside erodes more rapidly (Dorn 2004a). In other instances, a wide variety of agents may coat the rock and add some resistance to weathering; these include dust films, rock varnish, and silica glazes (Dorn 2004b).

When weathering disintegrates the granite bordering the cracks, it produces rotten stone (saprolite), which lies in place next to the hard core stone until the bedrock is uplifted by faulting. Then the debris is removed by wind, rain, and runoff water, leaving rounded and irregular shapes, including overhangs and cavities (Linton 1955; Palmer and Neilson 1962). Then the rock is susceptible to even more weathering. As Bryan (1925) noted, the niches at Tinajas Altas work inward from the rock face as successive flakes and chips fall away. Apparently the interiors of tafoni openings favor and accelerate weathering, perhaps by increased humidity, and the hollows grow in size (Goudie 2004a). Exposed granite must endure insolation and exfoliation, as well.

Wind does not produce the tafoni caves at Tinajas Altas—it would need to carry abrasive sands to do that—but wind does blow away loosened sediments, and rain or runoff water cleans away the crumbled rock debris also. As you can imagine, if a surface weathers even one ten-thousandth of an inch per year, or about one-tenth the thickness of a coat of paint, in ten thousand years, an inch (2.5 cm) of rock face will have eroded. Like inscriptions on centuries-old granite headstones, historical signatures on rock walls are by no means permanent.

But, many questions remain: how rapidly *is* the granite weathering at Tinajas Altas? What is the exact composition of this granite? Which weathering processes are busiest here? Why do tafoni form here and not in all granitic desert ranges? Did seawaters of the Proto-Gulf contribute to weathering or tafoni in what was then subsurface granite? And what new shapes of boulders and tafoni are to come? Further research is needed to answer our questions about the tafoni caves at Tinajas Altas.

Weathering pits are another interesting feature at Tinajas Altas. These pits are found on level flat surfaces and are sometimes mistaken for old bedrock metates. For all we know, people may have employed weathering pits to start their metates. These pits occur when water ponds in small horizontal hollows and promotes chemical weathering. Then wind, rain, running water, and even passing rodents clear away the minute rubble. Over thousands of cycles the pits deepen and widen (Goudie 2004b). The freshly loosened grains of granite are called grus or grit, and they make some pitches of the granite slopes treacherous to walk on.

The valley and stream channel above the tinajas being surveyed, 1998. (Photo courtesy of Austin Lenhart)

GEOLOGY AND WATER

Tinajas Altas displays fascinating rock formations that intrigue modern travelers, but over the centuries the main attraction at Tinajas Altas has been water found in the rock-basin pools—the tinajas—that lie in a steep, stair-step ravine. These natural depressions are eroded into bedrock by stream action, dissolution, or other weathering processes (Brown and Johnson 1983). In the hanging valley above the pools lies a dry, sandy wash among the granite peaks, which some call the Upland Valley. Rapid faulting raised the valley faster than running water could cut the rock to a steady grade.

The tinajas are created by the force of water working on rock. Running water carries sands and stones that scour and abrade the bedrock (corrasion), and turbulent water flicks out sections of jointed rock (hydraulic plucking). Waterfalls, even short-lived ones, pound and erode holes at their bases. Some of these waterfall potholes capture stones, rocks, and pebbles, and the eddy currents swirl or tumble the stones like abrasive grinding balls. The moving stones actually "hammer" chips and chunks out of the pool walls, further enlarging and deepening them. Anyone who has

Mapper Scott Lunt measuring a tinaja. (Photo courtesy of Austin Lenhart)

Coyote Water in the bed of Coyote Wash. Note the unusual waves of sand, which in the period from 1999 to 2006 were flattened by foot traffic from across the international border. (Photo by Luke Evans, March 1998)

heard and felt the rumble of rocks tumbling in a stream or river will appreciate the power of moving water. Likely all of these forces are at work here during the rainstorms that infrequently fill the streambed.

Though floods are rare, their erosive effects can be quite significant and obvious. Within a few hours one megastorm may markedly alter a landscape that had changed little over several centuries. The area's rocky watershed cannot absorb much rainfall, so when flash floods do occur, they funnel remarkable volumes of water between the narrow canyon walls. A line of flotsam left by one storm at tinaja number 3 indicated that the water at the lip of the basin may have run 20 inches (51 cm) deep and 30 feet (9 m) wide following a cloudburst. The threshold for rainfall to generate runoff in arroyos is generally more than three-quarters of an inch (1.9 cm; Shreve 1934), less on rock surfaces. At Tinajas Altas, rainfall captured by the upland valley may cascade and fall more than 400 vertical feet (130 m) before reaching the alluvial fan and sandy arroyo below the lowest waterfall.

Coyote Water, three miles east of Tinajas Altas, is another type of natural water storage. Coyote Water is located in Coyote Wash, the main channel of the broad Lechuguilla Valley. Unlike the bedrock tanks at Tinajas Altas, Coyote Wash is a channel full of coarse sand, but the size of the sand waves in its bed indicates that it flows forcefully if briefly and sporadically. The water has scoured a depression in the hardpan that underlies the channel, creating a subsurface basin for water trapped in the pores between the grains of sand (Bryan 1925). Plant roots give the water a bad taste but, like Tinajas Altas, Coyote Water has been an important waterhole historically (Bryan 1925).

STANDING WATER

Kirk Bryan (1925:133) described the tanks at Tinajas Altas in *The Papago Country, Arizona: A Geographic, Geologic, and Hydrologic Reconnaissance with a Guide to Desert Watering Places* and noted nine worth talking about, while our archaeological survey in 1998 listed fifteen. Counting tinajas along the stream channel may seem like a simple task, but the steep, craggy terrain makes it difficult to describe them in detail, and people puzzle over how long specific tanks hold water. Fifteen of the tanks seem to hold water for a

Tapping the sands at Coyote Water. (Photo by Luke Evans, March 1998)

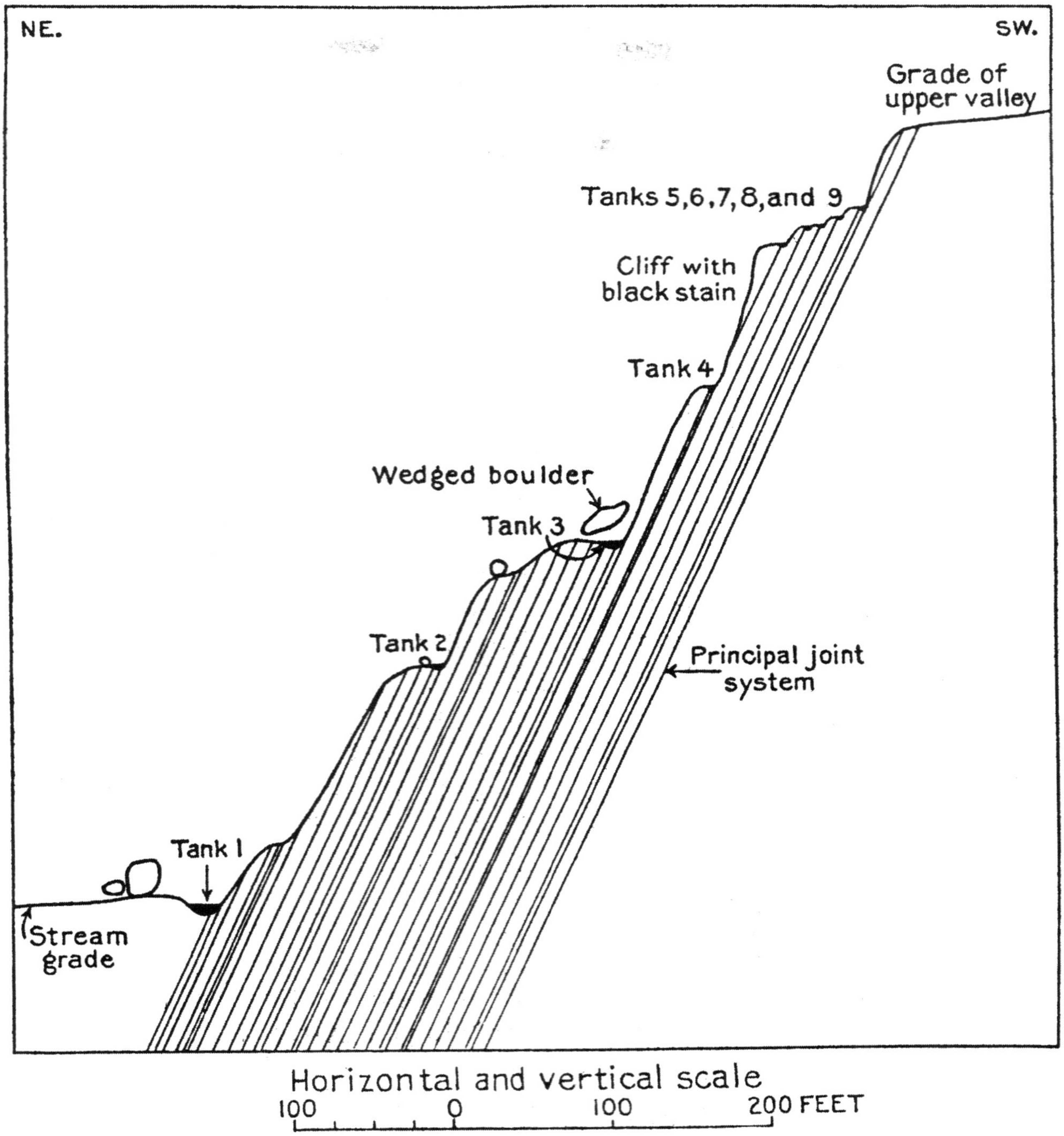

Kirk Bryan's profile of bedrock basins at Tinajas Altas (Bryan 1925:133)

few weeks after a rain, but fewer are useful to humans, and only five are reasonably reliable and accessible. Longevity, accessibility, and potability each play a role in human usage of the tinajas.

Rainwater fills the catchments in four ways: direct rainfall, runoff from the channel, runoff from the surrounding rock aprons, and seepage from above (Brown and Johnson 1983; Broyles 1996b). The catchments at Tinajas Altas sit in a narrow canyon with a relatively small watershed, but the only exit is the stream channel, so all tanks benefit from runoff, with the upper catchments filling first. Sand in the channel of the upper valley and in some of the tinajas holds water and feeds it into the tanks by seepage from fracture joints within the granite, especially to the lowest pool (Bryan 1925:134). Commonly too little rain falls to make the canyon flow, but some of the pools fill anyway because the impervious rock apron surrounding a pool "magnifies" the rain, much as a roof fills a rain barrel. Because the pools are small in surface area

No water—the dry sands of the bottom tank at Tinajas Altas. (Photo courtesy Gayle H. Hartmann, November 1989)

and most rainfall episodes here are slight (less than 0.5"; 1.3 cm), direct rainfall adds little to the pools. A half-inch of rain would add half an inch directly and might not cause water to flow in the canyon. However, that half-inch running off of the expansive slabs of bare rock may raise the water level several feet in a tinaja. It is not uncommon to visit Tinajas Altas and find full tanks but no sign of runoff in the arroyo above or below the tinajas.

But evaporation takes its toll of standing water, and "no tinajas in this region are truly permanent. At times every tinaja in the region has been known to go dry, either due to unreliable rains or from general drought" (Broyles 1996b:491). Total water volume in a tank depends on water depth and tank shape, but the longevity of the pool's water is controlled by two main variables: depth and evaporative rate.

In a desert, evaporation far exceeds precipitation. In summer, evaporation typically draws the water level down six-tenths of an inch per day (ranging from 0.8 to 2.0 cm/day; averaging 1.5 cm/day)(Brown and Johnson 1983; Broyles 1996b; Henry and Sowls 1980). In theory, an unshaded tinaja must hold a depth of at least 106 inches (270 cm) to be perennial, with intermittent ones lasting one to six months and holding between 24 and 106 inches (60 and 270 cm). Anything shallower than 24 inches (60 cm) is classified as ephemeral and is expected to hold water less than one month (Brown and Johnson 1983). The effects of evaporation are accelerated by leakage into the soil or through cracks, as well as the use of water by wildlife. Eight catchments (1, 3, 4, 7, 8, 10, 11, and 13) meet the necessary depth criterion to be classified as intermittent or perennial. In practical terms, however, shallower basins may hold water year-round, because tinajas are usually replenished every six months, the interval between summer and winter rains in the Sonoran Desert.

Short-lived or slow-moving runoffs may fill the tanks with sand, but eventually an intense storm will scour and purge them clean again with rushing water. When full of sand, a tank may still hold water, much like a sponge. About 25 percent of the volume of sand can be subsurface water—you may see places where coyotes have scratched holes in the sand to reach water, and you can readily imagine thirst-crazed pioneers desperately shoveling or pawing to find life-giving liquid. Water underneath sand lasts longer than in an open pool and generally tastes better. Standing surface water is an odd mix of bacteria, viruses, toxins, dead animals, dung, decaying plant matter, living—and excreting—invertebrate animals, salts, and unusual chemical and mineral concentrations. As you can imagine, the potability and taste of the water varies, and under normal conditions sensible people refuse to drink it. For persons pushed by thirst to the brink of death, the quality and taste of the water are ignored.

During Carl Lumholtz's 1910 visit to Tinajas Altas his guide told of a bighorn sheep that, when cornered, fell into the bottom tank and died there, perhaps deterring travelers from using the water (Lumholtz 1912:240). Raphael Pumpelly (1918:774), traveling along the Camino del Diablo with his family in 1915, met potability head-on. He wrote, "The next night we camped at the abandoned Tule well. Its water was both brackish and offensive, but facing thirst on the desert one can't be squeamish. Several months later a friend who had been over that route on a survey asked: 'How did you like the Tule well water?' 'Not much,' I answered. 'Naturally,' he said, 'for we found and left a man in it two years ago.'"

During wetter times in prehistory, the canyon may have had running water, and we can imagine the refreshing sound of trickling water and its refreshing taste, but when the water level is low or stagnant, a pool may have a pronounced odor

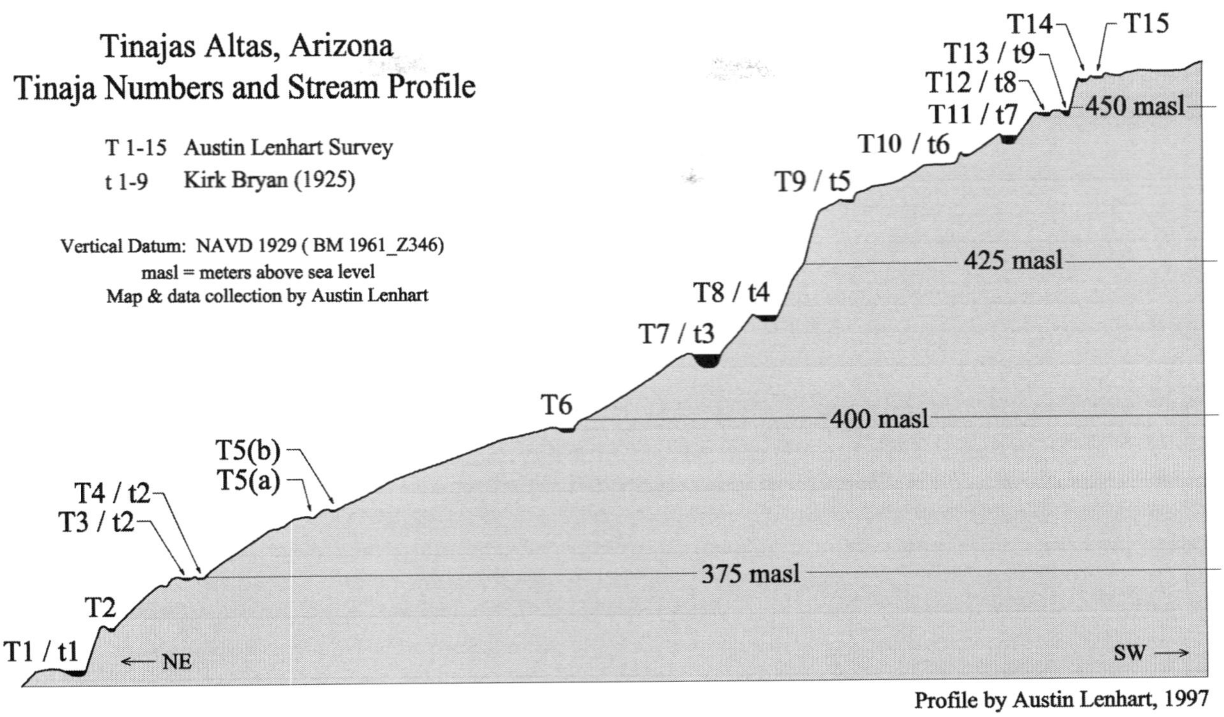

Tinaja numbers and stream profile of Tinajas Altas. The scale, both horizontal and vertical, is represented in meters above sea level (masl). T = tinajas counted by the 1998 archaeological survey; t = tinajas reported by Bryan (1925). (Map by Austin Lenhart)

and peculiar taste. Modern visitors report that water at Tinajas Altas tastes brackish, mineralized, gritty from debris, and rancid, particularly in summer when bees drown and float on the surface. One daredevil noted that you can almost get a few sips of it down if you hold your nose.

All water in the tanks is assumed to be contaminated and unfit for humans to drink except in life and death circumstances. Tinajas receive dead matter including corpses of animals and birds, fecal matter, and plants, all of which are attacked by invertebrates and bacteria, in effect making the pool a maceration tank where organic material is recycled and dissolved. This is a natural, healthy process, but it also means that the water potentially harbors a number of sickening organisms, such as *Giardia*, or harmful chemicals, such as hydrogen sulfide (Kubly 1990:120; US Environmental Protection Agency website April 2010).

Water quality for organisms living in the water depends on a number of measurable factors: temperature, pH, turbidity, dissolved oxygen, hardness, total dissolved solids, and trace minerals and chemicals. If you have ever tried to maintain a pet fish in a bowl or aquarium, you know how difficult it can be to balance all of the factors. Temperature and light affect water organisms in obvious ways as well as limiting dissolved oxygen and algae productivity. The shallower the pool, the wider the swing in daily temperatures; the shallowest pools may approach daytime-high and nighttime-low air temperatures. Turbidity is the amount of suspended matter (silt, plankton, dissolved organic compounds) in the water; ranging from cloudy to clear, it affects the ability of light to reach water organisms.

Dissolved oxygen is derived from mixing (falling water, surface agitation by wind) as well as photosynthesis (algae, blue-green algae). In stagnant waters, oxygen is depleted by respiring organisms, active chemical processes, bacteria as they consume dead organisms, and algae. Dissolved oxygen readings vary widely as organic loads fluctuate. Without sufficient oxygen, most species in a pool will die and then decay or rot.

Hardness measures the amount of calcium and magnesium in the water. Trace minerals (such as uranium or lead) and chemicals (toxins, pollutants) can harm or kill life in a tinaja. A test using an electrical current (specific conductance) can measure total dissolved solids. Because of low organic content and high mineral content, desert waters tend to be alkaline instead of acidic. Pure water has a pH of 7.0, with extreme acids, such as concentrated sulfuric acid, being 1.0 and extreme alkalis, such as sodium peroxide, being 14.0. Desert tinajas tend to fall within a range of 7.0 to 9.4 pH, but vary by season. By comparison, peat bogs run as low as 3.4 pH, stream water normally varies from 6.2 to 8.6, and

Table 1.1
Depths and Estimated Volumes of Tinajas Altas Basins

Tinaja Number[a]	Maximum Depth (cm)[b]	Shape[d]	Estimated volume (gallons)[d]	Exposure to sun (shading)	Duration in most years[e]
1 [B1]	210	basin	5996	exposed	perennial
2	—	cone		protected	ephemeral
3 [B2]	70	cone	1069	exposed	intermittent
4 [B2]	154	cone	4925	exposed	intermittent
5	—	cone		exposed	ephemeral
6	33	—		exposed	ephemeral
7 [B3]	120	cone	1419	very protected	perennial
8 [B4]	200	basin	3948	protected	perennial
9 [B5]	60	cone	Sand-filled in most years	exposed (subsurface)	intermittent
10 [B6]	75	cone	499	exposed	intermittent
11 [B7]	150	cone	2064	protected	perennial
12 [B8]	50	cone	318	protected	intermittent
13 [B9]	85	cone	205	protected	intermittent
14	30	cone	—	exposed	ephemeral
15	16	cone	—	exposed	ephemeral

[a] Tinajas are numbered from the bottom to the top; Bryan's numbers are in brackets: [B1].
[b] Depths are those measured by Austin Lenhart and crew.
[c] Shape was estimated as cone, basin (parabola), or cylinder.
[d] Volumes were calculated based on cone (V = 1/3 πr2h), although tinajas are irregular in shape and may contain volumes closer to parabolas or cylinders (Brown and Johnson 1983; Broyles 1997.
[e] Duration is based on observations and depth. All have large slick-rock catch aprons.

sea water averages 7.5 to 8.5 pH. An accepted standard for drinking water is 6.5 to 8.5 pH (US Environmental Protection Agency website April 2010).

Water quality at Tinajas Altas was measured by geologist Kirk Bryan (1925:135), who noted that when water in the tanks is replenished after rains it "is at first muddy, but rapidly clears, and then by the growth of algae and the decay of organic matter it becomes foul." During his first test, in October 1917, it was "particularly foul," but three years later, in September 1920, he found the water "sweet and pleasant to the taste." At the time of the October 1917 test, it was found to contain the following:

silica	19 ppm
iron	0.06 ppm
calcium	72 ppm
magnesium	9.1 ppm
sodium and potassium	9.2 ppm
carbonate radical	0.0 ppm
bicarbonate radical	244 ppm
sulfate radical	14 ppm
chloride radical	23 ppm
nitrate radical	6.0 ppm
total dissolved solids	312 ppm
total hardness as $CaCO_3$	217 ppm

Other water-chemistry studies for this and nearby desert waters include Kubly (1990), Rosenstock, O'Brien, et al. (2004) Rosenstock, Bleich, et al. (2005), and elsewhere in Bryan (1925).

But just because a catchment holds water doesn't mean that it can be used. At Tinajas Altas, the difficulty of reaching the upper tanks is painfully known. Thirsty travelers have arrived to find the lower tank empty, but they were too exhausted to scale the granite slopes to the upper tanks. For practical purposes, catchments 2, 8, 12, and 13, as counted by Austin Lenhart, are inaccessible without special climbing equipment or skills, or extreme desperation, and it is unlikely that they would hold water if the others were dry. Catchments 9, 10, 11, 14, and 15 are difficult to reach. In order to gain access to these upper tanks, it is necessary to climb a steep side-canyon, crawl through a window in the cliff, and then descend by means of the even steeper stream channel over slippery rock. This is a strenuous climb and would have been

Dart Tank. (Photo by Bill Broyles)

Not surprisingly, most prehistoric and historical activities (shown by bedrock grinding holes, rock art, and historical inscriptions) were associated with the lower catchments, particularly 1, 3, 4, and 6. Nonetheless, despite difficulty of access, there is some evidence of cultural activity in the upland valley, especially along the edges of the ephemeral stream. The upper catchments (10 through 15) have very limited signs of cultural activity, with only one shallow grinding feature at catchment 9 (Lenhart's designation).

risky even for travelers still strong; for those weakened by thirst, weariness, heat, and hunger, the attempt itself could be fatal. The lowest group (catchments 1, 3, 4, 5, 6, and 7) is accessible, but footing can be treacherous, especially when the slopes are wet or damp.

In the second half of the twentieth century the Arizona Game and Fish Department constructed water catchments in the Tinajas Altas region, ostensibly to serve wildlife. These include Borrego Tank, Dart Tank, Ewe Tank, and Lamb Tank in the Tinajas Altas Mountains and Betty Lee Tank, Geology Divide Tank, Coyote Peak Tanks, Cipriano Tanks in the

Borrego Tank. (Photo by Bill Broyles)

A coastal pozo. (Photo by Bill Broyles)

region. The newer catchments are 10,000-gallon fiberglass tanks buried underground and piped to a watering basin. At others, dams were constructed to raise the water level in natural potholes or pipes were installed to carry water from seeps to storage tanks that feed float-valved guzzlers. The US Fish and Wildlife Service has also built some catchments in the nearby Cabeza Prieta National Wildlife Refuge (Broyles 1997; Broyles, Evans et al. 2007). Such catchments have led to debate whether constructed tanks bring benefits for wildlife and the ecosystem. On the one hand they offer an increased water supply, but on the other hand, roads are constructed into pristine canyons and the tanks may or may not provide any actual species-level benefits to wildlife populations (see Arizona Game and Fish Department 1997; Brown 1998; Broyles 1995, 1998; Broyles and Cutler 1999; Cain et al. 2008; Rosenstock, Ballard, et al. 1999; Rosenstock, O'Brien, et al. 2004; Rosenstock, Bleich, et al. 2005 for more of this continuing discussion).

A Network of Waterholes

Because of the arid terrain and the limited food resources, human survival in the middle of the desert depended on thorough knowledge of the entire region. In practical terms this meant knowing routes throughout an area of about 15,000 square miles, the land from the Gila River to the Gulf of California, and from the Colorado River to the Sonoyta River (Altschul and Rankin 2008; Broyles 1996a, 1996b; Fontana 1965, 1974; McGuire and Schiffer 1982). This is an area larger than Connecticut, Massachusetts, and Rhode Island combined. Groups living along the Colorado might have made an occasional trip to Tinajas Altas to hunt desert bighorn sheep or to grind seed pods from the leguminous trees, but then returned to the relative bounty and safety of the river corridor. Groups living in the Gran Desierto west and north of the Pinacate volcanic region would have migrated seasonally, hunting sheep in one place, digging clams along the ocean beaches, and then grinding beans at another site. Their favorite sites were likely dozens of miles apart and quite seasonal. Water, too, would have been a primary consideration, and even if sandfood (Nabhan 1980) or other moist plants were available to assuage thirst, they also relied on scattered natural pozos along the salt flats and bases of coastal bluffs. These shallow, freshwater artesian upwellings vary from tub-sized to several thousand gallons (Ezcurra et al. 1988).

Human settlement in the American West depended on waterholes, springs, wells, streams, and rivers. Today

Murray Chappel waters a horse at the lower pool at Tinajas Altas, 1913. (Photo by E. A. Goldman, courtesy of Neil B. Carmony)

communities thrive with the help of aqueducts from distant rivers. The plots of many Western novels and movies revolve around water for livestock, crops, or personal survival. Old photographs and Western paintings of waterholes are etched in our memories. For example, Frederick Remington's painting "Fight for the Water Hole" portrays a desperate battle with US cavalry soldiers firing at marauders from the rim of a nearly dry waterhole. And, Edward S. Curtis's tranquil photo "At the Old Well—Acoma" depicts women dipping water from pools in a domestic scene that evokes what Tinajas Altas may have been like centuries ago. Books by Zane Grey (*Desert Gold*, 1969), Louis L'Amour (*Last Stand at Papago Wells*, 1957, and *Kid Rodelo*, 1957) and Jeanne Williams (*Lady Bought With Rifles*, 1976) are set in the landscape astride El Camino del Diablo and their plots involve water. Waterholes continue to enchant and entice us. They not only quench our thirst, but they also whet our imaginations.

If not for other waterholes in the region, Tinajas Altas would be just a curiosity instead of an international landmark. Few modern people would have ventured here to visit the place itself. It is not the Garden of Eden. It is not a destination but a waypoint on a long route. But it did lie invitingly along a track that crossed the vast desert wilderness—a *despoblado* in Spanish, an intimidating expanse without town or home or ranch. Before the advent of autos and airplanes, travel required time, endurance, self-sustaining skill, and an abiding motivation to reach a distant destination.

The history of Tinajas Altas was shaped by geography. It was on a path between a rock and a hard place, the passage between Scylla and Charybdis. Travelers wishing to take the most direct route from the interior of Mexico to Alta California funneled to the far northwestern corner of Sonora. They may have passed through Hermosillo, but unless they went through Tucson, their road led to tiny Sonoyta, a village advantageously situated on the banks of a small but perennial desert stream. From there they could head north, and in some seasons find water at Horse Tanks, Black Tanks, and then the Gila River near the modern town of Gila Bend. That route was longer, like taking the two 90-degree legs of a right-angle triangle.

The more direct route lay along the hypotenuse. Though shorter, it added uncertainty, not only about finding one's way but also finding water. A great sea of dry sand—El Gran Desierto—lay to the south, along with crippling Pinacate basalt flows and steep lava ridges too imposing for most walkers and certainly impassable for horses or oxen. To the north were rows of steep, rocky sierras—knife-edge granite mountains much like those at Tinajas Altas. Travelers navigated through wide, sandy valleys that lay between these long, narrow, northwest-trending ranges. Few usable passes cut through the sierras, which rear 1,000 to 3,000 feet (305 to 914 m) above the valley floors.

Historic travelers calculated three important factors: water, distance, and bandits. Humans require water and cannot easily survive more than a day or two without it. In the torrid summers in this region, where the thermometer can reach 120°F in the shade, a thirsty person can die within hours. The beeline distance from Sonoyta the Colorado River was about 135 miles (216 km), but travelers could easily add another 20 or 30 miles with detours and missteps. Afoot and on fresh legs they might travel two to three miles (3 to 5 km) per hour, and in an hour horses could cover five or six miles (8 to 10 km), hour after hour. Even so, the trip could fill three to six arduous days, all while one's water supply dwindled. Water weighs 8.4 pounds per gallon, and during hotter seasons a person requires two to three gallons a day. Livestock, of course, need much more but can glean some moisture from green plants and can go longer without drinking.

Hostile Indians and bandits made some routes dangerous, especially along the Gila River. Apaches raided as far southwest as Sierra Blanca, just north of Rocky Point, Sonora (Lumholtz 1912), and renegade Hia C'ed O'odham killed some travelers along El Camino Diablo, taking all their belongings including their gold pouches. They discarded the nuggets and dust, which were useless to them, but kept the handy pouches, as well as the victims' tobacco (Childs 1954). Palmer Valor, a Western Apache who was born near Fort Apache in Arizona's White Mountains, reported participating in two raids that went as far as the Gulf of California (Goodwin 1971).

By the luck of geography, the straightest line to California wound its way through a series of passes and valleys that took travelers past the spot that could serve them best: Tinajas Altas. Travelers could water themselves and their livestock at Agua Dulce or Agua Salada, the last pools on the Río Sonoyta, before jumping off into the dry, sandy desert that lay between them and the Colorado or Gila rivers. Along the way they might find that the summer monsoon or the winter storms had left puddles and pasture at Las Playas and had filled the rock basin at Tule Tank. These would be bonuses but could never be taken for granted.

If after stopping at Las Playas and Tule Tank travelers felt strong and confident in their provision of water, they might ignore Tinajas Altas and angle northwest for Pozos de en Medio, now called Coyote Water, and then Pozos de San Miguel near the Wellton Hills. They could make a good pace along this route in mid-valley if they could tolerate dry camps.

If they arrived at Las Playas with guides who knew the puzzling route, wayfarers could go northwest up the Tule Desert between the Sierra Pinta and Cabeza Prieta Mountains. Known as the Cabeza Prieta route, this way offered good options for water (Michler 1857-1859:115). Heart Tank—the fabled Aguaje de la Luna—sits on the west flank of the Sierra Pinta. It is usually perennial, but reaching it requires scaling boulders and arroyo banks and then scampering up a steep granite apron. Agile bighorn sheep walk to the tinaja, but it is probably impassible for livestock. From Heart Tank, the logical route is west to Cabeza Prieta Tanks and then north to Baker Tanks. Early auto roads followed this route (Bryan 1925). Baker Tanks sit along an arroyo in a wide opening between Baker Peaks and the Copper Mountains, and almost always contain water.

Travelers plodding in the most direct line to the Colorado River at Yuma Crossing aimed for Tinajas Altas, where they might find water in the lowest tank. They watered livestock there or dipped water from the upper tanks and carried it down in *botas* or buckets. If few other people had traveled this way for several seasons, they might even find browse for their animals and firewood for their own comfort. After leaving Tinajas Altas in cooler seasons, tired travelers might slide westward through Tinajas Altas Pass or Cipriano Pass and head directly for the favorable bend in the mighty Colorado where Yuma now stands. Following this route, they might be lucky enough to find pasture at Albino, a green-up area where several washes joined. Or, they might detour a few miles to what is now known as Spook Tank. From there they faced a do-or-die slog to the river.

If they chose to go north from Tinajas Altas directly to the Gila River, travelers would look for pasture at Pozos de San Miguel, where they might be able to dig for water in the sandy wash. They might be aware of Dripping Springs, farther along their route, but it is so secluded in the mountains that it is not only difficult to find but it requires more

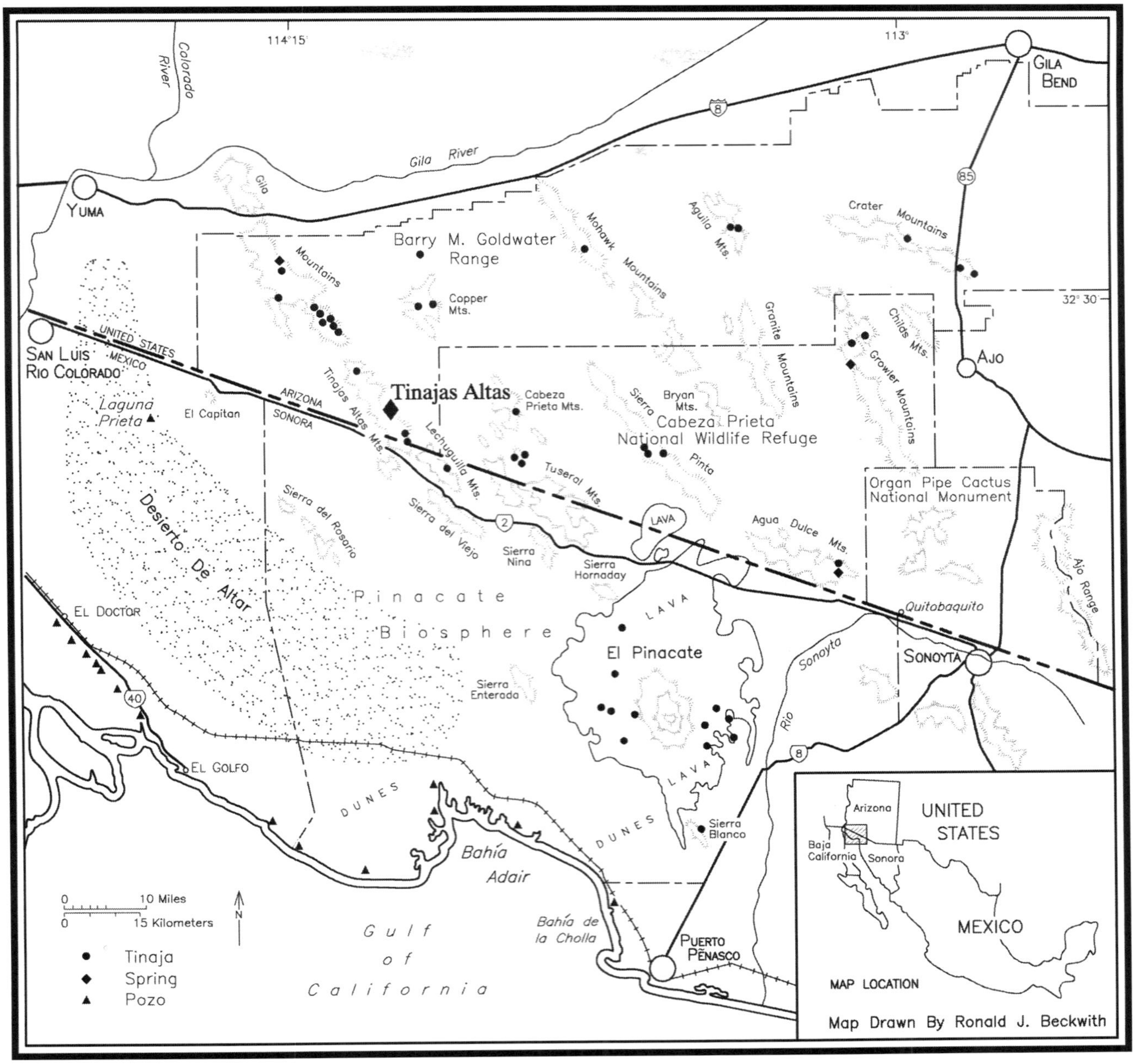

Natural waterholes of the region. (By Ronald J. Beckwith; after Beckwith in Broyles 1996b)

work to reach than they would expend getting to the sure water at the river.

If they were driving cattle to Mexicali or Tijuana by way of the crossing at San Luis Río Colorado, herders would bear west from Las Playas, across the gently rolling sands. They usually attempted this in winter or spring while the annual plants—evening primroses, mallows, mustards, and grasses, especially big galleta (*Hilaria rigida*)—were green and luscious. They might find water at a small tinaja on the southwest side of the Sierra Lechuguilla (Lumholtz 1912), or, in a pinch, detour 15 miles (24 km) round-trip to Tinajas Altas. But going to Tinajas Altas lost precious time and forced them to retrace their steps, so it probably was more common to send a *vaquero* up to Surveyors Tank or Tinajas Altas to fetch water for the cowhands than to take the livestock there. From Sierra Lechuguilla they headed westward, guiding on the low peaks of El Capitán, then proceeded to Laguna Prieta, which held permanent water in the days before groundwater pumping (Matson and Fontana 1996).

A few other tinajas, such as Surveyors and Raven, held water for a few weeks or months. But in its essence, the trek from Sonoyta to the Colorado was a race against thirst and death. And the vision that lulled them to sleep at night and guided their steps by day was water at Tinajas Altas. It

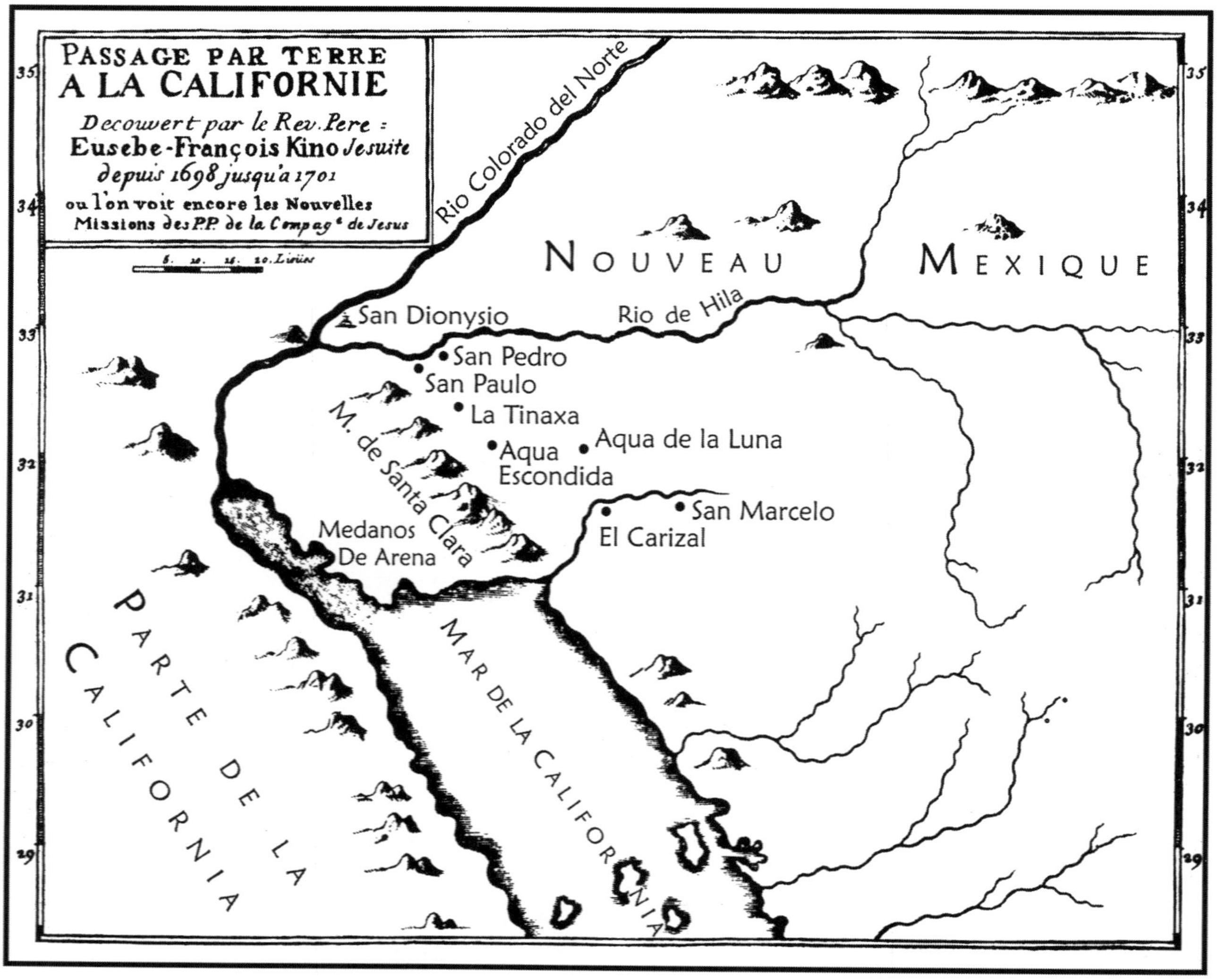

Portion of Kino's 1701 map, redrawn to feature selected landmarks (by Ronald J. Beckwith). Note the relative locations of Agua de la Luna, Agua Escondida, La Tinaxa [Tinaja], and the village of San Pedro (northwest of modern Wellton). The complete map is shown at the beginning of chapter 3. Map available in Burrus 1965: plates X and XI.

surely filled their every thought. Water became an obsession. Most of the tanks, Tinajas Altas included, are not noticeable from a distance—there are no stately cottonwood trees or green thicket banks, so travelers must have relied on guides or spoken instructions, followed tracks of prior travelers, or lucked onto coyote trails that converge toward places holding water.

Riddles of the Routes

Historical itineraries leave us with some confusion, especially since parties applied their own names for places. The question "Who drank where?" has led to a great deal of sleuthing, as we'll see, especially in chapter 3. Decoding padre and pioneer journals and calculating their rates of travel requires attentive map work and sometimes arduous field trips. Early travelers measured distances in time traveled—a league is the distance they could travel in an hour. Some permanent missions may have had accurate sundials or hourglasses, and Padre Eusebio Kino carried an astrolabe with which he could tell time fairly accurately if need be (Ives 1960b), but mainly the method was to estimate time by the positions of the sun, moon, and stars, much as a modern hiker would without using a wristwatch. Converted to linear distance, a league in this terrain runs between 1.25 and 2.5 miles (2 to 4 km), with 1.4 to 1.8 miles an hour being a reasonable range. Reading the historical records and plotting routes on maps, archaeologist Paul Ezell calculated that Kino traveled at an average of 1.83 miles (2.9 km) per hour, and Manje at 1.57 miles (2.5 km per hour, with an hour equaling a league. A party entirely on horseback could travel somewhat faster.

Based on Kino's previous trips in northern Sonora on established and well-used trails, cartographer Robert Lenon (1987:3) concluded, "The Kino legua was found to run from about 1½ to 2¼ mi.; the Manje legua has a bit more variance: from 1¼ to 2¾ mi., say, but normally considerably shorter than the Kino value. Hence, we have used, with good results, 2 miles for the Kino league and 1½ miles for Manje's version." Lenon goes on to say, "There are some indications, also, that both men may have used the term 'legua' to designate the 'distance made good' . . . in an hour of travel. It is natural to expect . . . that leguas tallied during the passage of rugged terrain are shorter than those recorded when the stock is fresh and the condition of the trail is good."

Further complicating matters, summer days would have more daylight than winter days, and none of the diarists seems to have recorded fractions of a league or to fret about accurately accounting for delays and rest stops along the way. Discussions over the length of a league fuel endless campfire debates.

The league problem can be seen readily in Padre Eusebio Kino's trek of 1699. Historian Herbert Bolton (1919b) believed that one of the watering stops on this expedition—the place Kino called Agua Escondida (Hidden Water)—was Tinajas Altas. Other scholars, such as Jesuit historian Charles Polzer, accepted Bolton's identification, but a number of individuals familiar with the region, particularly Robert Lenon (1987) and William K. Hartmann (1989), argue that Agua Escondida was farther east, probably Cabeza Prieta Tanks in the Cabeza Prieta Mountains. We may not be able to resolve exactly which route or routes Kino followed, but we can evaluate the ambiguous and occasionally contradictory documentary evidence of Kino's expeditions along the Camino del Diablo. Because none of the descriptions of Agua Escondida is complete enough to allow a certain identification, all the different legs of Kino's treks to and from Sonoyta and the Gila River must be examined. One strong clue that Tinajas Altas was not Hidden Water is offered by Lt. Michler, "Although no vegetation marks the place, still it [Tinajas Altas] is readily found" (Michler 1857-1859:114), that is to say, not hidden.

Kino's first trip along the Camino in February 1699 included Jesuit Padre Adam Gilg, Captain Juan Mateo Manje, an unspecified number of Indian helpers, and 80 horses (Karns 1954). They left Kino's mission headquarters at Nuestra Señora de los Dolores along the Río Dolores in north-central Sonora on February 9 and arrived at Sonoyta on February 16, where they left most of the horses.

On February 17, 1699, accompanied by three O'odham from Sonoyta as guides, Kino, Manje, and Gilg took the best mounts and headed west for 10 leagues to a place the party named Los Carrizales along the Río Sonoyta. Also known as El Carrizal, this canebrake was the place where most travelers left the Río Sonoyta and headed northwest across the Camino del Diablo. The waterhole upstream at the beginning of the canebrake came to be known as Agua Dulce, because the water was "sweet" or fresh there. The waterhole downstream at the end of the canebrake was called Agua Salada because the water was brackish (Ives 1989:26). Today, thickets of carrizo (*Phragmites australis*) mark the remnants of El Carrizal (Felger 2000), and it is here that the reader can appreciate the crux of the mileage issue. Burrus (1971:230, 291) reports that El Carrizal is about 25 miles west of Sonoyta, but the route on modern topographic maps is 17.6 miles, indicating that their "league" was about 1.7 miles on level ground of a riverside trail.

The next day, February 18, according to Manje's first version, the expedition headed north-northwest for six leagues to a place, probably Las Playas, where the guides assured them there was water. Finding none, they headed north another 14 leagues, completing the ride on a moonlit night. "After midnight, we were shown a waterhole high up in the crags and canyons where, climbing with difficulty, we drank but the horses could not, even though they were in great need, because the climb up the rocks was so rugged and slippery," Manje noted. "We called it the Waterhole of the Moon [because we arrived there with its light]" (Burrus 1971:391; translation by Thomas E. Sheridan).

Bolton (1936:412) believed that Aguaje de la Luna was Heart Tank in the Sierra Pintas. The identification of Heart Tank as Aguaje de la Luna is reasonably consistent with the distances and directions given by Manje. On modern maps it is about 37 miles from El Carrizal to Heart Tank, meaning that they averaged 1.85 miles per hour (37 miles divided by 20 leagues). The description also sounds like Heart Tank: a waterhole back in a rugged canyon reached by climbing over boulders and slippery slopes. Horses cannot clamber to the water. Because of the light-colored rock, the tinaja may be reached in moonlight without aid of a flashlight.

On February 19, Kino, Manje, Gilg, and the three O'odham left Aguaje de la Luna. In Manje's words,

> On the 19th day of February, we continued our journey, always in a northwesterly direction, until after three leagues on the trail, we turned to the west, in order to search for water for the horses, because they had not drunk in 36 hours. At four in the afternoon, after traveling about nine leagues, we came upon a small ranchería of poor Indians, who sustained themselves with roots and [lizards]. Among them we counted thirty souls of one or the other sex. Here we found good and sufficient water. The horses, in great need, drank there but there was very

Sonoyta River, near El Carrizal. (Photo by Bill Broyles)

Table 1.2. Names Applied to Watering Places Used by Early Spanish Explorers in the Tinajas Altas Region

Modern Name	Kino 1699–1702	Manje 1699	Sedelmayr/Ensign 1750	Anza 1774, 1776	Juan Díaz 1774	Garcés 1771–1774	Font 1776	Fages 1781–1782
El Carrizal (Agua Dulce; Agua Salada)	Los Carrizales (D)	Los Carrizales (D)	"the ranch of Captain Cipriano" (D)	El Carrizal (D)	—	El Carrizal (P)	El Carrizal (D)	El Carrizal (D)
Heart Tank	Aguaje de la Luna (D)	Aguaje de la Luna (D)	"a tank. . . from which only the men were able to. . . drink" (L)	Aguaje Empinado (D)	"a tank of water which is very high up" (L)	Aguaje Empinado (D)	Aguaje Empinado (D)	La Tinaja Alta (P) Several possible locations.
Cabeza Prieta Tanks	Agua Escondida (D)	"good waterhole" (L)	"three tanks of high elevation" (P)	Tinajas de la Purificación (L), Tinajas de la Candelaria (L)	La Purificación (L)	—	Tinajas de la Candelaria (D)	—
Baker Tanks	La Tinaja (D)	La Tinaja (D)	—	—	—	—	—	—
Near Baker Tanks	Aguaje de los Algives [Alquives]	—	—	—	—	—	—	—
Laguna Prieta	—	—	San Juan Nepomuceno (D)	—	—	—	—	—
Tinajas Altas	—	—	"a tank where we watered the animals . . . [with] water-tight baskets with some difficulty" (L)	—	—	Tinaja de Eusebio (named for Kino but he did not visit here)	—	—
Agua Dulce Spring? A tinaja in Sierra de los Tanques?	—	—	—	—	—	—	—	La Tinaja Alta (P) Several possible locations.
Coyote Water	—	—	—	Pozos de En Medio (D)	"an arroyo . . . in the middle of a valley" (L)	"an arroyo which has water when it rains" (L)	Pozos de En Medio (L)	—
Coyote Wash (east of Wellton Hills)	—	—	—	Pozos de San Miguel (L)	—	—	—	Pozos de San Miguel (L)
Spook Tank (in Arroyo San Albino)	—	—	—	Agua Escondida (D)	Agua Escondida/ Pozo Blando(D	Pozo Blando (D)		

Key: D = Definite, L = Likely, P = Possible

Heart Tank. The concrete lip was added by US Fish and Wildlife Service. (Photo by Bill Broyles)

> little pasture. Because we did not know whether we would find better pasture further on, we resolved to spend the night at this ranchería. (Burrus 1971:413–14; translation by Sheridan)

This apparently was the place Kino referred to as Agua Escondida on his return trip from the Colorado River the following year (Bolton 1919b:253), although neither he nor Manje gave it a name in their accounts of the 1699 journey. Some identify the watering place as Tinajas Altas (Bolton 1936; Polzer 1982), but we disagree and contend that it is Cabeza Prieta Tanks (W. Hartmann 1989; Lenon 1987). Cabeza Prieta Tanks are obscure and difficult to find. They are in dark, dull, loose rock, and walking up the canyon to them at night is difficult and dangerous, even by the light of a full moon. They are reached by navigating toward the distinctive Cabeza Prieta Peak, a regional landmark, and negotiating a series of canyons. The tanks themselves sit about a mile back in a rocky, boulder-strewn side canyon. The lower tanks are easy to reach, but the upper tanks require scaling several steep ledges. There is scant "pasturage," but there was once an Hia C'ed O'odham camp there, although it may have been outside the mouth of the canyon, not at the tanks themselves.

According to Manje, the party had traveled 3 leagues to the northwest and 9 leagues west on February 19 for a total of 12 leagues. The shortest viable foot distance from Heart Tank to Cabeza Prieta Tanks is 16.9 miles (27 km) by way of Smoke Tree Wash. Tinajas Altas, on the other hand, is 33.1 miles (53 km) from Heart Tank by foot, meaning that for this place to be Tinajas Altas, they would have needed to double their pace from 1.4 mph to 2.8 mph, which is unlikely given the terrain. So, we are confident that Agua Escondida is Cabeza Prieta Tanks.

From there, Kino's group went northward.

> On the twentieth of said month, after mass, we left in search of the Río Grande [Gila River], traveling to the northwest a point of the compass to the north, across spacious plains that were, above all, sterile with little pasture for the horses. We passed some hills with many veins of minerals of a gravelly metal, green and black and a variety of other colors, that appeared to

Herbert E. Bolton at Cabeza Prieta Tanks, probably February 1, 1928, for "it was in honor of the feast of the next day that Anza called these tanks the Tinajas de la Purificación" (Bolton 1931:60). (Courtesy of the Arizona Historical Society/Tucson; image no. 15998)

> be some type of silver. After traveling since the morning for a distance of 14 leagues, we camped in a dry arroyo where there were some tanks of water from which the horses drank. Among them there was one in the middle of a rocky hill that had very clear water in the form of a very large tinaja. For that reason, we named the waterhole La Tinaja. (Burrus 1971:414)

A second, though less detailed, version of this leg of Kino's trip (Burrus 1971:391) doubly convinces us that La Tinaja is the major set of tinajas now called Baker Tanks. In both of these accounts, the description, direction, and distance fits a march to Baker Tanks, 30.3 miles by foot from Cabeza Prieta Tanks. From the northern end of the Cabeza Prieta Mountains, travelers pass low hills, either the Devil Hills or outliers of the Buck Mountains, and then cross a broad plain while guiding on Baker Peaks. The open plain is the broad valley between the Copper and Mohawk mountains, and the northern portion of the Copper Mountains shows traces of ore—copper and other colorful streaks, hence their name. Hikers from anywhere south of Baker Tanks can locate the tanks by walking toward the mid-section of the rocky hills that resemble the sails of a full-rigged clipper ship on a desert sea.

At Baker Tanks a number of tinajas sit in an unusual conglomerate bedrock. The largest tinaja rests in a small box canyon with sheer rock walls except for the sandy beach at its downstream end, making it an easy spot to water livestock. These tanks are perennial in good years. Because of their proximity to Antelope Hill, a significant, multicultural Native American site along the Gila River, Baker Tanks were undoubtedly known to Kino's O'odham guides. Interestingly, Kino's 1701 map shows Agua Escondido and La Tinaja aligned as if on a fairly straight route to the Gila River, as can be seen on modern maps. Kino's map also indicates the relative location of Agua de la Luna from El Carrizal and then the westward jog to Agua Escondido. In some later maps this relationship becomes distorted.

Some have interpreted La Tinaja to be Drippings Springs (Bolton 1936), but the description, distance, and direction far better fit Baker Tanks. Dripping Springs is difficult to find, even today, for it is located back in an obscure canyon in the Gila Mountains, whereas Baker Tanks are obvious and easily accessible. And Dripping Springs was just that—low-volume seeps coming out of a rock wall and collecting in small rock basins; the flow was so paltry and unreliable that the Arizona Game and Fish Department built a catchment dam there in 1969 to capture the scant water before it dribbled away or evaporated.

Further evidence that Baker Tanks were La Tinaja comes in the account for the next day, February 21, when Kino and his companions reach the Gila River after only five or six leagues. Modern maps peg this distance at 6.6 miles at the closest point, roughly where modern Avenue 25-E intersects the river, and about 8 to 9 miles (13 to 14 km) to the favored camping spot at Laguna Salada. On the map, the distance from Dripping Springs to the Gila looks to be about 7.3 miles (11.7 km), but the bajada badlands and rocky, loose terrain forces travelers to loop far eastward before heading north toward the river and with difficulties that the diarists would have noted with complaint.

Baker Tanks (La Tinaja). (Photo by Bill Broyles)

The next day, they went a short distance and reached the Gila River.

> On the 21st of the said month of February, we dispatched two of the Indian natives who came with us as guides. [They carried] assurances to the people of the Great River that we were coming to see them. Later, we followed our course there, heading in a west-northwesterly direction. And at a distance of five leagues, we reached the much-desired river, where we found ourselves among 450 souls. Some of them were Pimas, the rest belonged to the Yuma nation (Burrus 1971:414–415)

After his visit, Kino and his troop returned to Mission San Xavier del Bac, Tucson, by way of the Gila River and then southward up the Santa Cruz River valley. Along the way he passed by a number of native villages and named them.

Kino retraced his steps a year and a half later, in September and October of 1700, when he left Tucson, followed the Santa Cruz northward, and descended the Gila to its junction with the Colorado. Then he returned to San Marcelo de Sonoyta, traveling along the Camino del Diablo. Manje did not accompany him on this trip; Kino's companions were ten Pimas from Dolores. The expedition lasted from September 24 to October 20, 1700. On this trip Kino convinced himself that Baja California was not an island, and that San Marcelo de Sonoyta was a key oasis along the road he hoped to blaze between the Pimería Alta and Baja California (Bolton 1936). Regarding the Camino del Diablo route, this is what Kino had to say:

> On the tenth [of October], leaving the Rio Grande [Gila River] and coming by the route by which we had entered in the month of February of the past year, we arrived in time to take our siesta at the tank of La Tinaja; and traveling in the afternoon about twelve leagues farther, we arrived an hour after nightfall at the watering-place which also the past year we had called El Agua Escondida, because it was among the rocks.

So they went from the Gila River to Baker Tanks to Cabeza Prieta Tanks. As discussed previously, Cabeza Prieta Tanks are 30.3 miles from Baker Tanks. A walk from the Gila River to Tinajas Altas would be minimally 27.2 miles, but there is no La Tinaja along that route nor would Tinajas Altas put them within 12 leagues of La Luna (Heart Tank).

Drippings Springs. The dam was installed in 1969 by the Arizona Game and Fish Department; in 1985 the department added a shade roof to inhibit evaporation. (Photo by Gayle H. Hartmann)

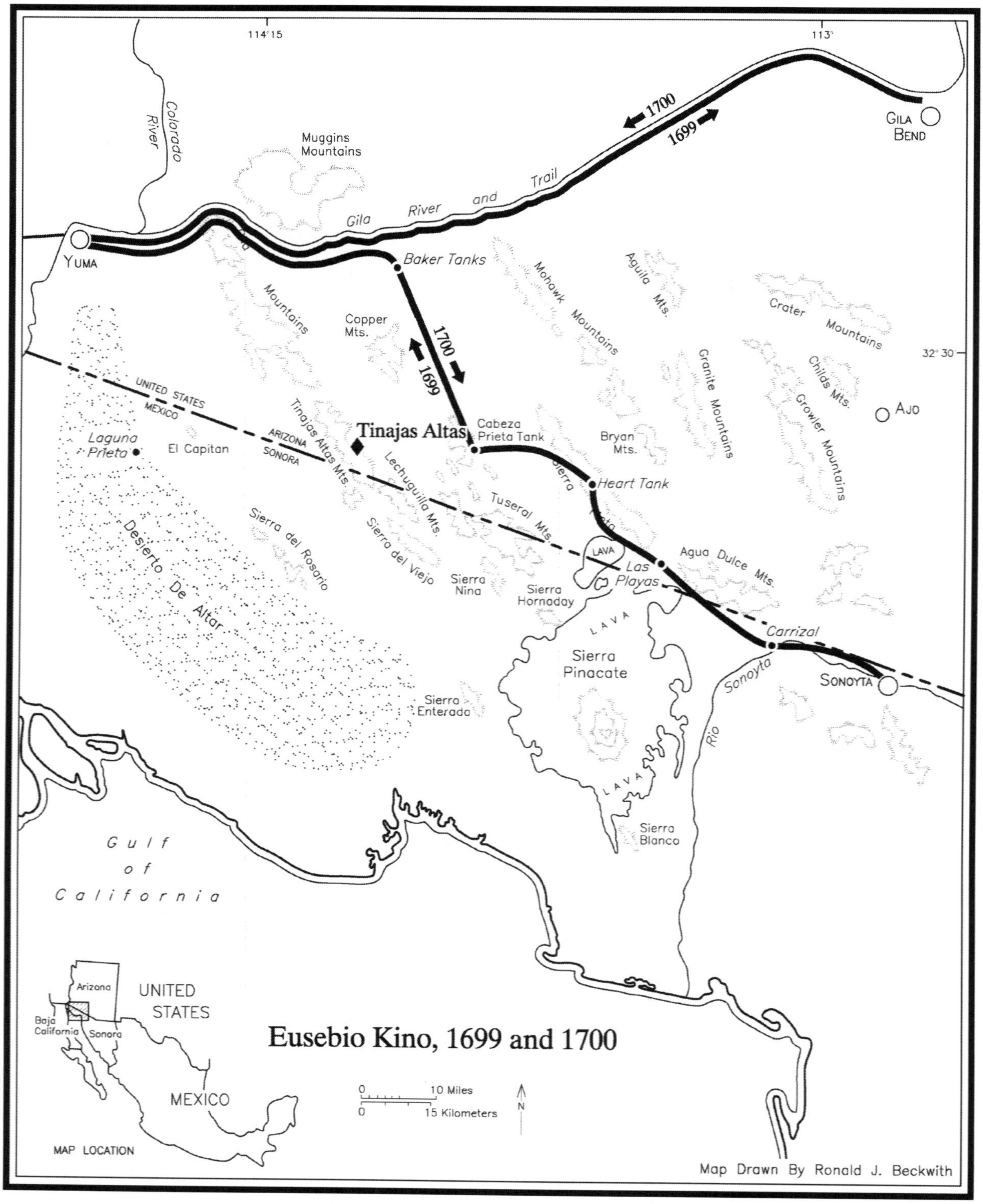

Kino's routes, 1699 and 1700. (By Ronald J. Beckwith)

On the eleventh, after dispatching most of the servants with the relay, that they might go on to take their siesta and wait for me wherever they should come across good pasturage for the pack-animals, I turned toward the west and ascended another hill, but I saw nothing more than the continuation of these lands with those of California, and the sands of the California sea. Overtaking the servants, and traveling today twelve leagues, we arrived before sunset at the tank of La Luna, for we had given it this name the year before because we arrived there

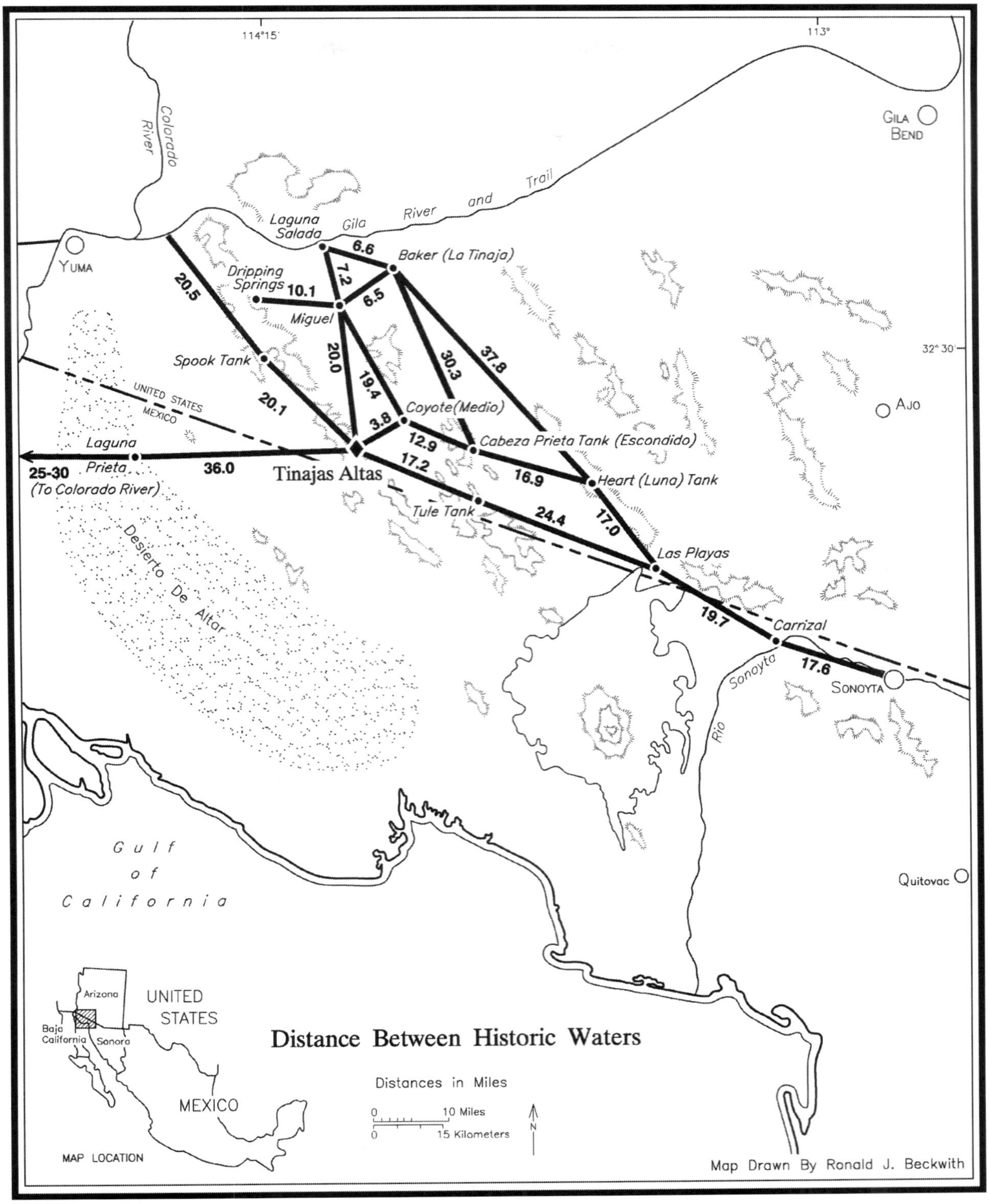

Distances between historical water sources. (Map by Ronald J. Beckwith)

at night by moonlight. And because this watering-place is among some rocks so high that the pack-animals cannot ascend to drink water, we determined to eat a morsel of supper there and then travel, and we did travel, three hours more by night, in order to reach the watering-place of Carrizal with more ease the following day.

We wonder which "hill" he surmounted to view the gulf, though it may have been Cabeza Prieta Peak, a two-and-a-half-mile walk from the tanks but along the same route, prob-

Table 1.3. Walking Mileages between Key Waterholes

Waterhole	BT	CP	CW	DS	EC	GR	HT	LP	PSM	RB	SO	ST	TA	TT
Baker Tanks (BT)	*	30.3	20.6	16.2	88.5	6.6	37.8	51.2	6.5	16.4	88.5	31.2	22.0	39.3
Cabeza Prieta Tanks (CP)	30.3	*	12.9	37.3	67.7	37.0	16.9	30.4	30.3	22.0	67.7	36.6	16.2	8.9
Coyote Water (CW)	20.6	12.9	*	25.5	76.5	26.5	29.8	39.2	19.4	9.5	76.5	23.7	3.8	14.8
Dripping Springs (DS)	16.2	37.3	25.5	*	102.0	7.3	50.7	64.6	10.1	17.7	102.0	15.0	26.8	40.2
El Carrizal (EC)	70.9	50.1	58.9	84.3	*	77.5	34.6	19.7	77.2	68.3	17.6	81.4	61.3	44.1
Gila River (GR)	6.6	37.0	26.5	7.3	95.1	*	44.4	57.8	7.2	19.4	95.1	20.5	27.2	41.3
Heart Tank (HT)	37.8	16.9	29.8	50.7	54.3	44.4	*	17.0	43.2	38.9	54.3	53.5	33.1	15.4
Las Playas (LP)	51.2	30.4	39.2	64.6	19.7	57.8	17.0	*	57.5	48.6	37.3	61.7	41.6	24.4
Pozo de San Miguel (PSM)	6.5	30.3	19.4	10.1	94.8	7.2	43.2	57.5	*	12.9	94.8	25.1	20.0	34.1
Raven Butte (RB)	16.4	22.0	9.5	17.7	85.9	19.4	38.9	48.6	12.9	*	85.9	12.6	9.1	24.2
Sonoyta (SO)	88.5	67.7	76.5	102.0	17.6	95.1	54.3	37.3	94.8	85.9	*	99.0	78.9	61.7
Spook Tank (ST)	31.2	36.6	23.7	15.0	99.0	20.5	53.3	61.7	25.1	12.6	99.0	*	20.1	36.8
Tinajas Altas (TA)	22.0	16.2	3.8	26.8	78.9	27.2	33.1	41.6	20.0	9.1	78.9	20.1	*	17.2
Tule Tank (TT)	39.3	8.9	14.8	40.2	61.7	41.3	15.4	24.4	34.1	24.2	61.7	36.8	17.2	*

Source: Compiled by Luke Evans.

ably Smoke Tree Wash. The summit stands at an elevation of 2,559 feet (780 m) with a clear view to the south, though other satisfactory ridge tops were nearer his route.

> On the twelfth, arising more than two hours before dawn, and setting out from the stopping-place [Heart Tank] at the rising of the morning star, after thirteen leagues of very good roads we arrived at ten o'clock at the good watering-place of El Carrizal, of the Arroyo de San Marcelo de Sonoydag [Sonoyta River]. I said mass, we breakfasted, and after eating dinner we took a very good siesta; and after eight leagues' journey farther we arrived at eight o'clock at night at . . . San Marzelo. (Bolton 1919b, 1:253–55)

Have we come to the correct conclusions? The evidence is strong, but new evidence may yet be found in a distant archive. Later maps, such as the 1859 or 1889 maps of New Mexico Territory, clearly show four major routes: from Sonoyta two trails go north to Gila Bend or to Mohawk Station via Don Diego Tanks; the next path to the west veers north at Las Playas and passes Heart Tank, Cabeza Prieta Tanks, and Baker Tanks; and finally, the most direct route follows El Camino

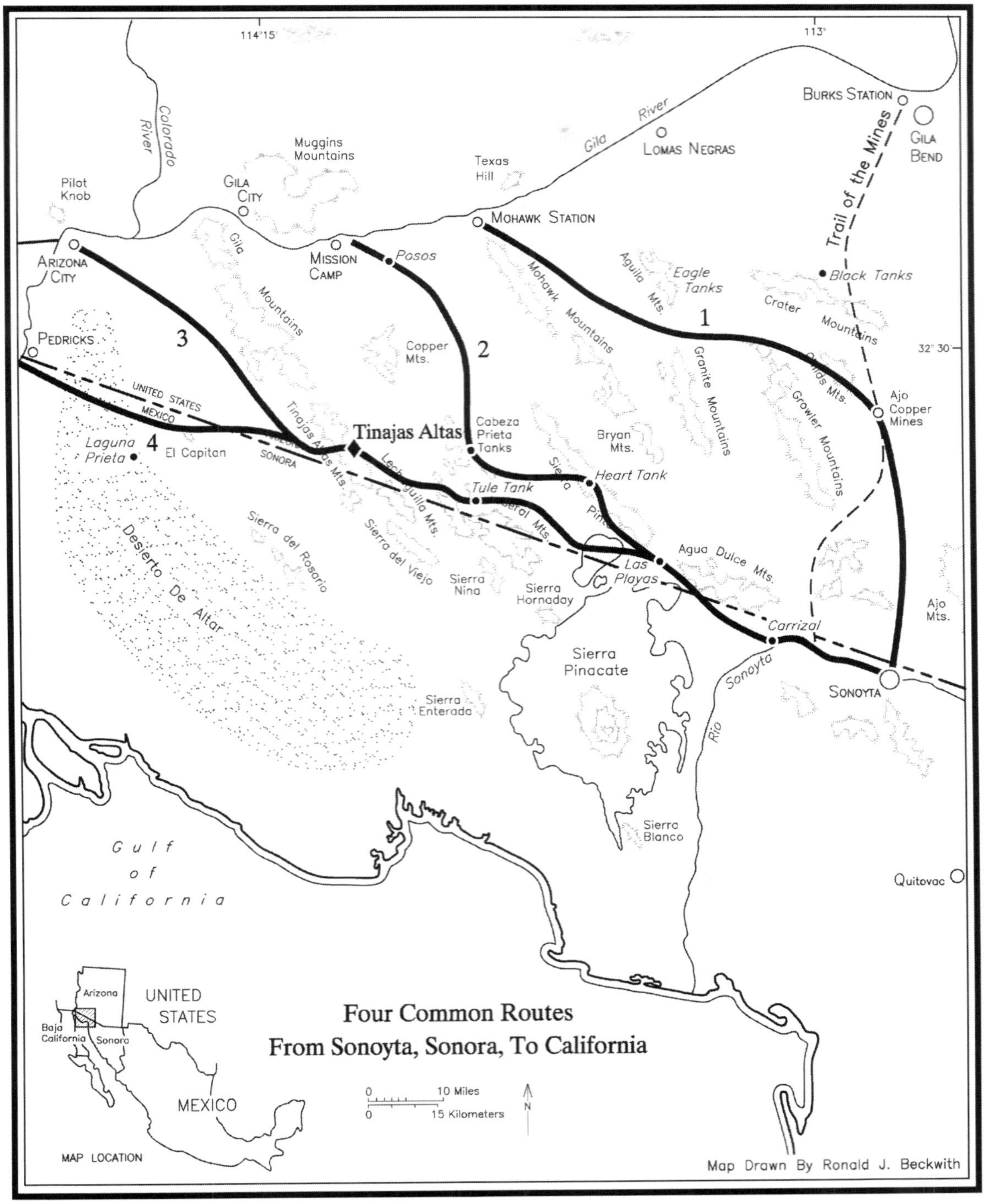

Common routes from Sonoyta, Sonora, to Yuma, Arizona. (Redrawn by Ronald J. Beckwith from Bureau of Topographic Engineers map, 1889, New Mexico Territory.)

del Diablo to Tule Tank, Tinajas Altas, and then to Yuma. The careful observer will also notice that the Gila Trail stage stations at Mission Camp and Mohawk Station are at or near the intersections of the northbound trails, indicating they also fed and bedded travelers coming from the border.

Some readers may be surprised to learn that Kino did not drink from Tinajas Altas, but the larger story of desert water is the extensive network of Native American trails linking useful waterholes and forming a complete transit and supply system. Our story will return to these desert waters—Agua de la Luna, Agua Escondido, La Tinaja, or as we now know them, Heart Tank, Cabeza Prieta Tanks, and Baker Tanks. And our story will come again to Tinajas Altas, the most famous desert water of them all.

Tohono O'odham women gathering saguaro fruit and other plants. (Photo by Edward S. Curtis, about 1907. Photo courtesy Library of Congress, Edward S. Curtis Collection, image no. LC-USZ62-115801.) Judging from several out-of-season clues in the photo, such as the apparent lack of fruit or flowers, Curtis likely staged this and his other photos of saguaro fruit gathering.

2 Native Peoples of the Tinajas Altas Region

Prehistoric to Present

Gayle Harrison Hartmann, Mary Charlotte Thurtle, and Gary Paul Nabhan

One of the most fascinating aspects of Tinajas Altas is the overwhelming sense of the past one feels when there. Old trails, ancient grinding holes, and bits of seashell or pottery discarded thousands of years ago inspire wonder and curiosity. As we look at the remnants of ancient cultures all around us we ask ourselves, who *were* the people who sat around campfires in this same spot a thousand or ten thousand years ago in this area called the Western Papaguería?

Desert regions are unusual from an archaeological perspective because much of the evidence of past human activity is still lying on the surface of the ground. This is contrary to wetter regions where soil deposition buries evidence of ancient peoples. Thus, in a desert much can be learned from just looking at the ground. Archaeologists conduct surveys by walking in a series of evenly spaced transects to record evidence and designate what they call "sites." A site is a locale where there is a concentration of archaeological evidence.

In the 1920s, Malcolm Rogers, an archaeologist with the San Diego Museum of Man, visited the site of Tinajas Altas and made collections of artifacts, later curating them along with his written observations (Rogers 1929). Archaeologist Paul Ezell (1954) noted the site while he was surveying in the region for the National Park Service and the Arizona State Museum (ASM). The site was first brought to the attention of the Arizona State Museum by Carolyn Ann Carr, then a University of Arizona graduate student in resource economics. She had taken several hiking and car trips into the region in an effort to promote a federal wilderness designation for the Cabeza Prieta National Wildlife Refuge. After visiting Tinajas Altas in 1971 and realizing it was a significant archaeological site, she provided ASM staff archaeologist James Ayres with sufficient information so that the site could be entered in the museum's archaeological site system. More recently, in the 1980s and 1990s, two small surveys were conducted in the vicinity in connection with proposed military actions near the site (Bruder et al. 1988; Altschul and Jones 1989).

A large-scale archaeological survey, covering 5,456 acres, was conducted between January and April 1998; two of us (MCT and GHH) were involved in that fieldwork. The survey's primary focus was Tinajas Altas itself but it extended as far north as Tinajas Altas Pass and included portions of the western and eastern fronts of the Tinajas Altas Mountains as well as extending into the Lechuguilla Valley to the east. During this survey, sponsored by the US Air Force, we recorded 30 sites, with the site of Tinajas Altas clearly being the most significant. One of our conclusions was that most, if not all, of the other sites relied on water available at Tinajas Altas (Hartmann and Thurtle 2000, 2001).

Thus, archaeologists have begun to develop a fuller understanding of the Tinajas Altas region (Ahlstrom et al. 2000; Ahlstrom 2000, 2001; Altschul and Rankin 2008; Hartmann and Thurtle 2000, 2001; Hartmann et al. 2008; McGuire and Schiffer 1982). In the pages that follow we try to present a picture of what drew native peoples to this waterhole and the surrounding arid region and how they used local resources to their benefit.

The Earliest People: Paleoindian or Malpais (11,000 BC or earlier to 8,000 BC)

A mystery, and an interesting controversy, has simmered since the 1960s regarding the first occupants of the area, and indeed of North America as a whole. On one side, substantial evidence supports the notion that the earliest inhabitants of the Americas, referred to as Paleoindians, were hunters of big game including mammoth and other now-extinct species. The beginning of this tradition has been dated to around 13,000 years ago, when the Bering land bridge connected Asia and North America. During the Paleoindian period, the climate was wetter and colder than it is now, and the wooded parklands or grasslands that supported large animals were more widespread. Most evidence for Paleoindians in Arizona comes from the eastern and southeastern part of the state and includes "kill sites" where large mammals, particularly mammoths, were killed and butchered (Huckell 1982). Artifacts of the Paleoindian hunters include distinctive "fluted" projectile points (such as the famous Clovis points), which were hafted to hand-held spears. Also common are tools suitable for skinning large animals and cutting meat and bone. The people of the Paleoindian period must also have exploited wild plants for food, fuel, fiber, and tools, but few stone tools provide evidence for those activities.

Although no typical Paleoindian kill sites have ever been identified in the Papaguería, a fluted point, possibly Clovis, was found in a lower level of Ventana Cave that also produced a radiocarbon date of 9,350 plus or minus 1,200 years BC (Haury and Hayden 1975:v). Another fluted projectile point, found by archaeologist Paul Ezell (1954:13),

Clovis and Archaic points. Top row: three Archaic points; left, Datil type; center, Gypsum type; right, San Jose type, all from Ventana Cave, Tohono O'odham Reservation. Bottom row: two Clovis points; left, Leikem site, southeast Arizona; right, Ventana Cave, Tohono O'odham Reservation. Clovis points were used by Paleoindian hunters (11,500 BC to 11,000 BC) to hunt large game such as mammoths. Archaic points were used between 4500 BC and AD 300 to hunt smaller game such as deer. Both were hafted onto spears. Archaic spears were often augmented with atlatls, to give the spears more distance and power. (Courtesy Arizona State Museum, University of Arizona, Tucson; image no. 2009-223)

probably came from Organ Pipe Cactus National Monument or from the Cabeza Prieta National Wildlife Refuge (Huckell 1982:19). Other finds supporting a Paleoindian presence in the region include three fragmentary Clovis projectile points and a patinated fluted point from the Barry M. Goldwater Range (Tucker 20 00a) as well as other fluted points from both southwestern Arizona and northwestern Sonora (Sanchez 2001; Gaines and Sanchez 2009; Sanchez et al. 2009).

Another interpretation of the earliest inhabitants of the region has been advanced by Julian Hayden, an indefatigable explorer and pioneer in archaeological investigation of the Pinacate lava region that straddles the international border, some 30 miles southeast of Tinajas Altas (Hayden 1998, 2011). He argued that the crude stone artifacts he saw on the surface must predate the elegant Clovis points made by Paleoindian peoples (Heilen 2004). Hayden (1976) believed that a Malpais complex, identified in southwestern Arizona and northern Sonora, was pre-Clovis and represented stone tools of peoples who were in the area as much as 20,000 or 30,000 years ago. Hayden found support in the discovery of archaeological sites in both North and South America that may represent occupations by hunters who crossed into North America from Asia not when the Bering land bridge was exposed in 10,000 BC, but during earlier glacial cycles. Hayden's interpretation has been questioned because of the ambiguous nature of some of the tools and the lack of solid dating (McGuire 1982:160–164).

Meanwhile, since the 1980s, archaeologists in other parts of the Americas have unearthed increasing evidence of a small

Archaeologists Malcolm Rogers (left) and Julian D. Hayden (right) whose interests focused on the human history of the deserts of the Southwest. Photo at Ventana Cave, Arizona, April 1959. (Irwin Hayden photograph. Courtesy Steve Hayden)

population of early hunters—without fluted points—before 14,000 years ago (Goebel et al. 2008; Waters and Stafford 2007). Perhaps Julian Hayden did turn up evidence of them but this remains unclear. In any case, everyone agrees that the wave of immigration around 11,000 BC dwarfed any earlier population.

Evidence for these pre-Paleoindian peoples is still hotly debated. The question of whether the very First Americans were makers of simple stone tools or elegant fluted spear points will not be answered here. But, we do know that by about 8,000 BC people all over North America changed how they obtained food, thus changing their way of life.

ARCHAIC TIMES: ATLATLS AND GRINDING STONES (8,000 BC–AD 800)

The extinction of mammoths and other large Pleistocene mammals was at least one cause of a shift from the Paleoindian hunting economy to what archaeologists call the Archaic lifeway. Hunting was still important, but the prey changed to smaller animals such as deer and rabbit, and wild plant foods played a greater role in the diet. The Archaic period is defined by stylistically distinctive projectile points that were mounted on atlatl darts or spears, as well as by other stone tools and a variety of grinding implements. By late Archaic times, the abundance of grinding tools such as bedrock mortars and metates is the crucial evidence supporting the notion that plant foods were increasingly important.

The projectile point styles of the Archaic period changed over time. Projectile point evidence suggests that human occupation in southwestern Arizona during the early Archaic period was extremely limited, increasing in intensity during the middle Archaic period. The early Archaic period is represented by what archaeologists call Jay or Mohave-style projectile points found at one site in East Pass of the Crater Range (Tucker 2000a) and one possible early Archaic projectile point recovered from White Tanks in the Tank Mountains (Shackley 1993, 2005). Points from the middle of the Archaic period have been recovered from sites along San Cristobal Wash (Slaughter et al. 2000), from sites in and near the Aguila Mountains (Olszewski et al. 1995: table 2; Tucker 2000a); and from several sites in the Sauceda Valley (Lyon 2000; Roberts et al. 2000).

In 1998, the Tinajas Altas archaeological survey collected four Archaic period points. One was a tapering stem point from the early or middle Archaic period recovered several hundred meters east of the tinajas themselves. Three others, probably late Archaic in age, were recovered closer to the tinajas (Hartmann and Thurtle 2001:495). Although this sample is small, it suggests that even during Archaic times when the climate may have been more temperate, the catchments at Tinajas Altas were a focal point in the region, with hunters both taking advantage of game animals that congregated at the water and relying on the water themselves.

In much of the Southwest, the hunting and gathering lifeway gave way to farming during the late Archaic period. Crops began to play a major role in the lives of people living along the rivers of eastern and central Arizona, such as the Salt, Gila, San Pedro, and Santa Cruz. With this shift

Evidence for ceramic vessels usually occurs in fragments called "sherds" by archaeologists. (Photo by Bill Broyles)

to agriculture, people spent more time in villages near their fields to accommodate the requirements of growing crops. In the arid Western Papaguería, however, Native Americans continued to depend heavily on hunting and gathering. The extreme climate and lack of water simply did not make farming a viable option. Substantial numbers of grinding holes of various sizes recorded in and around Tinajas Altas during the 1998 survey suggest that large quantities of seeds and beans were being ground there—some perhaps as early as Archaic times.

THE CERAMIC PERIOD: POTTERY DEFINES A NEW LIFEWAY (AD 800–1850)

A landmark event in the history of any people is the time when pottery is first made. This invention marks the end of the Archaic period around AD 200–300 in most of southern Arizona, but not in the Western Papaguería. Pottery was significant because it provided durable containers that could be used for everything from carrying water to cooking to storage. From the point of view of modern archaeologists and historians, pottery is significant because often it can be dated, providing temporal underpinnings that clarify past lifeways. Pottery is datable because the shapes of vessels, the sizes of vessels, and especially painted decorations change over time—usually on the order of a few hundred years or less.

ANCESTRAL YUMAN/PATAYAN PEOPLES: BUFF-COLORED POTS AND ROCK CIRCLES

In the Western Papaguería the ceramic-based lifeway started quite late—beginning around AD 800 and lasting until around AD 1850 when native life was disrupted by the roads, towns, and farms of the expanding Americans. The pottery makers were ancestral Yuman-speakers, commonly called Patayan, a term proposed in the 1930s by archaeologist Lyn-

Lower Colorado Buff ware vessel. Typically these were used as water jugs, cooking pots, and storage containers for perishables. (Photo courtesy of Arizona State Museum, University of Arizona, Tucson; image no. 2206-960-055)

don Hargrave, who was told by an interpreter that it meant "old people" in Walapai (Colton 1945:119). The ancestral Yumans with their Patayan pottery were likely descendants of the Archaic period people who preceded them.

Evidence for the Ceramic period is not extensive at Tinajas Altas compared to other parts of the Southwest. There are no large village sites and little painted pottery. Instead there are sparse scatters of broken pottery and bits of stone tools, and, occasionally, camp sites consisting of rock circles, rock-filled roasting pits, and bedrock mortars (Schroeder 1979:100). The rock circles, often called "sleeping circles," are widespread throughout the western Sonoran Desert. These circles, whose use might seem obvious from the name, are actually one of the greatest mysteries of the desert. It is uncertain whether they were really for sleeping or whether they were religious or ceremonial in nature. The most common suggestions for their use have been personal bedding areas in temporary camps or foundations for simple brush structures (Rogers 1966). Additional uses as a vision ring or game circle have been proposed for a rock circle within Seri territory (Bowen et al. 2004).

By far the most widespread evidence of the Patayan tradition is thin-walled, tan pottery, called Lower Colorado Buff Ware. This ware is relatively common throughout southwestern Arizona and southeastern California, and several types can be distinguished. Archaeologist Michael Waters (1982b) has produced a thorough typology and time-line for Patayan ceramics. His analysis, based on the pioneering work of Malcolm Rogers, divides the chronology into three periods: Patayan I, II, and III. The 17 pottery types recognized by Waters are distinguished by surface treatment, jar rim form, vessel form, and temper.

Hohokam and Trincheras: Neighbors to the East and South

Away from Tinajas Altas and outside the most arid western region, other ceramic-using cultures were prominent in the Sonoran Desert. Evidence of these cultures, in the form of small amounts of Hohokam and Trincheras pottery, was found at Tinajas Altas. Whether these pieces of pottery were left there as a result of visitation by people of more distant cultural traditions or whether they were traded into the area is unclear. The presence of this pottery does tell us that the prehistoric people who used Tinajas Altas were not isolated, but lived within a network that extended outside of the western desert.

The Hohokam were agricultural people who lived in central and southern Arizona. They built houses in shallow pits and produced both plain and decorated pottery, along with numerous other crafts of shell, stone, and clay. The Hohokam also constructed extensive irrigation canal systems within the major river valleys and built large structures called "ball courts" (earthen berms surrounding a flat, oval surface) where archaeologist believe they played a game similar to those played in prehistoric Mexico.

The Trincheras people were also farmers and pottery makers who lived in what is now the northwestern part of the state of Sonora, Mexico. In addition to villages near watercourses, on hills they constructed extensive systems of rock walls, called *trincheras* in Spanish. These hillside and hilltop sites served a variety of purposes including locations for crops, houses, and ceremonial activities. The canal systems and ball courts of the Hohokam and the rock walls of the Trincheras tradition attest to complex social organizations.

History Begins: Yuman-speaking Peoples

The history of a region begins with its first written records. In southwestern Arizona the first records were kept by members of the Coronado expedition, who briefly passed through the region in 1540. It wasn't until 1699, however, when the Jesuit priest Father Eusebio Kino and his military companion Captain Juan Mateo Manje journeyed as far west as the Colorado River, that much was written about the area and its peoples.

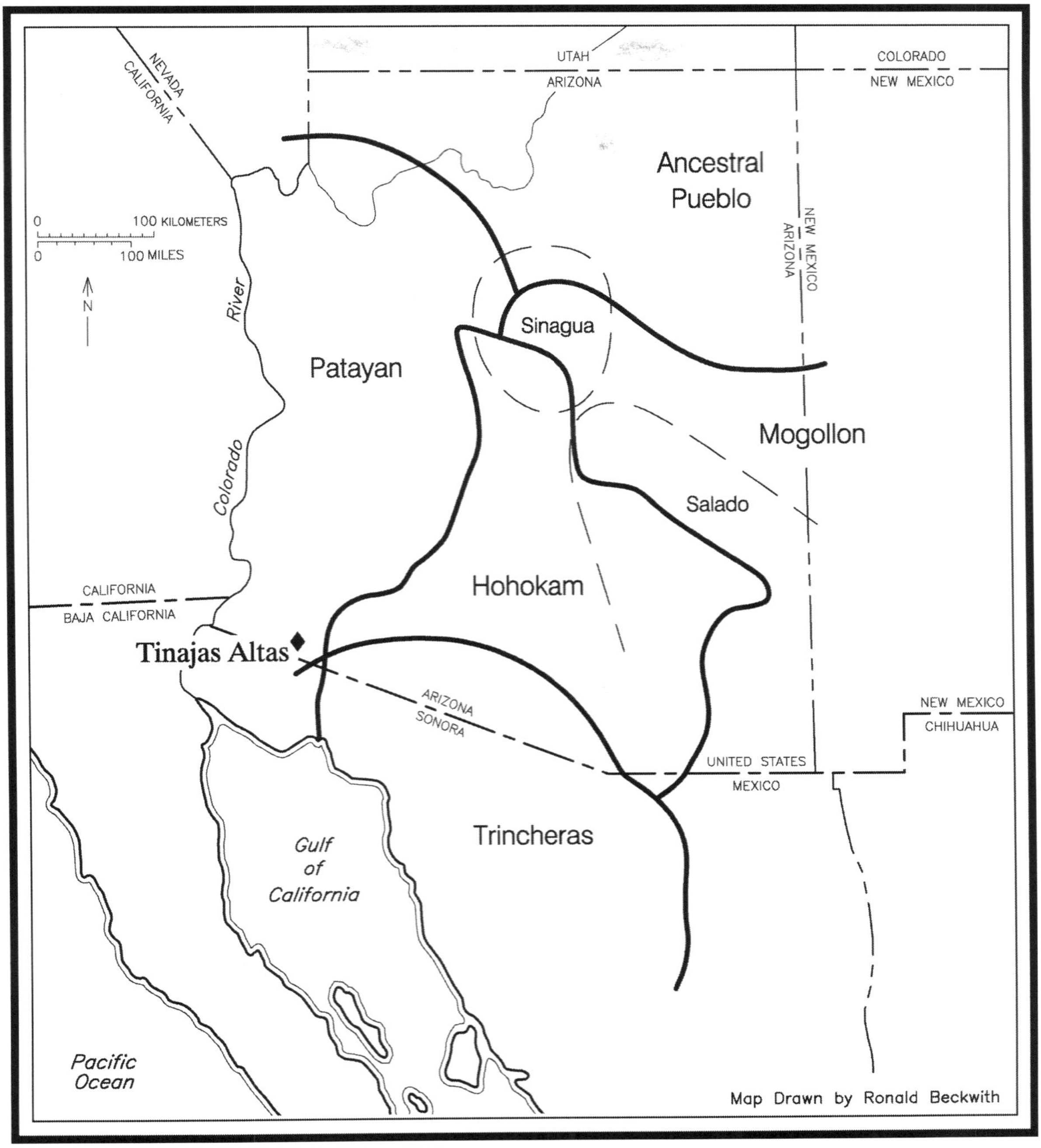

Major cultures of the Ceramic period in the Southwest, AD 800–1450. These generalized boundaries overlapped and changed over time. Some types of pottery were exchanged widely. (By Ronald J. Beckwith. After Reid and Whittlesey 1997, Walker and Bufkin 1986)

In historical times, the primary native inhabitants of southwestern Arizona were (and still are) Yuman-speaking groups. They have inhabited the valleys of the lower Colorado River, lower Gila River, and southeastern California since before the beginning of the early historic period. These Yuman speakers were agricultural people who relied on floodwaters from the Colorado and Gila rivers to provide moisture for their crops. The modern Yumas, or Quechan (Kw'tsan, Kwatch'an) as they prefer to be called, are the descendants of the people called ancestral Yuman or Patayan by archaeologists and historians. According to early records, there was considerable warfare among the Yuman peoples of the lower Colorado River during the early historic period, and on more than one occasion this conflict led groups to move to new locations along the Colorado or Gila rivers.

In this region, three groups of Yuman-speaking peoples were especially important during the early historic period: The Mohave occupied the northern part of the Colorado

A Hohokam vessel. (Photo courtesy of Arizona State Museum, University of Arizona, Tucson; image no. 2006-960-060)

A Trincheras vessel. (Photo courtesy of Arizona State Museum, University of Arizona, Tucson; image no. A-30474)

Valley, the Quechan lived near Yuma, and the Cocopah concentrated around the Colorado delta to the south. The Quechan and the Cocopah are the modern Native American groups living closest to Tinajas Altas.

The Quechan were observed in the Yuma area by Kino in 1701 and by Franciscan priest, Father Francisco Garcés in 1771–1776, and they have remained there to the present day. During the early historic period, their territory extended up the Gila River as far as Antelope Hill, just east of the modern town of Wellton (Spier 1933:23). A historical account provided by John Cremony (1868), an interpreter of Indian languages for the US Boundary Commission and later a captain in the California Volunteer Cavalry, speaks of Yumans coming to Antelope Hill to quarry milling implements in the winter of 1861–1862.

The presence of the Cocopah at the southern end of the delta was noted by Spanish explorer Juan de Oñate in 1605 and Garcés in 1771–1776. Cocopah living within the United States currently live west of Tinajas Altas on a small reservation near Somerton, Arizona, but some Cocopah also live in Mexico closer to the Colorado River delta. Spelling of the name varies; they are also known as the Cocopa, Cucapa, or Kwapa. The Cocopah were not recorded north of the present-day Arizona-Mexico border until 1873 (Williams 1974, 1975, 1983, 2004). They, too, may have visited Tinajas Altas.

The focus of Cocopah life was the river delta, so they rarely traveled *east* of the delta or as far as Tinajas Altas for food, but they did travel northeast to the Maricopa villages on the Gila River (Kelly 1977). The delta provided them with arable land, mesquite pods, fish, *nipa* (saltgrass grain), and small game. To exploit upland and mountain resources such as pine nuts, agave, paloverde, ironwood, and deer, Cocopah did travel *west* to the nearby Cocopah Mountains or farther afield to the mountains of Baja California, especially in summertime (Castetter and Bell 1951:202; Kelly 1977).

Also living west of the Colorado River are members of the Pai branch of the Yuman language family. Known as the Paipai, Akwa'ala, or Axua, they now live in the uplands of Baja California, but some of them formerly lived among the Cocopah. Prior to the 1830s, some of them apparently lived near Caborca, Sonora, but regularly visited the delta region (Williams 1975). They ventured back and forth between Caborca and the delta, but it is unknown whether they visited Tinajas Altas.

Still other Yuman speakers, the Kumeyaay and Kamia, lived to the west of the Colorado River. They ranged widely and engaged in long-distance travel and trade (Gifford 1931; Hedges 1975), so they may be among those who reached Tinajas Altas.

WESTERN APACHE

Surprisingly, the Apache may have visited the area at least a few times. The Western Apache are known to have raided the Quitobaquito, Yuma, and Ajo areas in the nineteenth

A Quechan (Yuma) home, about 1907. (Photo by Edward S. Curtis. Courtesy the Library of Congress; image no. LC-USZ62-ct02041)

Quechan woman, about 1880, carrying a Lower Colorado Buff ware jar on her head. (Arizona State Museum, University of Arizona, Tucson; image no. 8689)

century, and to have roasted mescal as far southwest as Seri encampments near Puerto Libertad (Seri elder Adolfo Burgos, personal communication to Nabhan 1997). James Barney (1943:19) recorded that Don Francisco Salazar, a miner and merchant from Hermosillo, told him of meeting Apaches on the Camino del Diablo around 1859: "At Tinajas Altas—in the old days—Apache were always skulking in the nearby hills and arroyos, ready to pounce." Hia C'ed O'odham elder Candelaria Orosco asserts that Western Apaches visited Tinajas Altas; she argues that it was Apache arrows that are part of the creation story of the tinajas, *not* arrows shot by "hero-gods" as Lumholtz (1912) refers to them.

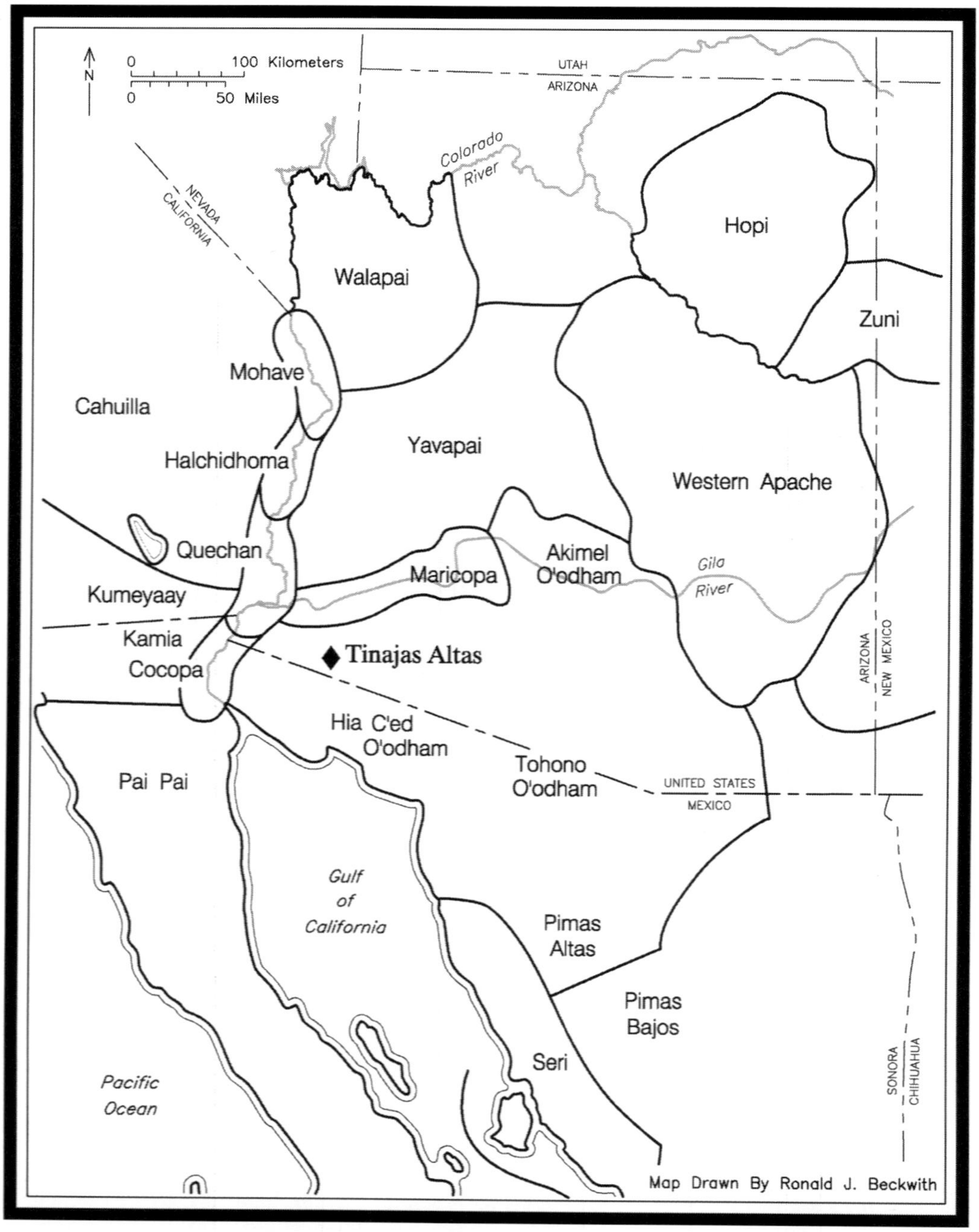

Key to eighteenth-century tribes of the Southwest. The lines suggest tribal boundaries, but the areas changed over time. In many cases tribes shared lands, or tribes moved seasonally or permanently to other areas. (Drawn by Ronald J. Beckwith, after Handbook of North American Indians, *vols. 8 and 9. Washington D.C.: Smithsonian Institution, and other sources)*

O'ODHAM

The O'odham , who are Piman speakers, are generally divided into three groups that reflect differences in lifeways, and to a lesser extent, language. The Tohono O'odham (Desert People) lived in the desert region between Ajo and Tucson, the Akimel O'odham (River People) lived along the Gila River, and the Hia C'ed O'odham (In-the-Sand People) lived in the arid western deserts. The Tohono O'odham and Akimel O'odham continue to live in independent reservations centered on their ancestral lands, although their current tribal properties are much reduced. Because they were much more mobile, the Hia C'ed O'odham were never awarded lands of their own by the US government and most now live in urban centers of southern Arizona.

The Tohono O'odham were adapted to living in the portions of the Sonoran Desert where agriculture could be

A Tohono O'odham village and homes, 1907. (Photo by Edward S. Curtis. Photo courtesy Library of Congress, Edward S. Curtis Collection, image no. LC-USZ62-ct02025)

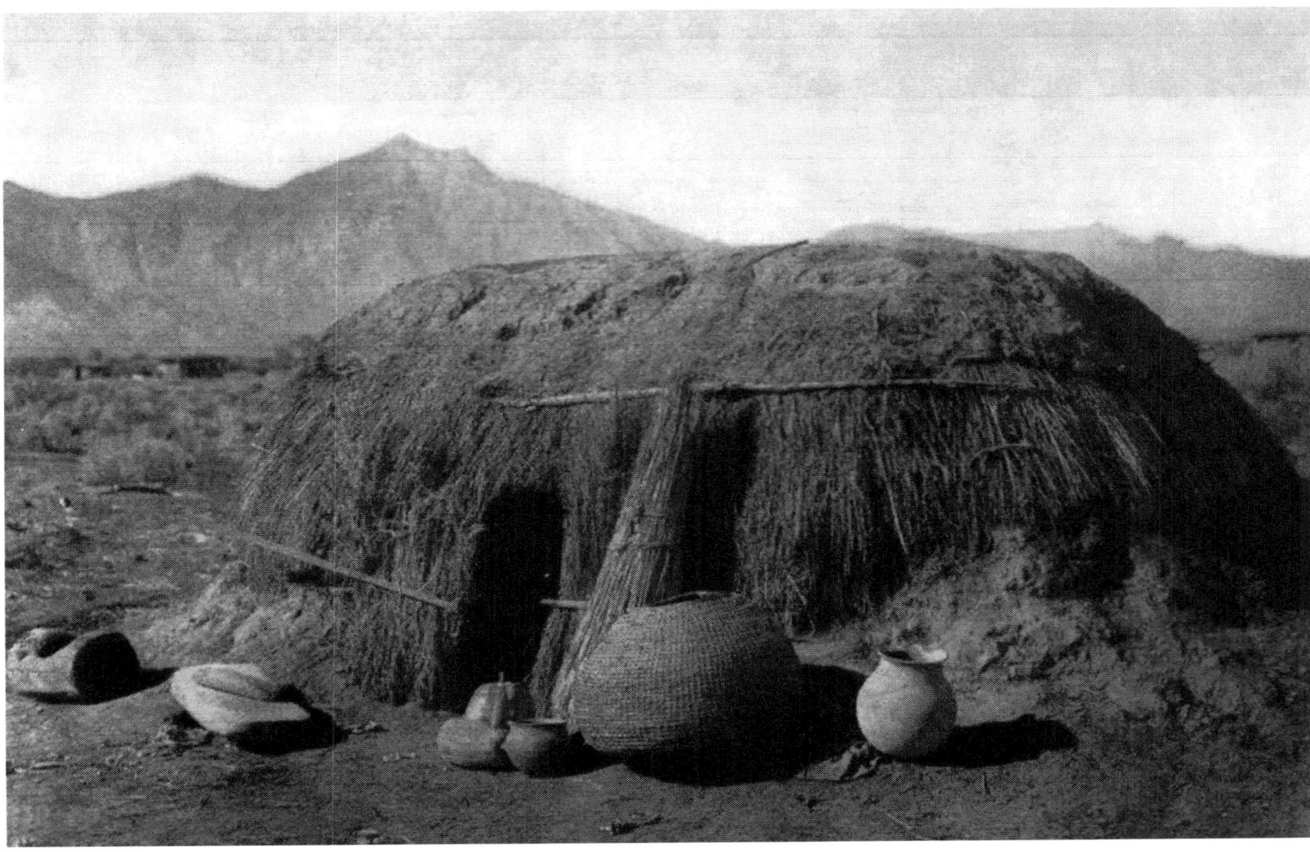

An Akimel O'odham home, called a ki. *(Photo by Edward S. Curtis, 1907. Photo courtesy Library of Congress, Edward S. Curtis Collection, image no. LC-USZ62-101255)*

Akimel O'odham granary, 1907. (Photo by Edward S. Curtis, about 1905. Photo courtesy Library of Congress, Edward S. Curtis Collection, image no. LC-USZ62-ct02009)

achieved only by precise timing in relation to the annual summer rains. As observed by Father Kino in the late 1600s, they were wide ranging and made long expeditions to the eastern shore of the Sea of Cortez where they collected salt. They traveled past modern-day Sonoyta, through the lava fields of the Pinacate Mountains, and across the dunes of the Gran Desierto.

The Akimel O'odham were adapted to living near readily available water. Their well-developed irrigation agriculture, along the Gila River, supported them with ease while providing surplus agricultural products for trade.

The Hia C'ed O'odham were the historic people of the Western Papaguería and were observed living in the greater Tinajas Altas region as recently as the late nineteenth century (Clotts 1915). These hardy desert dwellers have been known by a variety of names. They have been called the Sand Papago, the Areneños, the Pinacateños, the Soba, the S-o:bmakam ("Apache-like People"), the Hia Tatk Ku:mdam ("Sand Root Crushers"), and the O'otkol Ha-Kuadam ("[Iguana] Lizard Eaters").

How did these people sustain themselves in this difficult region? Among the favored food resources of the upland were the saguaro, mesquite and other leguminous trees, agave, cholla, and prickly pear. In the valleys annual plants and their seeds could be gathered. Among the most important parts of the primary food plants were the leaf bases and hearts of the agave, fruits or flower buds of various species of cacti, seeds of ironwood and foothill paloverde, and pods of western honey mesquite, berries of wolfberry and mistletoe, and seeds of mustards, peppergrass, and amaranth. Clearly, the Hia C'ed O'odham were primarily gatherers and hunters, but they did plant some crops at places like Quitobaquito, so agriculture seems to have been more important to them than previously thought (Eiler and Doyel 2008; Felger 2000, 2007; Nabhan et al. 1989).

The Hia C'ed O'odham were opportunistic and had two geographic centers. One was the shore of the Gulf of California, which usually provided abundant food resources, specifically fish and shellfish (Crosswhite 1981; Felger 2007). A second was the Sierra Pinacate, a varied and resource-rich region with a dozen or so major tinajas. Sandfood, a source of fluid and food, may be found in sandy areas at the head of the Gulf of California (Crosswhite 1981:53; Felger 2000; Nabhan 1985).

Alliances and Enemies

In the 1700s and 1800s many native groups lived within the larger region, and their alliances—and disagreements—were complicated. For example, the Hia C'ed O'odham had strong ties with the Gila River Pima and Maricopa (O'odham elder Fillman Bell, personal communication to Nabhan 1997), but there is also evidence of intermarriage with the Quechan, Mohave, Seri, and Tohono O'odham (O'odham elder Claude Passos, Quechan elder George Bryant, Seri linguist Mary Beck Moser, and O'odham elder Francis Manuel, personal communication to Nabhan 1997).

Although the Tohono O'odham were suspicious of the Seri of western Sonora, the Hia C'ed O'odham regularly traded with both the Seri and other O'odham groups. The Hia C'ed O'odham worked in Quechan, Cocopah, Mohave, Tohono O'odham, and Akimel O'odham fields as harvesters, according to many sources. The Hia C'ed O'odham traded their cactus products for cultivated crops (Hia C'ed O'odham elder Candelaria Orosco, personal communication to Nabhan 1997). During the nineteenth century, Tohono O'odham potters apparently supplied ceramics to the Hia C'ed O'odham.

"Areneños. Subtribe of the Papagos." Fishing on the coast of the Gulf of California. (By Arthur Schott, in Emory 1987:123a)

These relationships were confused further by the fact that tribal territories overlapped or were shared. For example, Antelope Hill, specifically, and the stretch of the Gila River north and east of the Gila Mountains, in general, were among the more dynamic areas in all of the Sonoran Desert, with at least four cultures—Quechan, Maricopa, Western Yavapai, and Hia C'ed O'odham—all using the floodplain at one point or other over the last four centuries (Schneider 1993a, 1993b). In addition, the Cocopah, Paipai, Kumeyaay, and Hia C'ed O'odham all used the resources of the Colorado River delta west and south of the Tinajas Altas Mountains (Kelly 1977; Williams 1983).

The Tohono O'odham also traveled widely and made traditional trips to salt flats near Puerto Peñasco. Because of the hazardous sand dunes south and southwest of the Tinajas Altas range, it is quite likely that for centuries, Native American traders utilized the Camino del Diablo and the gravelly passes just north of Tinajas Altas on their routes from northwestern Sonora to the lower Colorado and lower Gila Rivers. The building of the Southern Pacific railroad lines to Yuma in 1877, the paving of Mexico 2, and the completion of US Interstate 8 through the Gila Mountains in the mid-1960s removed the need for risky, arduous travel on desert pathways. In less than a century—a mere moment in archaeological time—the old ways had ended.

The Users of Tinajas Altas in Historical Times

During the 1700s and 1800s the region around Tinajas Altas probably was visited by native peoples coming from different directions and so it was in some sense a "free trade zone." For example, the historical account of Captain José Romero (presented in chapter 3) specifically refers to the Lechuguilla Valley as a Quechan trail to northwest Mexico.

*Flower of sandfood (*Pholisma sonorae*), whose underground tuber was an important food. (Photo courtesy Mark Dimmitt, Arizona-Sonora Desert Museum Collection)*

To the Quechan, from their farming villages on the Colorado River floodplain, the Tinajas Altas Mountains and the Gila Mountains to the north formed an eastern horizon of hope, sometimes piled high with storm clouds when the river valley itself had yet to receive any rains. Both the Quechan and the Cocopah also considered Tinajas Altas culturally significant, a place that connected them through trade with the O'odham communities to the east. In the early 1800s, when the Maricopa moved east to the middle Gila River, they frequently came to trade with the Cocopah by way of a trail with two springs along it, one of them called "lofty water," which ethnohistorians assume to be Tinajas Altas.

To the Hia C'ed O'odham, who used many of the tinajas in the vast riverless area east of the Colorado River

Table 2.1. Tribal Names

English names	Native peoples' names for themselves	Spanish names for the tribe
Apache	N*nee* [person] or *nde*e [man]; includes recognized subtribes and complex bands	Apache
Cocopah	Cocopah (US), Cucupá (Mex) (Kwapa)	Cocapa, Cucapá
Maricopa	Pee-Posh (Pi-Pa-Kwes, Pi-Pa-s, Pipaash) [The People]; now includes Halchidhoma, Halyikwamai, Kahwan, and Kavelchadom	Cocomaricopas, Opa
Mohave	Hàmakhá·v [People who live along the water]; makhá·v	Amacaua, Jamajá, and others
Paipai	Akwa'ala (Axua)	—
Tohono O'odham, Papago	Tohono O'odham [Desert People]	Papavi Otam, Pápago
Akimel O'odham, River Pima	Akimel O'odham [River People]	Pima, Gileños
Hia C'ed O'odham, Sand Papago	Hia C'ed O'odham [Sand People], (Oks Hihika)	Areneños
Seri	Comcáac, Komkaak [The People]	Seri, Heris
Yavapai	Yavapai; subgroups: Tolkepaya [Western People], Kewevkapaya [Southern People], Wipukpaya, Yavepe [Eastern People]	Yabipai, Cruzados, Tacabuy, Tehuas, and others
Yuma	Quechan (Kw'stan) [Those who descended]	Yuma

Sources: Based on Ortiz 1983 and other sources.

delta, Tinajas Altas was among the more reliable sources of water that served as stepping stones along their seasonal migrations from one ancient settlement to another. They typically named most every tinaja, seep, or spring within their reach, often referring to the landmarks above them, but they seldom named entire mountain ranges. In contrast, the Yuman-speaking peoples named entire ranges, and they associated particular peaks, mesas, or ranges with episodes from their migration epic. The tanks at Tinajas Altas are near the southernmost end of the long, linear range, and for the Yuman-speaking peoples living along the rivers west and north of the Tinajas Altas and Gila Mountains, these ranges were poetically named "cumulus clouds straddling the rocks like a fleecy blanket"; the Tinajas Altas range was known as "the tail end." We wonder if "the tail end" refers to the range being the southernmost section of a longer string of mountains extending northward to the Gila River, or if it indicates the extent of their tribal migration and a southeastern boundary of their homeland.

From conversations with Hia C'ed O'odham elders, we know that as recently as the late nineteenth century a group of Hia C'ed O'odham seasonally resided at Tinajas Altas; unfortunately, we don't know their names. We do, however, know the names of a few O'odham who visited there. Tomaso, who lived with Tom Childs Jr. and his wife near Ajo and who served as a guide for scientist-explorer Raphael Pumpelly and his family to Tinajas Altas in 1915, claimed to have been the only O'odham he knew alive to have been to the tanks, and he had not visited there previously for 40 years (Weight 1968). Victoriano Garcia said that he had gone to Dome by way of Tinajas Altas by burro, before "the white people ate all of them up" (Thomas 1991). Francisco Ortega, perhaps the uncle of Tom Childs Jr.'s wife, Marta Garcia, related that his grandfather had gone to Tinajas Altas [he called it Oo'oowac] perhaps in the late 1800s (Thomas 1953). William Merrill, whose father John was a contemporary of Tom Childs Jr. (1870–1951), reported that Tinajas Altas was "a tank way up in the big mountains . . . hard to get up there." In addition, he claimed that his uncle Augustine used to live out that way (Zepeda 1985). Finally, Francisco "Tulpo" Suni, the last resident of Cabeza Prieta National Wildlife Refuge who died in 1999, had been there (Margaret Ross personal communication to Nabhan 1997).

WHAT'S IN A NAME: TINAJAS ALTAS AND O'OVAK

The first transcription of the Hia C'ed O'odham name for the waterholes we know as Tinajas Altas was published by Lumholtz (1912). Tinajas Altas, according to Lumholtz, was

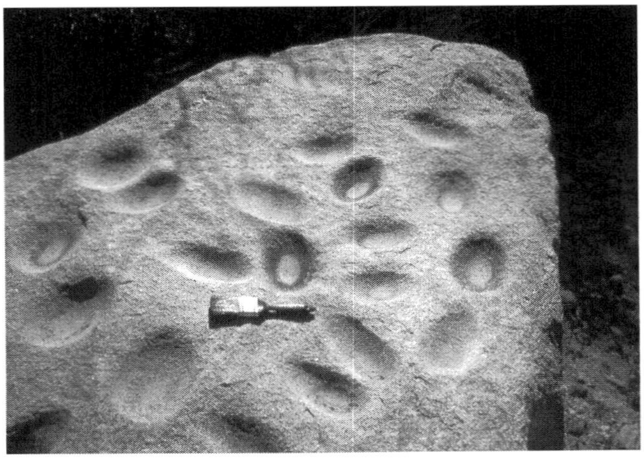

A boulder with grinding depressions. Native peoples used these in concert with hand-held rocks (known as manos) presumably to grind small seeds such as spring annuals and grasses. (Photo by Gayle H. Hartmann)

Mortar hole. In deeper holes, pods of mesquite, ironwood, and other plants were pounded with a stone pestle or hardwood stick. (Photo by Juel and Madeleine Rodack, March 1978. Courtesy Gayle H. Hartmann)

an important Sand Papago (Hia C'ed O'odham) camp called O'ovak, from *o'o* "arrows" and *vuak* "were shot." Lumholtz relates this name to a Hia C'ed O'odham legend, relating that two "hero-gods" shot their arrows from each side of the range; one bowman's arrows made it all the way across, but where the arrow of the other fell, pools came forth.

O'odham dictionaries shed further insight into the nuances of this place name (Alvarez and Hale 1970; Mathiot 1973; Saxton and Saxton 1969). The complexities of translation, coupled with differing human memories, illustrate how challenging it is to definitively record and translate Native American place-names. For example, Mathiot's 1973 dictionary includes *u'u* "arrow," and *wua* "shoot," as in *gatwua* "gunshot," and *vaak* "to get an object to sink in." *Vaak* is perhaps related to *ivua* "to throw an object down." The term *vuhak* can mean "to grow from the ground," "to shoot upward," or "to shoot outward." An O'odham elder, Mr. Moore, told ethnographer Robert Thomas (1953:n.p.) that the term *Oo'oowac* referred to "bullets," and Thomas added that he thought "originally the word [was] for arrowhead, but the meaning of bullet has been transferred to it." Thomas' notes also that the Akimel O'odham knew of Tinajas Altas, "On the way back to Ajo I got in a conversation with Mr. Moore. . . . He said that one time he was traveling from up on the Pima Reservation to Yuma with an old Pima, who pointed south and told him that way to the south was a place called Oo'oowac."

Although most of Thomas' O'odham consultants translated Oo'oowac as "arrowhead sunk," he recorded another variant. Victoriano Garcia (Thomas 1953:n.p.) "knew it took three days to get to Dome [from Ajo], but sometimes they would stay at a chupak [*ce:po*, "tinaja"] for a couple of days. He had been to Oo'oowac (arrowhead). He said it was the same place that they called Oo'oo'woopa, close to Yuma. He said it was right next to the Mexican border south of Dome."

Our best guess at the etymology of *Oo'oo'woopa* would be *'u'u* (arrow points), and *voop* (to pull out), or *voopod* (to lay object down) (Mathiot 1973). Candelaria Orosco, a Hia C'ed O'odham elder who lived to at least 104 years of age and tells of camping at Tinajas Altas, preferred the interpretation "arrows sunk" (Nabhan and Kahn 2000). Fillman Bell, a Hia C'ed O'odham, believed that O'ovak or Au'auvak might mean "place where the water is drawn forth." She said that

Fillman Bell, about 1970, with her daughter Mary. (Photo courtesy Oscar and Fillman Bell)

'u is to "draw forth or towards you," as in pulling of saguaro fruit toward you or the drawing of water. *Vak*, "moist place" or "place with tules," is the same element found in the place

Clouds over the Tinajas Altas Mountains.
(Photo by William K. Hartmann, October 1971)

names Quitovac and Quitobaquito. It seems clear, in general, then, that the O'odham name is associated with some legend involving arrows and perhaps an implication of "drawing forth" or "springing up" of water from the rocks.

There are just as many nuances with the Quechan names for the Gila-Tinajas Altas mountain ranges as there are with O'odham place-names for Tinajas Altas. The name given for these ranges by former tribal councilman George Bryant is Avik Amook To'or (Nabhan, field notes April 25, 1997). *Avik* can be translated as "rock," "peak" or "mountain range." The clouds that occasionally straddle the ridge tops are referred to by the term *amook*, which is also used to refer to "sheep," "wool fleeces," and "wool blankets." Quechan elder Lorey Cachora preferred the transcription Mokwintaorv for this place name (Nabhan, field notes April 26, 1997).

Quechan elder Preston Jefferson offered Ahpar as a place name for the northernmost Gila Mountains, and Yua Pad "tail end" for the more southerly mountains, presumably the Tinajas Altas range (Jacquie Gage Kahn, field notes January 13, 1997).

In his ethnohistory of the Maricopa, Spier (1933:44) offers considerable commentary with regard to the trade route from Sacate on the middle Gila to the Cocopah homeland on the Colorado River delta, passing by a "spring" called *xamilkwilau'* "lofty water":

> From the Mohawk Mountains the course was southward across a big mountain containing a spring [Baker Tanks?], called *kuhwi'tc*, thence southwestward across a sandy desert. In this desert was a second spring, called *xamilkwilau,'* "lofty water" [Tinajas Altas?], distant from the first by a day, a night, and the morning of the second day. From there they went westward to another spring [Laguna Prieta], *xacapa'kwinyako'Rac*, "little wells"; thence west to the Cocopah, who lay two days journey downstream from the Yuma.

In a footnote, Spier adds that Tinajas Altas is almost certain to have been the objective since it was the principal camping place in the region. From Tinajas Altas the Maricopa must have used the only practicable route to the Cocopa, that followed by Lumholtz . . . westward to Laguna Prieta in the desert, thence southwesterly to the vicinity of Colonia Lerdo (Spier 1933:44).

In spite of Spier's and other historians' assumptions, our fresh research, reported in chapters 1 and 3, suggests that Kino did not visit Tinajas Altas, but instead followed another, more easterly route to the Colorado River.

The Resources at Tinajas Altas: Water, Plants, and Animals

Water was the limiting factor for human occupancy in the western Sonoran Desert between the Gila River, Colorado River, and present-day US-Mexico border. In that entire region prehistoric waterholes and springs collectively offered less than 200,000 gallons of water when full, though above average rains could, of course, replenish the tinajas (Broyles 1996b). Under prehistoric or early historical subsistence conditions, this amount might be hardly enough to support a few hundred humans in addition to use by desert wildlife.

From many of the oral histories with O'odham elders, it is clear that any seasonal migrations from Dome, Quitobaquito, the Sierra Pinacate, or Ajo were carefully staged so that travelers could resupply themselves with water at Tinajas Altas or other tinajas along the way (Bell et al. 1980; Zepeda 1985). In discussing intertribal trails extending out from the territories of the Yuman-speaking tribes, ethnobiologists Castetter and Bell (1951:58) write that one trail took off to the east from Yuma, followed along the western edge of Sierra de las Tinajas Altas, crossed this range through Tinajas Altas Pass, proceeded to Agua Dulce, Sonoyta, Quitovac, and terminated in Caborca. For most of these intertribal routes,

Tohono O'odham woman carrying water in an olla, 1909. (Photo by Carl Lumholtz 1912:120)

they add, "the waterholes on these trails attest their former importance in the large amounts of potsherds, charred wood, etc., which strew their borders."

Stumps near Tinajas Altas show evidence of both recent and historical cutting of limbs of ironwood and mesquite, as well as entire trees, presumably for firewood, although both are rare at the site today. However, the hundreds of bedrock mortar grinding stones suggest that some legumes, probably ironwood and mesquite, were likely ground on a large scale. It is impossible to know how many grinding stones were in use simultaneously, and therefore what volume of legume foodstuffs were processed at the same time, but these trees served as sources of food, medicine, construction, utensils, and fuel (for cooking, heating, and ceremonial bighorn sheep bone cremations).

Native American consultants interviewed near the end of the twentieth century listed at least 52 plant species among those currently present at Tinajas Altas that were useful or significant (Nabhan and Kahn 2000; see Felger 2007, Felger, Rutman, et al. 2007; Nabhan et al. 1989). This inventory only scratches the surface of the rich lore associated with the plants of cultural importance growing at Tinajas Altas (see Felger 2007; Rea 1997, 1998). During a January 14, 1997, outing to Tinajas Altas, for example, Hia C'ed O'odham elder Frank Jim offered the following commentaries on uses of the plants he encountered at Tinajas Altas:

- *Echo* [*Krameria grayii*], ratany. Its roots were used [as dye] in making baskets; as a medicine, it is good for cuts and sores. You boil the root then wash the wound. Or if you get hurt inside, you boil it and drink the tea.
- *Nunuije:j* [*Ambrosia ambrosioides*], canyon ragweed. If your feet hurt, you pick the leaves, and boil them in a big kettle of water, then stick your feet in the water. This plant and greasewood [creosotebush, *Larrea divaricata* ssp. *tridentata*] were also used for "dry sweats" for arthritis.
- *Tobawes* or *tobos* [*Encelia farinosa*], brittlebush. The sap is used as a chewing gum.
- *Vipinoi* [*Cylindropuntia ramosissima*], diamond cholla. The fruits are gathered with sticks, then brushed with greasewood [creosotebush] to dethorn them before eating. They're much smaller than *babidaj* [*Carnegiea gigantea*] which was also eaten.
- *U:pad* [*Acacia greggii*], catclaw acacia. Catclaw wood is used to make *toka* (field hockey) sticks.
- *Jiawuli* [barrel cactus]. The fruit were not eaten but when they got real thirsty they cut the top off of the plant and with an ice pick they mashed the meat and let the juice rise up. They would chew it to save their lives. I had to do it twice myself like that or I would have died (see appendix 1).

We know that historically Tinajas Altas was strewn with horns from wild desert sheep (Hackenberg 1964), and that other documents refer to it as a bighorn hunting camp for the Hia C'ed O'odham, who ritually cremated the bones of bighorn and piled the horns high in heaps away from their camps or villages (Hayden and Friends 1993). But large mammals were not the only animals hunted or captured in the area. As mentioned in chapter 1, when Juan Mateo Manje traveled through the region in February 1699, he encountered, at Cabeza Prieta Tanks 16 miles east of Tinajas Altas, "thirty naked and poverty-stricken Indians who lived solely on roots, lizards [perhaps mistranslated as "locusts"] and other wild food." It seems probable that these were Hia C'ed O'odham. Although during Manje's brief stay these O'odham appeared to be impoverished, they knew how to utilize a number of reptiles and a great variety of mammals for food. Oral histories suggest that desert iguanas, chuckwallas, desert tortoises, rabbits, and rodents probably provided as much meat as the occasional bighorn sheep.

However, these animals should not be considered solely as meat sources. They also provided hides for sandals, fur for blankets, and bone for tools, among other uses. In addition, many of the animals listed were important in songs and stories, although some were taboo (Felger 2007; Rea 1998; Spier 1933).

THE EVIDENCE FOR NATIVE ACTIVITIES AT TINAJAS ALTAS

In recent decades, thousands of people have visited Tinajas Altas with the result that the site has been greatly compromised. Portable grinding stones, decorated pottery, and projectile points, all of which are always popular as collectors' items, have very likely been removed in large numbers. Though of little or no monetary value, these items are typical keepsakes taken by tourists both as remembrances of their trip, and perhaps as a way to touch their deeper human past. A historical cemetery thought to be the result primarily of forty-niner casualties and, at one time, holding approximately 65 graves or memorials has also completely disappeared. Although some of this damage came as a result of a late 1940s Bureau of Animal Industry camp at Tinajas Altas (monitoring free-ranging cattle for hoof-and-mouth disease), we conclude that, during the twentieth century, the increasing number of visitors, including recreational users and representatives of various government agencies, have altered or removed many cultural remains from the Tinajas Altas area.

Once items are removed and taken to shelves and closets in distant homes, they lose much of their informational value. Thus, we may never know the full story of Tinajas Altas. Nonetheless, a great deal was learned during the archaeological survey conducted between January and April 1998 and we present our major findings here (for more details see Hartmann and Thurtle 2001, and Hartmann et al. 2008).

Ceramics

As with most studies based on surface evidence, pottery is the most useful indicator of the temporal and cultural affiliation of the makers. Based on the observations of archaeologists Malcolm Rogers (1929) and Paul Ezell (1954, 1955) the ceramic evidence points overwhelmingly toward Patayan or Yuman use during the late prehistoric period and into historical times. Data gathered during our research support this conclusion. Of the ceramic fragments recorded by our survey, an estimated 86 percent of Native American ceramic vessels (107 out of an estimated minimum of 124 vessels, represented by 998 sherds) documented during the survey were Lower Colorado Buff Ware or related wares, that is, Patayan/Yuman. O'odham and probable O'odham wares were second in number, at nearly 10 percent of the total (minimum of 12 vessels). Only a few sherds of Hohokam Buff ware from central Arizona (three sherds representing three vessels, a little over 2 percent) and Trincheras Purple-on-brown from northern Sonora (two sherds representing two vessels, 1.5 percent), were identified. One-third of the potentially diagnostic sherds were found at the Tinajas Altas site itself, and the remainder from within a few miles (Gregonis 2000).

To create a sample of sufficient size for analysis, we combined the handful of sherds identified as to type or temporal phase from the Tinajas Altas survey with the collection from the site made by Rogers in 1929 (Waters 1982a). The results show that Lower Colorado Buff Ware comprises 91 percent of the sample. The fact that Rogers, in 1929, noted nearly 60 times as many diagnostic sherds as we did reaffirms that a great quantity of ceramics has been removed from the site by visitors.

What do the statistics tell us, then, about the people who dropped pottery at Tinajas Altas? Information from several sources indicates the region in which the most common type (Tumco Buff) was manufactured extended from Blythe, California, on the north to the Colorado River delta on the south (Waters 1982b). Evidence for its manufacture has also come from near the shoreline of ancient Lake Cahuilla in California (Schaefer 1994). It is the most abundant type found in Imperial County, California (Schaefer 1994), and is distributed as far eastward as the eastern end of Yuma County. The temporal range for this type is also quite broad, spanning nearly 700 years from the beginning of the Patayan II period (AD 1000–1500; Waters 1982b:563) into historical times (Schaefer 1994:84). The other buff wares also have a wide range of spatial distribution, having been manufactured on the west side of the ancient Lake Cahuilla shoreline (Pendleton 1984:289) and distributed from the west coast of California to Phoenix, Arizona.

Interestingly, the types that are primarily associated with Arizona, *combined,* constitute less than 22 percent of the total assemblage. Also, the southern extent of the distribution of wares of the three most represented types (Tumco, Colorado, and Black Mesa Buff) is the Sierra Pinacate, some 30 miles to the south and east of Tinajas Altas.

From the ceramic data, then, it would appear the Native Americans who utilized Tinajas Altas interacted throughout a wide sphere, and possibly had an affinity to peoples occupying or traveling to regions to the west and south. At the same time we should note that Lower Colorado Buff Ware is poorly understood, and we conducted no sourcing studies. Thus, the source of the Patayan pottery from Tinajas Altas remains uncertain. Although late Patayan wares are represented slightly more (63 percent) than those wares associated with the early Patayan period, no dominant period of use is indicated by the ceramic assemblage.

Rock Circles

Of the 30 sites recorded by the Tinajas Altas survey, 12 contained rock circles, for a total of 24 circles. These circles all consist of cleared areas bounded by local granite cobbles. They range from subrectangular to oval to circular, with two being portions of circles. Much scholarship, most notably by Hayden (1967, 1976), has been devoted to both the temporal and cultural affiliation of these features. Hayden believed the circles were of considerable antiquity, with both the rock-outlined and non-rock-outlined varieties dating from 20,000 to 7,000 BC. He based his argument on the presence of artifacts found in the vicinity of the features, many of which he believed were pre-Clovis in age. He asserted that the substantial amount of desert pavement, desert varnish, and caliche formation within and surrounding the features indicated their great age. However, Hayden's perspective on desert varnish dating and overall chronology appears inconsistent with more recent archaeological work (e.g., Beck et al. 1998; Watchman 2000). A lack of clearly diagnostic artifacts and direct dating methods has, so far, limited our ability to date rock circles reliably.

To better understand the age relationships among the circles we recorded, we examined the presence and absence of ceramics at sites containing circles, the size and shape of the circles, and the degree to which the associated rocks were partly buried or were standing on slight pedestals caused by erosion. About two-thirds (69 percent) of the circles without ceramics have

Rock circle. (Photo by Bill Broyles)

average diameters of less than 10 feet (3 m), whereas only a little over one-quarter (28 percent) of those with ceramics are this small. This distribution might indicate that older or preceramic circles were smaller, at least in our sample. Circles ranged in size from 2.8 square meters to 58.5 square meters. There is, however, no systematic difference in embedding, pedestaling, or qualitative appearance of the rocks to support the contention that they are older. Moreover, one site in the area offers evidence to the contrary. At that site pottery is associated with all circles; its circles range from the smallest to the largest and include all shapes. This suggests that circles of all types were constructed during the Ceramic period (AD 800–1850). Sites with single circles tended to lack ceramics, perhaps because they served as short-term camps where artifact use was minimal. We speculate that the Tinajas Altas site may have had many more circles of varying sizes before it was disturbed.

None of the four Archaic period projectile points was found in close association with a circle. Thus, our survey found no evidence to support Archaic period age (8,000 BC–AD 800) for any of the circles, though an Archaic age for some circles cannot be disproved. Our general hypothesis, then, is that differences in circle size and assemblage content may be more functional than temporal and that all circles recorded on our survey are likely of Ceramic-period age.

Rock Art

Rock art, primarily in the form of pictographs, is an abundant and impressive component of the Tinajas Altas site. Pictographs are designs painted on rock surfaces, while petroglyphs are pecked into rock surfaces. In a detailed recording of the Tinajas Altas rock art, Hartmann and Kolber (2000) compared it with rock art that has been identified as Western Archaic, Hohokam, Patayan/Yuman, Apache, or O'odham. Their conclusion was that the rock art at Tinajas Altas appears to be stylistically most similar to rock art located at other small sites in the arid region of southwestern Arizona, the Pinacate region of northwestern Sonora, and at the site of Indian Hill in Anza-Borrego State Park in southern California. At Indian Hill, the style is associated with the late prehistoric and historical Kumeyaay population—Yuman speakers and makers of Lower Colorado Buff Ware (Hedges and Hamann 1998). Although it cannot be proven whether the Tinajas Altas rock art was made by Yuman speakers (Quechan, Cocopah, Yavapai, Kamia/Kumeyaay), Piman speakers (Hia C'ed O'odham), or possibly their prehistoric ancestors, we favor the notion that at least some was created by Hia C'ed O'odham . Our conclusion is supported by two lines of evidence: documentation supporting a Hia C'ed O'odham camp at Tinajas Altas and rapid change of one rock art feature, suggesting that this feature, at least, is relatively recent. When we compared our 1998 photo to one taken by Malcolm Rogers in 1929, we could see that the design had deteriorated significantly (Hartmann and Kolber 2000).

Bedrock Grinding Features

Bedrock grinding features are very common at the Tinajas Altas site as well as at several sites nearby. As we have previously noted, their abundance and depth attest to the importance of food processing at these locales. Not surprisingly, a strong association exists between bedrock grinding features and watering holes in the southwestern Arizona desert, emphasizing the importance of locating subsistence activities near reliable sources of water in this harsh, arid region.

In our analysis of the grinding features, we divided them into two distinct types: 35 deep bedrock mortars were found only at the site of Tinajas Altas, and 622 shallow, oval depressions were found at Tinajas Altas as well as other smaller sites. The deep mortars are very similar in diameter (averaging 26 cm) but vary considerably in depth, ranging from 5.5 to 16 inches (14 to 41 cm), with an average of 12.6 inches (32 cm). The numerous oval depressions are all relatively uniform in size and shape. On average these

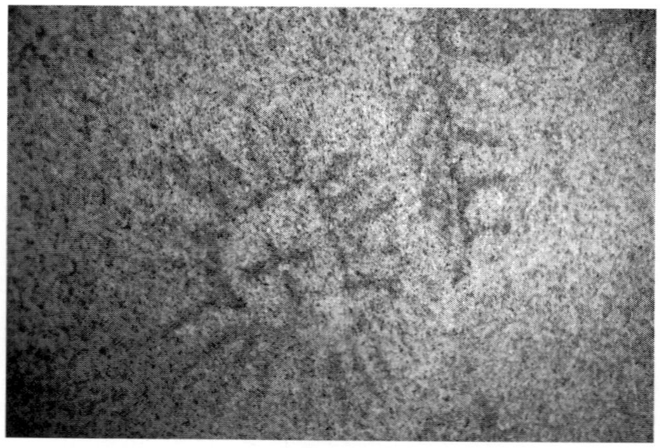

A pictograph near Tinajas Altas. (Photo by Gayle H. Hartmann)

A petroglyph near Tinajas Altas. (Photo by Bill Broyles)

Juanita Ahill, a Tohono O'odham, using a pestle to crush mesquite pods. Bedrock mortars at Tinajas Altas were undoubtedly used for the same purpose. Archaeologists have documented 35 of these bedrock mortars and 622 shallow grinding depressions at or near Tinajas Altas. (Photo by Helga Teiwes, 1975. Courtesy of Arizona State Museum, University of Arizona, Tucson, image no. 42540)

Shallow grinding depressions at Tinajas Altas. (Photo by Juel and Madeleine Rodack, 1967. Courtesy Gayle H. Hartmann)

depressions are very shallow, averaging a little over 0.8 inches (2 cm) deep; they average 6.7 inches (17.8 cm) in length and 3.8 inches (9.8 cm) in width.

What was processed in these mortars? As noted earlier, mesquite trees are present but not abundant in the Tinajas Altas area, but there are numerous references, such as Lumholtz (1912) and Nabhan et al. (1989), to the grinding and consumption of mesquite by historical inhabitants of this desert region. Thus, we assume that the Tinajas Altas mortars were used to process products of leguminous trees such as mesquite, ironwood, and paloverde.

Native processing of mesquite has been described by many authors (e.g., Bell and Castetter 1937; Castetter and Bell 1942) who relate how mesquite pods were generally pulverized in deep mortars using stone or wooden pestles. Archaeologist Julian Hayden (1969:156) describes the next step of this process by noting that "the roughly crushed material was taken to the villages for grinding into 'pechita' or flour, on individually owned metates." The seeds, which are considerably harder, are often discarded because they are so difficult to grind. When the seeds are used, they are ground separately either in a mortar or a metate (Felger 1977; Rea 1997).

It seems likely that the two kinds of grinding holes (deep mortars and shallow, oval depressions) were used in some combination to process various pods, beans, and seeds. Exactly how the two types of mortar holes were used is less clear. Perhaps the deep mortars were used for the initial crushing of pods and the smaller depressions were used with handstones for further grinding into flour. A study of bedrock mortars in the southern Sierra Nevada of California indicates that shallow mortars were used as "starter mortars" in the initial crushing of acorns (McCarthy et al. 1985:317, 336). Although acorns have certainly not been available in the vicinity of Tinajas Altas for thousands of years, the shallow depressions may have served to grind other seeds such as amaranth, goosefoot, or grass, all found in the pollen samples collected from two of the mortars.

SETTLEMENT, SUBSISTENCE, AND TRAVEL PATTERNS

What does the evidence from our survey, as well as earlier work, really tell us about the lives of the native peoples who used the desert resources at Tinajas Altas? What did their camps look like? What did they eat and how did they get their food ready to eat? How far did they travel and for what purposes?

Evidence for settlement must be inferred largely from the rock circles and associated artifact scatters. Only two sites recorded on the survey, Tinajas Altas itself and a smaller site in Tinajas Altas Pass, appear to have been anything other than very short-term campsites or resource-processing locales. The presence of bedrock grinding features at both those sites, as well as rock art at Tinajas Altas, strongly suggests that these two sites were extended-stay campsites, where food processing was a chief activity.

The only evidence for campfires and agave roasting pits, which would commonly be part of long-term campsites, consisted of bits of charcoal at a single small rock circle. It is worth noting, however, that some small, rock circles (not "sleeping circles") in the Growler Valley proved, after excavation, to be the surface manifestations of small roasting pits (Tucker et al. 2000b). In addition, during his 1929 visit to Tinajas Altas, Rogers (1929) recorded "two old hearths about 1 foot underground . . . with nothing but charcoal in them." Thus, buried hearths or other cooking features, present in the early twentieth century, may still be present or may have been destroyed or eroded in recent decades.

Interestingly, although agave is relatively common in the canyons of the Tinajas Altas Mountains and is known to have been used extensively by desert dwellers (Haury 1975; Doelle 1980), we found no certain evidence of its use. No definitive roasting pits were recorded. Two artifacts, a fragment of a tabular stone tool found at the Tinajas Altas site and a digging stick purported to be from Tinajas Altas, may be connected with agave processing; they are on display at the Pioneer Museum in Wellton, Arizona. Digging sticks have long been associated both with construction of roasting pits and separating the agave crown from its roots (Doelle 1980:90–91).

Further evidence relating to plant processing was obtained from pollen analysis of two mortars—a portable mortar and a bedrock mortar from the Tinajas Altas area (Holloway 2000). These mortars contained substantial plant pollen including pollen from goosefoot or amaranth, grass, sunflower, Mormon tea, and a small quantity of corn. Although some of these plant remains may be attributable to windblown pollen, some inferences can be drawn from these samples. First, people were likely carrying corn or corn pollen with them and were processing it or

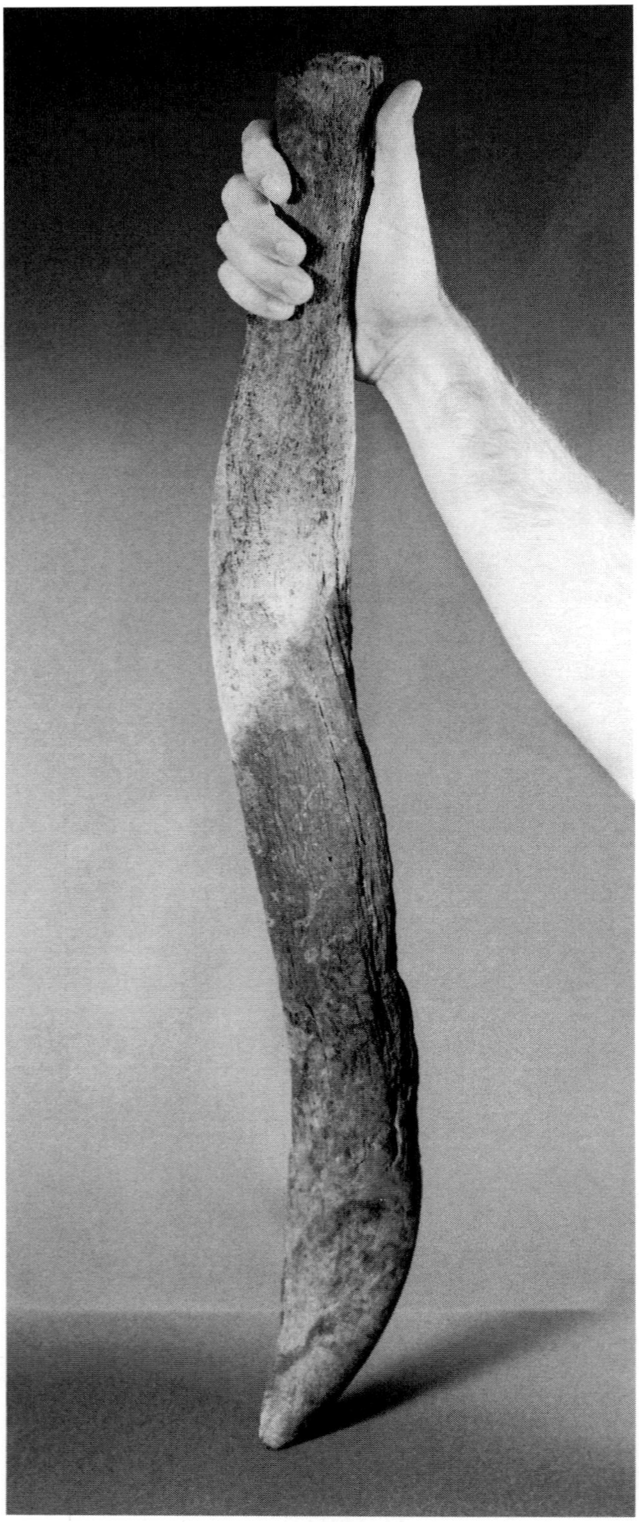

A wooden digging stick, probably made by a Yuman-speaker. These sticks were hewn out of hard woods such as mesquite or ironwood. This one was found by Thomas R. Van Devender south of Wellton, Arizona, in 1975. (Photo by Helga Teiwes. Courtesy Arizona State Museum, University of Arizona, Tucson, image no. 35849)

Segments of an ancient trail to Tinajas Altas. (Photo by Bill Broyles)

using it at the camp. It is also possible that people who carried the mortar came from an area where corn was grown. Second, plants belonging to the Asteraceae family (perhaps sunflowers), grass, Mormon tea, and some kind of chenopodium or amaranth, may have been processed in the mortar.

Recent research and ethnographic reports indicate the use of several of these plants by native people of this region. As part of our Tinajas Altas study, Nabhan and Kahn consulted with several individuals of Hia C'ed O'odham, Tohono O'odham, Cocopah, and Quechan ancestry. Plants recorded as having been "useful or significant among the O'odham (with most being used by other cultural communities as well)" include canyon ragweed, brittlebush, wooly plantain, Mormon tea, and wolfberry (Nabhan and Kahn 2000:64). At least one member of the sunflower family is recorded as having been used by the Yumans (Castetter and Bell 1951:187), perhaps desert sunflower (*Geraea canescens*), which is seasonally abundant in the area.

A tantalizing glimpse of Hia C'ed O'odham life in the early twentieth century notes their use of both desert plants and corn. Anthropologist Edward Gifford (1933:262) reported in his ethnography of the Cocopah that a small band of Papago [Hia C'ed O'odham] lived near Somerton. They "lived on wild foods, except vegetables from Yuma in exchange for gulf shells. [They] came in regalia, marched, and danced with rattles at Yuma house. [In return, they were] presented with maize and other foods which they put in sacks and took away on horseback." Hia C'ed O'odham reportedly traveled seasonally to Akimel O'odham communities along the Gila River to help harvest crops and were paid for their labor with a portion of the harvest.

We also must consider the possibility that hunting was another important activity that drew people to Tinajas Altas. In addition to the 10 projectile points collected, an arrowshaft was discovered in a small cave. Faunal remains collected by the survey consisted of a small quantity of calcined (burned) artiodactyl bones. Some of these were identified as bighorn sheep (Strand 2000). Historical records reflect the importance of bighorn sheep hunting around the tinajas of southwestern Arizona. For example, Anza noted in 1774 (referring to Cabeza Prieta Tanks) that "certain Papago [probably Hia C'ed O'odham] always camped during the height of the dry season to hunt mountain sheep which came there to drink" (Castetter and Bell 1942:67). In addition we know that the horns of desert bighorn were ceremonially stacked near trails leading to waterholes, and that bighorn bones were burned to "quiet the wind" (Hayden 1985; Hayden and Friends 1993; Lumholtz 1912).

Information on ethnographically documented animal life in the wider traditional homeland was also collected by Nabhan and Kahn (2000). The animals they cite as important for food include desert tortoise, desert iguana, chuckwalla, bighorn sheep, pronghorn, various species of woodrat and mice, jackrabbits and cottontails, and ground squirrels. With the exception of pronghorn, evidence in the form of tracks, scat, or visual sightings of all of these animals was noted in the Tinajas Altas area.

Foot trails are numerous in the project area; 39 segments or groups of closely related segments were recorded. The shortest extends 36 feet (11 m) and the longest, a north-south-trending trail, extends 2,427 feet (740 m) along the east base of the Tinajas Altas Mountains. Ninety-five percent of the trails contain associated artifacts within or immediately adjacent to the tread. Potsherds of Lower Colorado Buff Ware are the most common artifact type found in association with trails.

Trail systems throughout the larger region are well documented. For example, in the Sierra Pinacate, Hayden (1967) documents an extensive trail network that connects the various tinajas and campsites. Examples have been cited of Native American routes that likely extended through Tinajas Altas Pass (Gifford 1933) as well as along the route of El Camino del Diablo. An extensive network has also been documented in the desert regions of southern California. For example, a significant long-distance trail extends along the Colorado River from Pilot Knob near Yuma to locations in southern Nevada (Altschul and Ezzo 1994: fig. 5.6). A trail corridor also extends from Yuma to the California coast, with branches extending both to the north and the south from this main route (Davis 1961).

Amazingly, long trips across this region utilizing desert trails and taking advantage of water in natural catchments are still recalled by a few elderly Hia C'ed O'odham. The following excerpt refers to a trip in the early 1900s from the vicinity of Yuma to Quitobaquito on the present international border by way of the Camino del Diablo (Bell et al. 1980:55):

> We followed the old trail that drops into Cuk Do'ag (Pinacates) and came through to Sonoita, Mexico. Here we stayed because we are afraid of the [A]nglos. We drank rain water in the ce:cpo (rock tanks) all the way from Yuma to Cuk Do'ag.

Clearly, long-distance travel was an integral part of the broad-based subsistence strategy required for survival in this arid region. Historical sources suggest that journeys were conducted in a context of ever-shifting intra- and intertribal alliances (White 1974). Tinajas Altas was undoubtedly one of several nodes in this network.

O'odham or Yuman

Archaeological evidence of historical Tohono O'odham (formerly called Papago) use within the area surveyed is limited to a total of 28 sherds from a probable 17 vessels. Nearly all examples come from two sites, Tinajas Altas and Surveyors Canyon, about three miles south of Tinajas Altas. Only the latter site had substantial numbers of Tohono O'odham pottery at the time of our 1998 survey. All ceramics from Surveyors Canyon were identified as Papago Plain, which dates from the late 1600s to the 1960s (Gregonis 2000). Thus, the Tohono O'odham presence in the area may have extended over the last four centuries but was apparently minimal, or, just as likely, this pottery was traded to another group, perhaps the Hia C'ed O'odham.

As we have seen, the historical record, with its written documentation, does provide considerable evidence for the principal use of the area by the Hia C'ed O'odham. The most comprehensive discussion of the early historical Hia C'ed O'odham presence in the Tinajas Altas area is provided by Ezell (1955). He cites numerous accounts, beginning with Oñate in 1605, through Kino and Manje in the late 1690s, to Lumholtz and Gifford in the early 1900s, to support his contention that Piman-speaking people inhabited this region.

Tinajas de los Papagos, Sierra Pinacate. Padre Kino visited here several times and called the small O'odham village at these tanks Pitaqui. (Hornaday 1908:180)

Routes to the Colorado River and delta headed west across the open desert, leaving the mountains behind. Photo taken from the ridgeline of the Tinajas Altas range. (Photo by Gayle H. Hartmann, 1998)

Tohono O'odham women harvesting saguaro fruit. Note the clay olla, harvesting pole, and carrying baskets, called kiho. *(Photo by Edward S. Curtis, circa 1905. "Resting in the harvest field—Qahatika [Kohatk]." Photo courtesy Library of Congress, Edward S. Curtis Collection, image no. LC-USZ62-115801)*

Further evidence that Hia C'ed O'odham camped at Tinajas Altas comes from an interview recorded by H. V. Clotts in 1915 with merchant M. G. Levy. Levy asserted that around 1875 to 1885 "there were about 250 or 300 of these people [Hia C'ed O'odham] living around the tinajas of Cabeza Prieta, Tinajas Altas, Tinajas del Tule" and other locations in the riverless area of western Arizona and Sonora (Clotts 1915:76). Levy, who traded with the Hia C'ed O'odham while in residence at Quitobaquito, Ajo, and Sonoyta, said the indigenous families persisted in seasonally camping at the waterholes in the region until they lost access to these water resources through a combination of drought and enticement away from them by Anglo ranchers who wanted to use these water sources for their own livestock. He also noted that after moving "they became victims of some fatal stomach disease with the result that only a few families remain." He concluded, "There is not now, according to the best information a single Indian living between Quitovaquito and the Colorado River, except a few at Wellton or Adonde on the S.P.R.R." (Clotts 1915:76). As noted earlier, Lumholtz (1912:240) wrote that, at some time in the past, the Hia C'ed O'odham had an important camp at Tinajas Altas, which was called O'ovak, but no one was living there when McGee visited in 1900 and 1905, or Lumholtz in 1910, and most appeared to have moved to towns such as Dome, Ajo, and Gila Bend shortly after the coming of the railroad in 1877 (Broyles, Rankin and Felger 2007; Vivian 1965).

Ezell (1955:370) also discusses the strong connection that the Hia C'ed O'odham had with Yuman speakers. He endorses Garcés's contention made in 1771 that "these people resemble Yumas in their customs, although in reality they are Pimas." Ezell notes that there is considerable evidence that these Pima speakers had friendly contact with Yuman groups and comments that "it should not be surprising if Yuman cultural elements had been taken over by them" (Ezell 1955:371). He remarks that there is some evidence for Yuman groups using the area, or at least traveling through it. The Cocopah are mentioned specifically. He further mentions that there is good evidence that the Hia C'ed O'odham made their own pottery and that pottery from their camps resembles Palomas Buff.

Ezell's observations support some of the "big picture" conclusions we reached regarding cultural and temporal affiliation at Tinajas Altas. The earliest documented use consists of four projectile points of Archaic age, proving at least some use in the ancient, Archaic time period. Overwhelmingly, however, the prehistoric evidence points to dominant use in the Ceramic period (AD 800–1850) of the region by the Patayan, those enigmatic gatherers-hunters-farmers who occupied a vast area centering on southwest Arizona and southeast California. The substantial presence of Lower Colorado Buff Ware from the Patayan I, II, and III periods indicates usage *throughout* this period. The ceramics associated with rock circles also support Ceramic period use, as opposed to Archaic period or earlier use. The rock art also may suggest a prehistoric Patayan affiliation, but at least some of it appears to be relatively recent. This conclusion is based on its similarity to rock art in California that is associated with the Kumeyaay and is likely to be very late prehistoric or even historic in age (Hedges and Hamann 1998:37).

Although prehistoric and historical Yuman speakers may have been the dominant people to use Tinajas Altas, Piman-speaking O'odham clearly visited there as well. The archaeological evidence most supportive of a Hia C'ed O'odham presence is limited to calcined artiodactyl bones, which have been previously documented as a Hia C'ed O'odham trait (Hayden 1985). Abundant historical documentation exists of Hia C'ed use of the area. We must not, however, discount that if Ezell (1955) is correct and the Hia C'ed O'odham had Yuman traits and made Lower Colorado Buff Ware, then much of the material culture may be theirs as well. Rock art that appears to be Patayan/Yuman could actually be Hia C'ed O'odham. In support of a limited Tohono O'odham presence is a small quantity of ceramics dating from the late seventeenth century to the mid-twentieth century. As we mentioned earlier, however, this pottery could have been traded to others. Still, as far as we know, there is no other documentation of Tohono O'odham pottery this far west.

Three broad scenarios sketch how Tinajas Altas might have been perceived by the users themselves (Hartmann and

Thurtle 2001:513). First, a "wayfarer" scenario treats Tinajas Altas as a water resource that helped travelers get from one populated area to another. The documented Hia C'ed O'odham trip mentioned earlier would be an example of this scenario as would the annual trips reported by Lumholtz (1912) during which Hia C'ed O'odham traveled to Yuma to "barter with the Indians there, and maize, tepari beans, and squashes were exchanged for baskets and sea-shells." Second, a "desert resource" scenario treats Tinajas Altas as a location with certain resources, such as spring annuals or bighorn sheep, that would be exploited on a short-term basis by short-term occupants from more settled communities (Doelle 1980). This scenario is documented by Castetter and Bell (1942:67), who refer to "certain Papago" hunting mountain sheep at water catchments "during the height of the dry season." Third, a "seasonal camp" scenario suggests longer-term use on a seasonal basis by more mobile groups, which fits the documented Hia C'ed O'odham use of other tinajas in the region (see Doelle 1980).

The harshness and remoteness of Tinajas Altas has made the first scenario seem natural to most modern visitors, but it was clearly not the primary one because of the extensive evidence for rock features and food processing (though some evidence for camping and food processing might be expected even as a result of very short-term stopovers). Most of the visitation is likely closer to scenarios 2 and 3, which represent a continuum. Scenario 3 well represents Hia C'ed O'odham use as it was perceived by late-nineteenth-century Anglo travelers, but the early Anglos may have misperceived the Hia C'ed O'odham as spending much of their time wandering from camp to camp. In reality, they had semi-permanent camps such as Quitovac, Suvuk, and Quitobaquito, where they practiced some farming (Doelle 1980; Eiler and Doyel 2008; Felger 2007; Lumholtz 1912).

Over millennia, visits to Tinajas Altas were most likely made by both Yuman and Piman speakers and their predecessors, and focused on plant gathering and processing, some hunting, and the creation of rock art. The visits could have varied from short stops primarily for water to longer stays for resource acquisition. Tinajas Altas was clearly an important water station on a trail network connecting the Gulf of California and the Sierra Pinacate with points farther north and east, such as Quitobaquito, as well as communities along the Gila and Colorado rivers. Resource acquisition was probably closely tied to the time of year, with plant gathering more likely in the early spring and hunting more likely during the summer, when game animals may frequent waterholes.

For much of its known human history, then, Tinajas Altas was a place where cultures overlapped, where goods or songs were traded while individuals were en route to other settlements. Its waters seem to have been a resource for visitors from various communities. The hundreds of bedrock mortars suggest its importance for subsistence, and the rock art hints at a greater spiritual significance.

A Tohono O'odham woman carrying a kiho. (Photo by Edward S. Curtis, circa 1905. Photo courtesy Library of Congress, Edward S. Curtis Collection, image no. LC-USZ62-ct02019)

At Tinajas Altas, prehistory blends into history, and cultural affiliations are blurred. The significance of Tinajas Altas, however, is clear: perennial water in this region of heat and aridity is of profound importance and has attracted humans for millennia.

Padre Eusebio Kino's 1701 map. Map available in Burrus 1965: plates X and XI.

3 First Europeans to Forty-Niners

1540–1854

Thomas E. Sheridan and Bill Broyles

Beginning in 1540, Spanish soldiers and European missionaries persuaded Indian guides to lead them across the scorched landscape of the western Arizona-Sonora border country. These travelers had no interest in the region itself; that came later, when Spanish and Mexican prospectors searched for alluvial deposits of gold in desert washes. Instead, the Spanish sojourners followed the ancient Native American trails in order to proselytize the Yuman-speaking Indians living along the Colorado River and to establish land routes between Sonora and both Lower (Baja) and Upper (Alta) California. The Colorado River and California were powerful destinations long before the Gold Rush of 1849 drew thousands of argonauts from North and South America.

Following portions of a trail system that came to be called the Camino del Diablo, the route roughly follows the Arizona-Sonora border west from Sonoyta to Yuma, and it became an alternative to the Gila Trail, the route following the Gila River. It was a way to shave distance and time, and perhaps to reduce the danger of Apache attack by riding from one tank or well to another across a landscape that had been explored and utilized by O'odham, Quechan, Pee-Posh (Maricopa), Cocopah, and other Native American sojourners for centuries, even millennia, before the Europeans arrived (see Broyles, Rankin, and Felger 2007). The starting point was Sonoyta, a crossroads village on the Sonoyta River where a perennial stream allowed residents to farm and raise livestock.

The Mystery of Melchior Díaz

The first non-Indian to follow those Native American trails may have been Melchior Díaz, one of Francisco Vásquez de Coronado's captains (see Flint and Flint 2005; Winship 1990). After conquering Cíbola, a Zuni community in west-central New Mexico, Coronado dispatched Díaz in the fall of 1540 to rendezvous with Spanish sea captain Hernando de Alarcón, who had sailed up the Gulf of California and ascended the Colorado River in a vain attempt to supply Coronado's expedition. Spanish knowledge of the Southwest was rudimentary at best in the sixteenth century; both Coronado and Alarcón thought Cíbola in western New Mexico and the Gulf of California were much closer together than they really were.

When Alarcón reached the vicinity of modern Yuma and found no Spaniards, he left letters buried beneath a tree marked with a cross. Meanwhile, Díaz had headed south from Cíbola to Corazones, a settlement in the modern Mexican state of Sonora probably on the Río Sonora. Then he and 25 soldiers traveled in a northwesterly direction into the arid heart of the Sonoran Desert, traversing northwestern Sonora on his way to search for Alarcón on the Colorado River.

His route is uncertain. Walt Wheelock (1988a, 1988b, 1989) suggests that Díaz's party followed Seri trails along the Sonoran coast. Herbert Eugene Bolton (1949) believed that Díaz, who was driving a flock of sheep, followed the Río Magdalena to Caborca, Quitovac, and Sonoyta, crossing what is now the international border to travel along the Camino del Diablo and then directly west to the river, a route that would have taken him across the low dunes and past Laguna Prieta. A more likely route is described by geographer Ronald Ives (1960a, 1989) who argued that Díaz headed north up the Lechuguilla Valley all the way to the Gila River, because Ives thought the route across the dunes was too difficult. The dune route was, however, later used effectively by cowboys and rustlers driving large herds of horses and cattle, so Ives' objection may not stand; besides, as we will see, that route was used regularly by the Cocopah. Some have discussed and discarded Wheelock's route along the coast (W. Hartmann 1989; Ives 1936; Sauer 1937), but Mexican historian Manuel Rojas (1986) suggests that this coastal route, which he calls *La Vereda Mesteña* (The Mustang Trail), was used in the mid-nineteenth century.

Whatever route Díaz took to get around the Pinacates, it is clear that he visited the Quechan on the Colorado River. After crossing the Colorado, Díaz and his men journeyed southwest to Volcano Lake below Cerro Prieto just south of the modern city of Mexicali. Somewhere in the area, Díaz threw a lance at a dog bothering the sheep. The lance stuck in the ground and Díaz ran into it on his horse, the butt end piercing his bladder. For 20 days, Díaz's soldiers carried their mortally wounded commander back toward Corazones so he could be confessed by a priest, but he died somewhere along the trail and was buried. Ives concludes that the litter-bearers retraced their route down the Lechuguilla Valley and that Díaz died somewhere between Sonoyta and Caborca. If Bolton and Ives are correct, the Díaz expedition may have been the first Europeans to pass near, if not actually drink from, Tinajas Altas.

Because no diary was kept of the expedition, however,

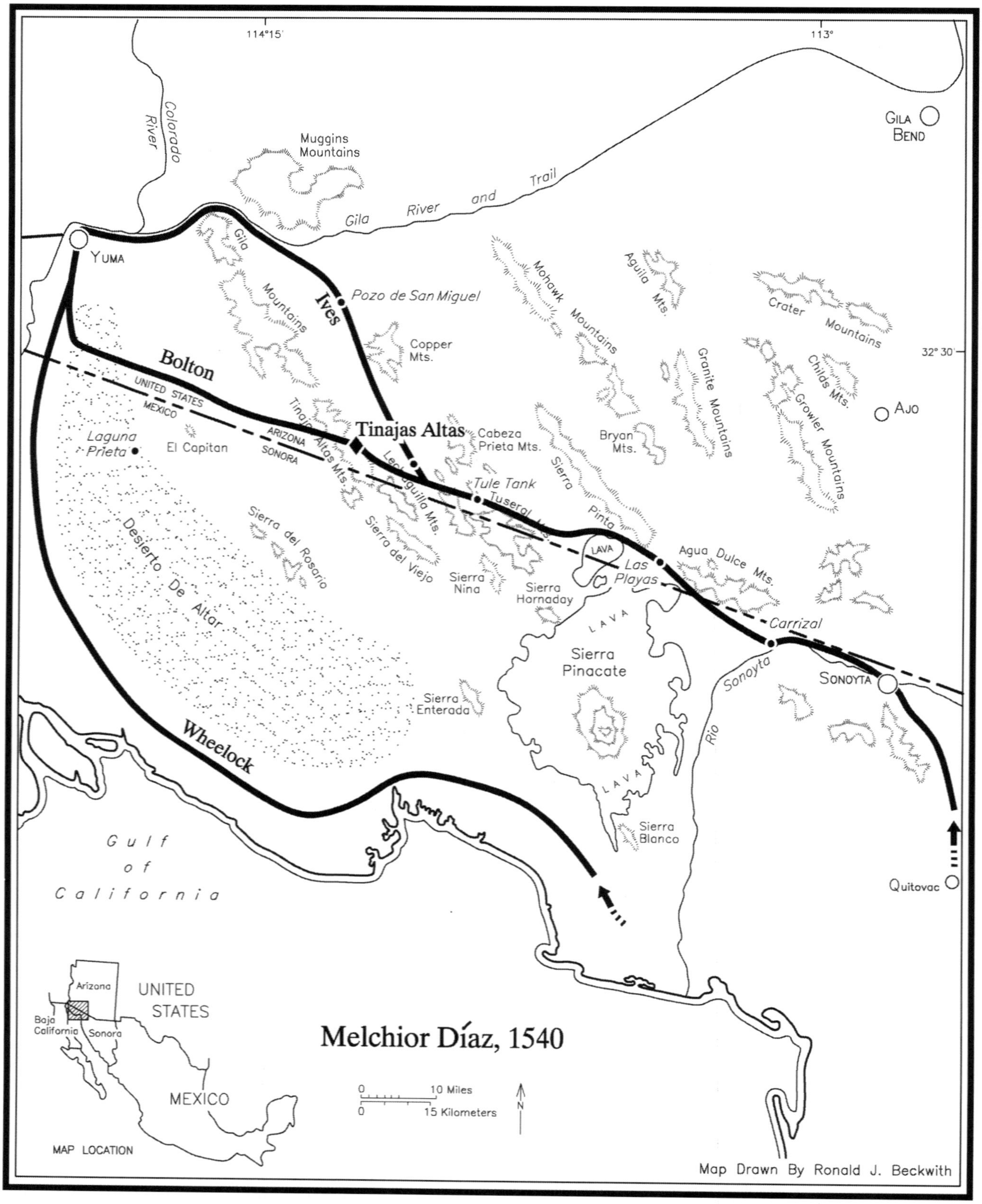

Possible routes taken by Melchior Díaz, 1540. (By Ronald J. Beckwith)

we have only a vague outline of the journey. Not until a half-century later do we encounter first-person, day-by-day accounts of travel across the region. Juan de Oñate led an expedition from Santa Fe, New Mexico, across northern Arizona and down the Colorado River to the Gulf of California in 1604–1605. His report supplies interesting information about the Native American groups living along the Colorado River as far south as the Gulf, but he and his troops apparently did not stray far from the Colorado itself (Bolton 1919a).

The Jesuit Explorer Eusebio Francisco Kino

Jesuit explorer Eusebio Francisco Kino and his companions were the first non-Indians to leave detailed records of the dry basin-and-range country the Camino del Diablo traverses and of the waterholes that sustained them en route. Kino, the Jesuit missionary who evangelized the Pimería Alta, made a series of expeditions from the Altar Valley of northwestern Sonora to the Colorado River between 1699 and 1702. Kino did so for a variety of reasons—to prove that California was not an island but a peninsula (Bolton 1919b, 1936; see Brusca 2007:423, n. 2); to establish a land route between Sonora and the missions of Baja California; and to advance the Spanish frontier to the Gila and Colorado rivers, where, in the words of Kino's frequent traveling companion, Captain Juan Mateo Manje,

> a small city can be founded which will serve as a haven, defense, and refuge for converting the other nations, namely the Hopis, Apaches, and the peoples of the north, northwest, and west, as far as the South Sea (Pacific Ocean), and a refuge for the navigators of the Philippines, with well-founded hopes of minerals. (Burrus 1971:150)

Kino, who regarded the village of Sonoyta as the hub of his work in the Papaguería, stated "it would be a halfway station on the road to the Colorado" (Bolton 1936:442). In Kino's words, "This post and ranchería of San Marcelo is the best there is on this coast," with fertile land, irrigation ditches, a perennial stream, good pasture, "and everything necessary for a good settlement" (Bolton 1936:442). Moreover, it was centrally located, being roughly 50 leagues to Yuma, 50 to Tucson, 50 to Piman settlements on the Gila River, and 50 to Caborca. It's no wonder that he invested so much time and energy in the Sonoyta region.

Kino pursued his dreams of northward expansion with unparalleled tenacity. His explorations of one of the most arid regions in the world followed ancient Indian trails converging on the Colorado River, with the contributions of his O'odham and Yuman guides often being overlooked by Kino's admirers. Nonetheless, Kino's accomplishments—as explorer, cartographer, diplomat, and promoter—continue to justifiably amaze and excite generations of scholars and desert rats alike.

Statue of Padre Eusebio Kino on horseback, Tucson, Arizona. (Photo by Bill Broyles)

In chapter 1 we traveled along with Padre Kino on his 1699 and 1700 journeys. He had already established his reputation as a tireless explorer. On that 1700 trip to the Colorado, he returned to his mission headquarters at Dolores, Sonora, "having traveled in going and returning three hundred and eighty-four leagues in twenty-six days, without our pack animals becoming weary and without any mishap" (Kino, *Favores Celestiales*, Parte II, Lib I, Caps. 10–11, quoted in Bolton 1936:443). That is roughly 770 miles, or 30 miles a day.

His next foray along the Camino came a year later, when he attempted to establish a route by land from the Pimería Alta to Baja California. The expedition lasted from November 3 to December 8, 1701. No other Europeans or Euroamericans accompanied Kino on his trek, so his account is the only one available for the journey. Kino wrote,

> On the thirteenth [November 1701] . . . , we set out for Carrizal, where we found many people, with a little new house which they [the Indians] had prepared for me, and with new messages from the Yumas and Quiquimas. . . .
>
> On the fourteenth we set out for the tank of La Luna, twenty leagues' journey. We arrived by moonlight half an hour after night fall, and although this tank is between rocks so sharp that the horses could not ascend to drink, we saw where on the return we could open a way for them, and afterwards we did open it.
>
> On the fifteenth we set out for the tank of Agua Escondida. We took a siesta halfway, where there was good pasturage, and at two in the afternoon, after a ten league's journey, we arrived at the tank, which we found somewhat scant. We therefore determined to set out as quickly as possible to arrive on the following day so much the earlier at the good tank of La Tinaja, and having set out at nightfall, a good shower came upon us which, however, by the darkness made us lose the way; notwithstanding, we found it afterwards when the storm cleared, and, breaking our sleep a little, we got up very early in the morning, and, on the sixteenth, traveling five leagues more, we said mass at a good stopping-place with water and pasturage. We breakfasted, and passing by the tank of La Tinaja, after a fifteen leagues' journey we arrived early at the Rio Grande de Hyla [Gila], and at its first ranchería of San Pedro. (Bolton 1919b, 1:310–311)

After visiting the River Yumans along the Gila and Colorado rivers and crossing the Colorado on a raft, Kino returned to the Pimería Alta via the same route. In his words:

> On the twenty-fourth [November 1701] we arrived at San Pedro de las Yumas; on the twenty-fifth at Agua Escondida; on the twenty-sixth at midday at the tank of La Luna, where all the afternoon we opened an impassable road through very sharp stones and rocks where animals had never been able to ascend to drink water, so that today all ascended to drink. And afterwards, an hour before sunset and in the night, we traveled five leagues more, so that the horses might have good pasturage. (Bolton 1919b, 1:320)

This passage is perplexing, for it is difficult to see how livestock could clamber up the steep granite apron to Heart Tank. There is a low ledge at Heart Tank that diagonally rises northward across the apron from its base and then zigs back to the tinaja. It is the easiest route for hikers, but it would be difficult access for even a Grand Canyon mule. In other respects the location described fits Heart Tank. However, the "sharp stones and rocks" comment much more closely matches Cabeza Prieta Tanks, whose approach winds through about a mile of irregular basalt cobbles with several hundred yards of talus and loose hillside stones. Today a cleared foot trail passes through those rough sections, and petroglyphs line both sides of the trail as one nears the tinajas. The difficult footing at Heart Tank involves large canyon-bottom boulders and a steep, slick granite slope, not sharp stones. We won't pretend to answer this riddle, except to say that perhaps Kino wrote his notes sometime later and flipped the events. We do point out, however, that subsequent travelers did not report or use Kino's trail at Heart Tank.

From Heart Tank, Kino and his troop marched to Sonoyta without incident.

> On the twenty-seventh, having risen very early, after thirteen leagues of good road we arrived before midday in time to say mass, eat, and take a siesta at Carrizal, and in the afternoon, after eight leagues more, at San Marcelo del Sonoydag. (Bolton 1919b, 1:320–321)

In February 1702, Kino, Father Manuel Gonzalez, and 12 Indian servants launched another expedition that took them along the Camino. The expedition left Dolores on February 5 with 80 horses and mules. They reached San Marcelo de Sonoyta on the afternoon of February 15. After resting for three days, they set out from Sonoyta. In Kino's words:

> On the nineteenth [February 1702], Sexagesima Sunday, we went from San Marcelo to Carrizal.
>
> On the twentieth, the friendly natives having given us three infants and a sick adult to baptize, we set out, and after about fifteen leagues' journey we arrived near the tank of La Luna.
>
> On the twenty-first, passing near this tank, and letting the horses drink, we arrived at the plains and the pastures near Agua Escondida.
>
> On the twenty-second we set out directly for the tank of La Tinaja, and reached it after twelve leagues of very level and straight road.
>
> On the twenty-third and twenty-fourth we were detained by a very great cloud-burst, which

La Tinaja (Baker Tanks, Los Pozos). (Photo by Bill Broyles)

> much increased the stream of that tank, and we saw that it passed by some very sightly rocks which appeared to be very fine tanks made by hand and with very great art, and it seemed to the father rector [Gonzalez] that this tank with much reason should be named Aguaje de los Alquives. [Alquives is a misspelling or miscopying of "aljibes" meaning "cisterns."] (Bolton 1919b, 1:338)

The "sightly rocks" cisterns may be the freshly filled intermittent tinajas slightly upstream from the main pool at Baker Tanks. Those pools are in colorful conglomerate rock and may look as if chiseled by human hands. Or, the group may have followed the flooded arroyo and found pools downstream between Baker Tanks and the Gila River; the fact that on the next day they went four leagues and reached "the opening" where they could see California Alta implies that on the 25th they traveled north from Baker Tanks to within about eight miles of the Gila River's bend through the northern tip of the Gila Mountains near modern Dome. The other possible location for La Tinaja would be Dripping Springs in the Gila Mountains, but the distances do not fit Dripping Springs, nor does the description of artful cisterns. And, traveling from Baker Tanks to Dripping Springs and then the Gila River would lengthen the walk by 16.9 miles (27 km) for no apparent gain, since they had just been blessed by a cloud-burst.

> On the twenty-sixth we set out for the Rio Colorado. After going four leagues we arrived at the opening where it may be said California Alta begins, because its meridian passes through the middle of the head of the Sea of California. As a rain-storm again threatened, we halted near there, and the natives brought us, from various parts, quantities of very fine fresh fish and other gifts, today as well as the following day, when we were detained by the rain.
>
> On the twenty-eighth we set out for San Dionisio and the confluence of the rivers [near modern Yuma]. (Bolton 1919b, 1:338–39)

Kino, Gonzalez, and their companions returned via the same route, but Kino has little to say about that trek, for by then it was routine. "On the seventeenth [March 1702],

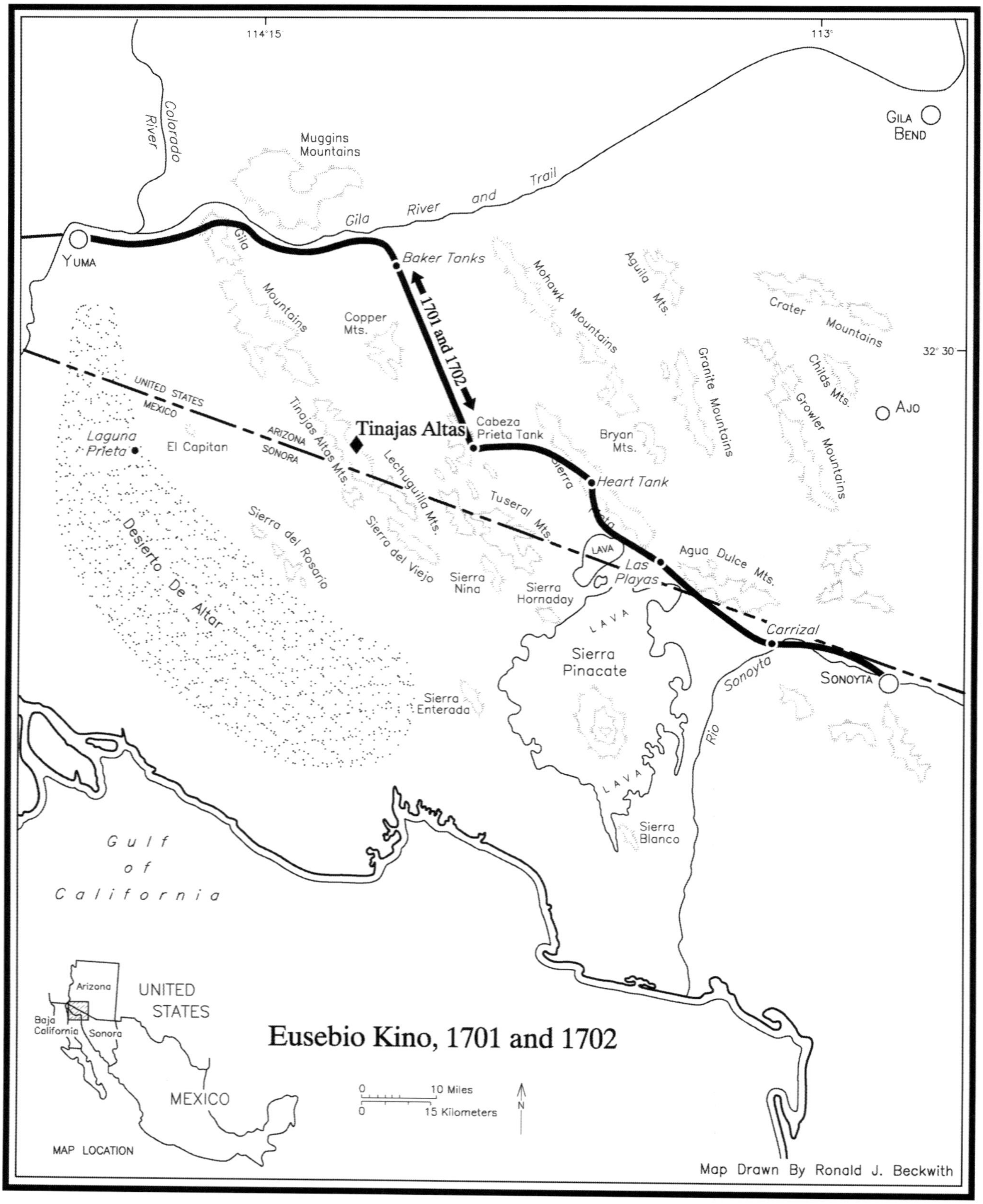

Map of Eusebio Kino's routes, 1701 and 1702. (By Ronald J. Beckwith)

we arrived at San Pablo of the Yumas; on the eighteenth at the Alquives; on the nineteenth at the plains of El Agua Escondida; on the twentieth at the tank of La Luna; and on the twenty-first at Carrizal," he wrote (Bolton 1919b, 1:345). By then, the Camino had apparently become a routine commute for the indefatigable missionary as he built his case that California was not an island. This was the last time Kino traversed the Camino to the Gila and Colorado. His last great expedition in October-November 1706 used El Carrizal as a jumping-off point for the Pinacate, which Kino called Sierra Santa Clara.

It is fascinating to share Kino and Manje's diaries, now more than 300 years old, and realize that we can retrace their steps in a landscape that looks much the same today as it did then. Whether or not either of these great pioneers ever drank the waters of Tinajas Altas itself, we still don't know for certain, but it doesn't seem likely. We do know they traveled back and forth through this region relying on Native American guides and on the water from the various tinajas. In addition, after careful analysis of their work, we may have a better idea of the route they used most frequently—from the Sonoyta River to Aguaje de la Luna (Heart Tank in the Sierra Pintas) to Agua Escondida (Cabeza Prieta Tanks in the Cabeza Prieta Mountains), to La Tinaja (Baker Tanks south of Baker Peaks) and on to the Gila River.

Anyone who walks these routes today comes away astounded at the difficulty of travel and the distances that Kino and Manje apparently considered part of a day's work. This is especially true now that some of the routes on the Cabeza Prieta National Wildlife Refuge that were formerly open to motorized travel must be traversed on foot, giving modern explorers a small taste of what travel was actually like during Kino's and Manje's time.

Jacob Sedelmayr and the Last Jesuit Attempt at Expansion

Kino's efforts to expand the Jesuit mission system northward languished until the 1740s, when a Bavarian Jesuit named Jacob Sedelmayr mounted a series of expeditions to the Gila and Colorado and even tried to reach the Hopi Mesas (Donohue 1969; Dunne 1955). In November 1750, one of those peregrinations took him across El Camino del Diablo. One published account of that trip is Jesuit historian Peter Masten Dunne's translation of the "Derrotero" or "Log" written by the "ensign" [probably *alférez*] of the soldiers stationed at Caborca. The account is terse and a bit confusing, but by now readers will immediately recognize Heart, Cabeza Prieta, and Baker tanks:

> November 22. We left the ranchería going west and at five leagues distance came to Sonóita. Here is the ranch of Captain Cipriano where there is an abundance of water for all and broad acres for raising crops. There were many Indians here, too, who were gentle and tractable. They assisted the Father Visitor and were obedient to the soldiers. A part of the Indians came about us and so the Father Visitor begged two of them as guides which were immediately furnished. There is a good house here made up of three rooms and a kitchen. There are ruins of a smaller house where Father Kino used to come to say Mass, so far as these Indians were able to remember.
>
> November 23. We left this spot heading in the same direction [west] and after eight leagues we came upon a watering place, though it was salty. But there was abundant pasture and here we made camp.
>
> November 24. We left the place at the break of dawn and after ten leagues journey we made camp without water or feed.
>
> November 23 [*sic*; 25?]. We left this spot going on in the same direction and at a distance of seven leagues we came upon a tank of water from which only the men were able to have drink and this with the greatest difficulty [Heart Tank]. We continued on without stopping, covering another nine leagues, and came to another spot in which there were three tanks of high elevation [Cabeza Prieta Tanks]. To get drink from them it was necessary to take the water from the first to the second, and then to the third. From this the animals were able to drink, although with great difficulty. Water had not been had for the two preceding days.
>
> November 26. We went on in the same direction for three leagues. We came upon a tank where we watered the animals *en coritas* [in watertight baskets] with some difficulty and following along in the same direction we traveled about five leagues. Then we made camp without water or pasture.
>
> November 28. We left this spot and traveled about sixteen leagues to the Río Gila and all this distance without water. The river is an abundant stream and its banks are fringed with trees. But it is without pasture, though there is much land for the sowing of wheat. The Indi-

> an tribe is large, little friendly, very naked and much given to theft. (Dunne 1955:68–69)

As Dunne himself points out, there are inconsistencies in the account, but William K. Hartmann (1989:61) contends, and we concur, that although the ensign labeled two entries "November 23," one must have been copied out of sequence and in actuality be the 25th, perhaps the "25" mistaken for "23." If so, the account would fit. On the 23rd they traveled 8 leagues from Sonoyta to Agua Salada, the salty watering place and pasture. On the 24th they traveled 10 leagues and made a dry camp, and then on the 25th they went 7 more leagues to Heart Tank, before continuing 9 leagues to three high tanks, such as they would find at Cabeza Prieta Tanks, 17 miles (27.2 km) from Heart Tank.

We are, however, still left to wonder at the name of the tinaja, three leagues distance from Cabeza Prieta Tanks, where they drew water in watertight baskets, and we would like to know where they were on the 26th. In November, following winter rains, a number of natural basins may hold water, but the distances do not jibe with La Tinaja (Baker Tanks). This sort of historical sleuthing may seem tedious, but it is part of trying to understand the complexity of the trail and waterhole system available to not only the Spaniards but to the Native Americans who lived in the region.

Information about their return trip may hold the key. After a battle with the Quechans in December, Sedelmayr and his escort took a different route home. Described as a "shortcut," it took them from the vicinity of Yuma through the sand dunes west of the Gila Mountains to Laguna Prieta. This route is of considerable interest to us, because it raises the possibility that they visited Tinajas Altas both on this return trip and the previous November 26th. According to the ensign (Dunne 1955:72–73), after stopping at Laguna Prieta,

> Then we went on another two leagues and made camp without water, wood or pasture. We broke camp and traveled in the same direction [east] for six leagues and we came to the trail which we had taken at our setting forth. Thence at two leagues distance we made camp at the place near the waterhole above referred to. Here we spent the night and next morning continued on the trail until we reached Sonoyta.

The ensign locates the waterhole referred to above as 10 leagues east of Laguna Prieta. The walk from Laguna Prieta to Tinajas Altas is about 36 miles, so covering it in ten hours would have required a stiff pace. According to the ensign, they struck their original trail eight leagues from Laguna Prieta and camped near the waterhole two leagues later.

Sedelmayr is less specific than the ensign, but his notes better fit the leg from Laguna Prieta to Tinajas Altas. When they reached Laguna Prieta, he gave thanks to God and wrote, "Because of this oasis one can get there [the Colorado] more easily and bypass the tribes at the river's mouth. I gave it the name of San Juan Nepomuceno." He added, "Continuing through the dunes we hit our original road through the mountains and finally we returned to San Marcelo del Sonoytac some 50 leagues from San Juan Nepomuceno. The Indians there had loyally guarded the horses and supplies" (Sedelmayr 1996:30). Travel from Laguna Prieta to Sonoyta by the Camino del Diablo route covers about 105 miles, and his fifty leagues average a believable 2.1 miles per hour. And this route would have brought them to, or at least near, Tinajas Altas.

These two short accounts of the return trip strongly suggest that Sedelmayr's party *had* stopped at Tinajas Altas on November 26 when they watered the animals "en coritas." They would have crossed the range at Tinajas Altas Pass or Cipriano Pass to travel up the western flanks of the Gila Mountains to the Gila River, as well as stopping there on his return trip. If this interpretation is correct, and discounting the unsubstantiated notion that Melchior Díaz had been there, then on November 26, 1750, Sedelmayr and his party were the first Europeans to take advantage of the water at Tinajas Altas. And he deserves credit for pioneering this more western route, though again the Europeans were following an existing Indian trail used by Cocopahs to reach the Gila River.

The Anza Expeditions and a Land Route to California, 1774

The widespread Indian revolts of the mid-eighteenth century put an end to Jesuit dreams of expansion (Donohue 1969; Sheridan 1999). Then, in 1767, the Jesuits were expelled from all Spanish dominions and replaced by the Franciscans in Sonora and the Pimería Alta. At the same time, however, Bourbon reformers led by Visitador General José de Gálvez were forging ambitious plans to counter their European rivals in North America and expand the Spanish empire once again. Key to those designs was the occupation of North America's Pacific coast, which the Spaniards called "California." Spanish officials were alarmed by England's search for a Northwest Passage to the Orient. They also were aware that Russian sea otter hunters were establishing outposts in Alaska and advancing down the North American coast. After mounting the largest military campaign in Sonoran colonial history to crush Seri and Pima rebels (Sheridan 1999), Gálvez sent expeditions by sea and by land from Baja California to occupy Monterey Bay and solidify Spain's

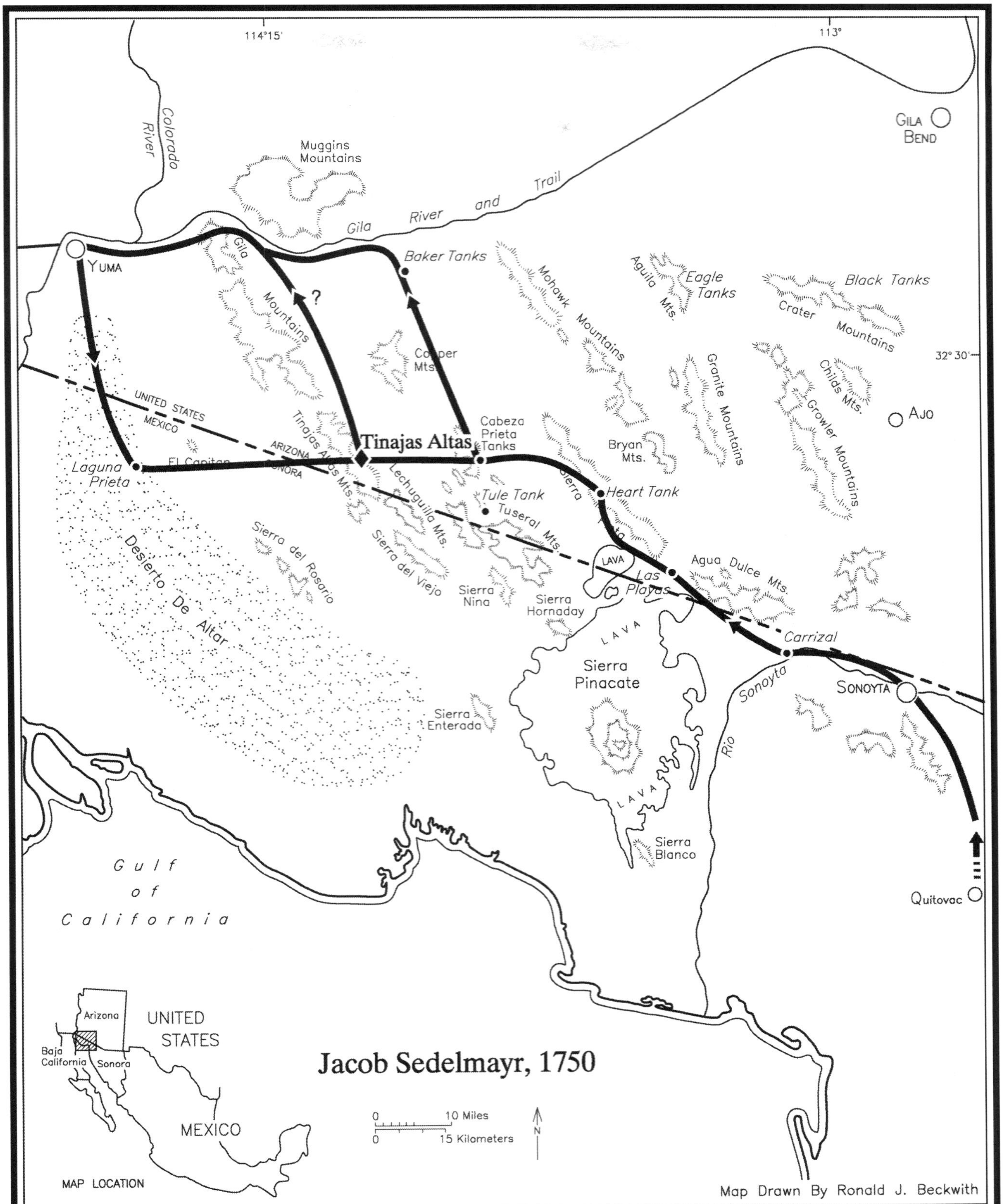

Jacob Sedelmayr's possible routes, 1750. On his way to the Colorado River he may have taken the eastern or the middle route. (By Ronald J. Beckwith)

One of the pools at Cabeza Prieta Tanks. (Photo by Bill Broyles)

claims to the Pacific coast. Ships had to sail north against prevailing winds and currents, however, and the missions of Baja were notoriously poor. Gálvez and his advisors therefore concluded that the new settlements of Alta California had to be supplied over a reliable land route from either Sonora or New Mexico (Weber 1992).

Two Franciscans—Silvestre Vélez de Escalante and Francisco Atanasio Domínguez—attempted to establish such a route across the Great Basin from Santa Fe to Monterey. Snow in the Sierra Nevada turned them back. In the words of historian David Weber (1992:256), their explorations "demonstrated the impracticality of reinforcing New California from New Mexico." Their failure made blazing a trail between Sonora and California that much more critical.

The man entrusted with establishing that trail was Juan Bautista de Anza, a second-generation Sonoran frontiersman of Basque ancestry who was captain of the Tubac presidio. Anza's father of the same name, captain of the presidio of Fronteras, had proposed a land route between Sonora and California in 1737, the year Apaches ambushed and killed him. Anza the Younger was therefore steeped in the dreams as well as the often harsh realities of the Sonoran frontier.

Then, in 1771, Franciscan missionary Francisco Garcés followed Kino's route to the Gila during one of his visits to the River Yumans, whom he hoped to missionize. In January 1774, Anza mounted a small expedition that included 21 soldiers, two priests including Garcés, and a Cochimí Indian named Sebastián Tarabal, who had fled the new California mission of San Gabriel and taken refuge with Quechans along

The main pool at Laguna Prieta, a small freshwater lake fed by semi-artesian springs and surrounded by sand dunes. (Photo by Julian D. Hayden, 1953)

the Colorado River. Their purpose was to determine whether a larger expedition of colonization bound for Alta California could traverse the Camino del Diablo, thereby avoiding the danger of Apache attack on the more northerly Gila Trail. The party left Tubac on January 8 and arrived in Sonoyta on January 28. The next day they reached El Carrizal. Anza kept a running tally of the distance he traveled from Tubac, listing Carrizal at 96 leagues. Anza kept two diaries of his expedition, one for the stretch from Tubac to San Gabriel, another for the journey from San Gabriel to Monterey and back to Tubac. Anza then combined the two into a complete diary of the trek. Bolton's translation of Anza's complete diary account of crossing the Camino del Diablo follows:

Sunday, January 30 [1774]. Because the next watering place is far away and very difficult to reach for watering the riding animals, I decided to march to it in two divisions, both setting out in the afternoon. For this purpose I left the pack train here at El Carrisal [Carrizal] with a corporal and seven soldiers. With the rest of the riding animals, the cattle, and the equipment that is necessary to use in reaching it, at twelve o'clock today I took the road which leads to it, going north-northwest for about three leagues and then three more to the northwest, with numerous hills always on one side or the other;

then going through a good pass, after nightfall camp was made in a flat, although there was no pasturage here, as was true of all the country traveled over this afternoon. For this reason the animals ate only some mesquite sprouts.

Monday, January 31. At half past seven in the morning, setting out in the same direction, we took the road and began our march for El Aguaje Empinado. Having traveled about a league we reached the skirt of a medium-sized range on the right, which is of bare rock, and for that reason has neither pasturage, tree, nor bush, either large or small. The ground over which the road runs is fairly good, and having traveled over it five or six leagues in all, I arrived at the canyon where the water was, but to reach the place where it is found it was necessary to go on foot, or rather more on my hands than on my feet. From Tubac to El Aguaje Empinado, 108 leagues. [Aguaje Empinado is Heart Tank. *Empinado* means 'steep'; *empinada* means a 'steep climb.' Again, we wonder about Kino's claim to have cleared a path to Heart Tank, for subsequent users do not seem to use it.]

I saw at once that there was not enough water for all our animals, and that if it were given to all of those which I was taking ahead, it would leave the pack train exposed to death, this being the part of the train that can advance least rapidly because of the load which it carries. For this reason I decided to go on to the next watering place and leave this one free. So I set out for it about two in the afternoon, and having traveled three leagues west-northwest, in addition to the six already covered, I camped for the night among some small hills where there was found a little pasturage, although it was bad.

Tuesday, February 1. At eight in the morning we set out on the road on which we had spent the night. Going northwest, immediately we climbed some small hills which must be hollow, because when the horses walked over them there was a sound similar to that of an underground cavern. Having passed these hills, which lasted for a league, and then traveled two more over much better terrain, the water was reached.

This watering place [Cabeza Prieta Tanks] is found in the midst of a number of hills whose opening is toward the east. It has six good tanks of very good rain water. In the first the animals drink. It is easily filled from the tanks above by climbing up and emptying them through the natural channel in the living rock which connects them all. In them is a large supply of water, and as much more could be supplied if they were cleared of the rocks and sand with which they abound. All the animals were well provided with water and when pasturage was sought it was found at a short distance.

These tanks, the one farther back [Heart], and another which is ahead of us [Baker], are inhabited by the Pápagos in the height of the dry season and as long as the water gathered in them lasts, through their desire to hunt mountain sheep there. These animals, as to skin and other characteristics, are like large deer, differing only in their horns, which are thicker than those of the largest ox, but in shape and position are like those of common sheep. They live among the cliffs that are highest and most difficult to scale. They are native to dry and sterile regions. Their flesh is better than that of deer. They multiply very slowly, and almost never run in the level country because of the impediment of their horns. These horns the Indians are careful not to waste. Indeed, whenever they kill the sheep, they carry the horns to the neighborhood of the waterholes, where they go piling them up to prevent the air from leaving the place. Those who, like ourselves, do not practice or do not know of this superstition, they warn not to take one from its place, because that element would come out to molest everybody and cause them to experience greater troubles.

Wednesday, February 2. I named these tanks La Purificación [Cabeza Prieta Tanks]. While here I learned that the pack train had reached El Empinado [Heart] greatly fatigued and with some of the mules worn out. I therefore decided to await the pack train here, and to send the relay muleteer to it with the few mules which I brought to carry my necessary baggage and other indispensables, and they set out today. From Tubac to El Aguaje de la Purificación, 114 leagues.

Thursday, February 3. In the middle of the day the pack train arrived, indeed completely worn out, and for this reason I thought it best not to march until tomorrow.

Friday, February 4. About half past eight in the morning we set out west-northwest on the road which leads to some wells. Going over good terrain, passing hills on both sides, and traveling five leagues, the wells were reached. I immediately had them opened [dug out], and it was seen that the water ran in quantities sufficient for all our animals, and that there was pasturage. Camp was therefore made at these wells for the night and they were named Pozos de en Medio. From Tubac to Los Pozos de en Medio, 119 leagues.

Saturday, February 5. At half past seven we set out along a made road toward the west-northwest, and having traveled seven leagues, continually passing hills on both sides, we reached the watering place which, on account of its long distance from the road, we inferred must be the one which the Jesuit fathers called Agua Escondida [actually Spook Tank in this case]. Here with much difficulty we succeeded in watering the horses and mules only, for, because the flow decreased and of the great inconvenience, the cattle could not be watered by nightfall, so this was left until tomorrow." From Tubac to El Agua Escondida, 126 leagues.

Sunday, February 6. At eight in the morning I sent the messenger named on a good horse, and he again promised to meet me on the road the next morning. An hour before this we had commenced to water our horses, in order to be free in the middle of the day to set forth to reach the rivers. Indeed, it was not possible to remain here any longer, because of the scarcity of the water and the still greater scarcity of the pasturage.

And so at two in the afternoon we set out over a traveled road toward the south, to round the little range [Vopoki Ridge in Gila Mountains] which contains the Agua Escondida. On its skirts we traveled in this direction about a league and a quarter, then, leaving it behind at the right, the road turns to the northwest, in which direction we traveled five leagues more, until nine o'clock at night, when camp was made in a place that had plentiful pasturage.

On the march of this day and that of yesterday we have had on our left the sand dunes through which the Jesuit fathers [Sedelmayr] made their way to the Rio Colorado and which cost them the failure of so many of their journeys. This transit, so difficult, unless we are mistaken, they never would have made except with the idea that only by crossing the Colorado near where it empties into the Gulf they would find the passage which they sought.

Monday, February 7. Shortly after sunrise we set forth over the road on which we had camped for the night, always skirting the sand dunes described on our left, although for a distance of about two and a half leagues some small patches of them are crossed before reaching the Gila River. (Bolton 1930, 2:26–36)

Bolton (1930, vol. 1) argued persuasively that Anza's first expedition passed through a gap in the O'Neill Hills and headed northwest through the Tule Desert to Heart Tank (Aguaje Empinado, Kino's Aguaje de la Luna) on January 31. The party then crossed the Tule Desert to Cabeza Prieta Tanks (Tinajas de la Purificación) on February 1, where they waited for the pack train. Cabeza Prieta Tanks did have evidence of a Hia C'ed O'odham settlement dating to the late nineteenth century (e.g., Lumholtz 1912), and stacks of bighorn sheep horns were recorded at the canyon's entrance to the tinajas (Hayden and Friends 1993). Two days later, February 4, they crossed Cabeza Prieta Pass and camped at a spot along Coyote Wash, the large arroyo in the center of the Lechuguilla Valley, where they dug for water in the sand. They called the shallow pits Pozos de En Medio, now Coyote Water.

On February 5, they traveled through Tinajas Altas Pass or Cipriano Pass, ignoring or missing Tinajas Altas just to the south, until they found water in a canyon on the west side of the Gila Mountains in Arroyo San Albino, between Vopoki Ridge and the Gila Mountains at what is now called Spook Tank. One map shows an "S. A. Tank" (S. A. for San Albino, we assume) north of the Albino pasture. We suspect that Albino was a campsite along the trail where it crossed Arroyo San Albino, and S. A. Tank was up the arroyo at Spook Tank. Anza thought they had reached Kino's Agua Escondida but he was mistaken—this was not Cabeza Prieta Tanks. Instead, he was following a route west, not east of the Gila range. On February 6, Anza and his men retraced their steps south out of the canyon and turned northwest toward the Gila, sand dunes visible to the west. A map hand-drawn by Ronald Ives shows Anza's "Agua Escondida" as Spook Tank. And Bolton's own Anza map (Bolton 1930: vol. I, following page 529) clearly shows the leg from Cabeza Prieta Tanks to Pozos de en Medio to Spook Tank, as well as bypassing Tinajas Altas. However, Bolton's Kino map (Bolton

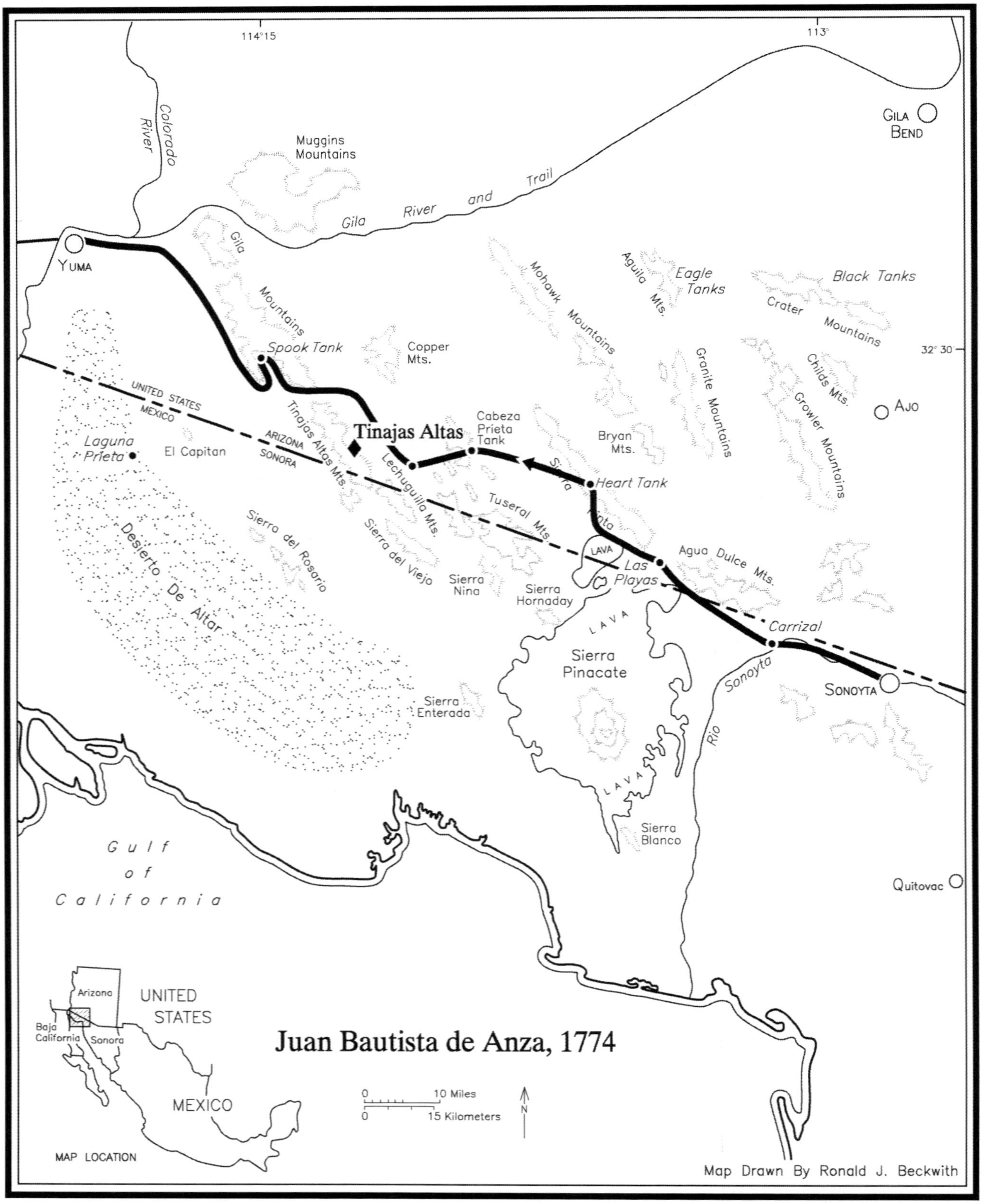

Juan Bautista de Anza's route, 1774. (By Ronald J. Beckwith)

Spook Tank (Anza's Agua Escondida). (Photo by Bill Broyles)

1919b, 1936) has misled many readers, because it erroneously conflates Agua Escondida (Spook Tank) with Tinajas Altas, La Tinaja with Los Alquives (Dripping Spring), and Aguaje de la Luna (Heart Tank) with Tinaja del Tule (Tule Tank). These pairs of tanks are not synonymous, though this Kino map shows them as being the same, leading to considerable confusion. We can't explain why Bolton was accurate on one map and not the other.

The diary of Fray Juan Díaz, a Franciscan accompanying Anza, is less ambiguous and reveals how reliant Anza was upon his O'odham guides. According to Díaz:

> January 30 [1774]. Being informed by one of these natives of the difficulty we should have to find water for the pack train and the saddle animals on the next march, the commander decided to divide the company, leaving the pack train to follow on the next day. The rest of us went forward in the afternoon to divide the journey, which is one of fourteen leagues. Having traveled eight of these leagues northwest and north-northwest, we came to the entrance of a valley formed by two rocky sierras, and halted near the one on the right, where we spent the night without water and very little pasturage.
>
> January 31. Having celebrated the sacrifice of the Mass we continued on our journey, keeping very close to this sierra; and having traveled six leagues to the northwest, with great difficulty because of the heaviness of the road and the little spirit which the saddle animals had by now, we arrived at a tank of water which is very high up in the same sierra on the right [Heart Tank]. Seeing that it was very difficult to get water to the horses, mules, and cattle, and that in the vicinity there was no pasturage, of

which they were in such need, the commander decided to go forward over a part of the next march, hoping to find some pasturage for the night and to arrive more quickly at the watering place the next day. With this in view we left this sierra and crossed the valley, and having traveled four leagues to the west and west-northwest, we camped at an arroyo which was dry but had some pasturage.

February 1. At seven o'clock in the morning, going northwest, we continued the journey, entering the sierra which is on the left through some hills, and having traveled three leagues we halted near some tanks of water which we called La Purificación [Cabeza Prieta Tanks]. In them was sufficient water for travelers, and cattle and horses can drink with ease. Here in this place we decided to await the pack train, because in the vicinity there was some pasturage, of which the riding animals had come so greatly to need.

February 2. We remained in this place, two Masses being celebrated for the success of the expedition.

February 3. Two Masses were celebrated, the same as yesterday. Because the pack train arrived at midday greatly fatigued, the march was not continued until the next day.

February 4. At eight o'clock in the morning we set out toward the west-northwest. Having reached an arroyo which is in the middle of a valley, with little difficulty we found plenty of water in its sands [Pozos de en Medio, now called Coyote Water]. According to the account given us by a native Pápago who was guiding us, it has water for the greater part of the year, although in small quantity. We made camp in this place, notwithstanding it was very short of pasturage, in order not to experience greater difficulty in going forward. In a sierra ahead of us, which runs from north to south, they say there is a tank of water [Tinajas Altas] that might serve in case there is none in this arroyo.

February 5. In the morning, after Mass, we set out from this place, and having crossed the sierra that lay in front, through a very easy pass [Tinajas Altas Pass or possibly Cipriano Pass], we traveled eight leagues in order to reach Agua Escondida, or Pozo Blando [Spook Tank], as these natives call it, where we halted. This water, according to what these Indians tell us, is permanent although very scarce. It is in a deep arroyo, which makes it necessary to travel a quarter of a league, going into a sierra of considerable height to reach the water. By departing from the direction of west-northwest which we have followed, and turning west by west-northwest four leagues before reaching this watering place, this march can be made with greater ease and for the following journey three leagues will be saved, because one can come out to another watering place on the skirts of this same sierra [Albino?], in whose neighborhood there is much pasturage. (Bolton 1930, 2:254–57)

Díaz's account clearly eliminates Tinajas Altas as Anza's Agua Escondida. The Spanish version of the Indian name "Pozo Blando" ("Soft Well") describes a waterhole in soft arroyo sand, not a bedrock basin. Again, this version fits Spook Tank, which lies in a "deep arroyo" within "a sierra of considerable height" and sits on the upstream edge of a sand-filled arroyo. On the other hand, the tank mentioned by the O'odham guide when they were at Pozos de en Medio ("the tank . . . in a sierra ahead of us, which runs from north to south") was likely a reference to Tinajas Altas, But since they had found "pozos" (water in shallow hand-dug wells) in the valley, they didn't bother going to Tinajas Altas. Instead, they crossed perhaps at Tinajas Altas Pass or Cipriano Pass, and moved along the west slopes of the Gila Mountains. We're unsure where the other watering place and pasturage was, but it may have been Albino. The O'odham guide obviously had a sophisticated knowledge of water sources in the region—not only where they were but in what seasons they held water.

The account left by Francisco Garcés, the Franciscan explorer who plunged into the Indian worlds of the Pimería Alta and the Colorado River and was also along on this expedition, adds some additional interesting details about Anza's first expedition. He provides a description of the mountain range where the high tank Anza called Aguaje Empinado was located: "part is black and afterward white" (Bolton 1930, 2:316). Bolton correctly argues, "This exactly describes Sierra Pinta as it loomed before us that January day in 1928. Most of the way to the tank the range is of dark colored rock; then, some four or five miles before the tinaja is reached the dark rock suddenly ends, forming a clearly marked line, where the range changes to light colored rock. Hence the name Pinta" (Bolton 1930, 1:88–89). This strongly reinforces Bolton's argument that Heart Tank is Anza's Aguaje Empinado and Kino's Aguaje de la Luna.

Garcés also provides a clear description of Pozos de en Medio as "an arroyo which has water when it rains, pools soon being formed in it." He goes on to say, "The preceding watering places I saw during my return from the last journey [in 1771?]. To the west-southwest of this one, in the sierra which runs to the north, is the Tinaja de Eusebio, of which I make mention in my diary; and there are other small watering-places in that sierra, but none of them are convenient for this journey." This description indicates that Tinaja de Eusebio is Tinajas Altas, and the name Tinaja de Eusebio suggests that Garcés associated Tinajas Altas with Eusebio Francisco Kino, even though Kino never visited it. Garcés also reports that the expedition crossed the Tinajas Altas range to reach Pozo Blando, "which on my last journeys I thought was Agua Escondida" (Bolton 1930, 2:317). This matches the location of Spook Tank, which by then he had learned was not Kino's Agua Escondido.

ANZA'S SECOND TRIP

Anza's second trip to California was a much more elaborate affair than the first because its purpose was to establish a permanent colony in Alta California. The expedition began in San Miguel de Horcasitas with 177 individuals on September 29, 1775. When it departed from Tubac on October 23, it consisted of 240 people—men, women, and children—165 pack mules, 340 saddle animals, and 302 head of cattle (Bolton 1930, 4:1–3). On their way to California, they did not follow the Camino del Diablo route, but rather followed what later became the Gila Trail, heading north along the Santa Cruz River to the Pima villages along the Gila River. With what he knew firsthand of the small quantity of desert water in the tinajas, he wisely chose to take the river route, which offered far more water for people and stock. Then they descended the Gila to the Colorado and crossed the Mojave Desert to Mission San Gabriel before heading north to found a presidio on San Francisco Bay.

When Anza returned to Sonora, however, he again followed El Camino with a much smaller party that included the Franciscan Pedro Font, seven soldiers from Tubac, six muleteers, two cowboys, and four servants but no cattle and far fewer horses. On May 16, 1776, Anza noted in his own diary of the expedition that the party reached Laguna Salada and the village of San Pedro, which Bolton (1936: 414, 436) identifies as along the Gila River five miles west of Wellton. At most stops, Anza kept track of the accumulating distance from "the river or lake of San Francisco"; Laguna Salada was 249 leagues. Then the group headed southeast, crossing one of the driest deserts in North America at nearly the driest and hottest time of the year. In Anza's words:

> Friday, May 17 [1776]. Since the next watering place was quite distant, and it was not easy to reach it before eleven o'clock, when the sun overcomes men and animals, because it is extremely hot, I decided to start for this place in the afternoon, and did so at three o'clock, going southeast for about two leagues over good terrain. This continuing, we marched to the east-southeast for six more leagues and arrived at half past eleven at night at the Pozos de en Medio, which afforded some water for drinking, and where we halted for the rest of the night, having traveled eight leagues in the same number of hours. By digging, these wells afford water for a large number of saddle animals even in the dry season like the present. They have plentiful pasturage, which is likewise found all along the road which we have followed [presumably along Coyote Wash]. For this reason and because it is free of stones it is always better than the one which up to now has been traveled to the junction of the rivers [according to Bolton, incorrectly, the route through Tinajas Altas Pass]." From the river named to the Pozos de en Medio, 257 leagues.
>
> Saturday, May 18. At five we resumed our march, going east-southeast over the road which runs from these wells to the Tinajas de la Candelaria [Candlemas], where we arrived at half past nine, having traveled five leagues. Going to examine them I found that they had plenty of water even for more animals than those which we had with us, which they took to water twice. The sun being now low, at six o'clock in the afternoon we continued our march, which was made for three leagues to the east-southeast over the regular road opened on our first journey. Having finished this distance and the crossing of some hills, and leaving these and the road which we had followed, we turned to the east for about nine leagues when we arrived at Puerto Blanco. From the river named to Puerto Blanco, 266 leagues." (Bolton 1930, 3: 186–87)

Going five leagues "east-southeast" from Pozos de en Medio would put them at Cabeza Prieta Tanks, 12 miles by a fairly direct route. One might ask if, instead, it is Tule Tank along the route employed by forty-niners and modern visitors, but Tule is a single waterhole southeast of Coyote Water, not east-southeast. The obvious discrepancy

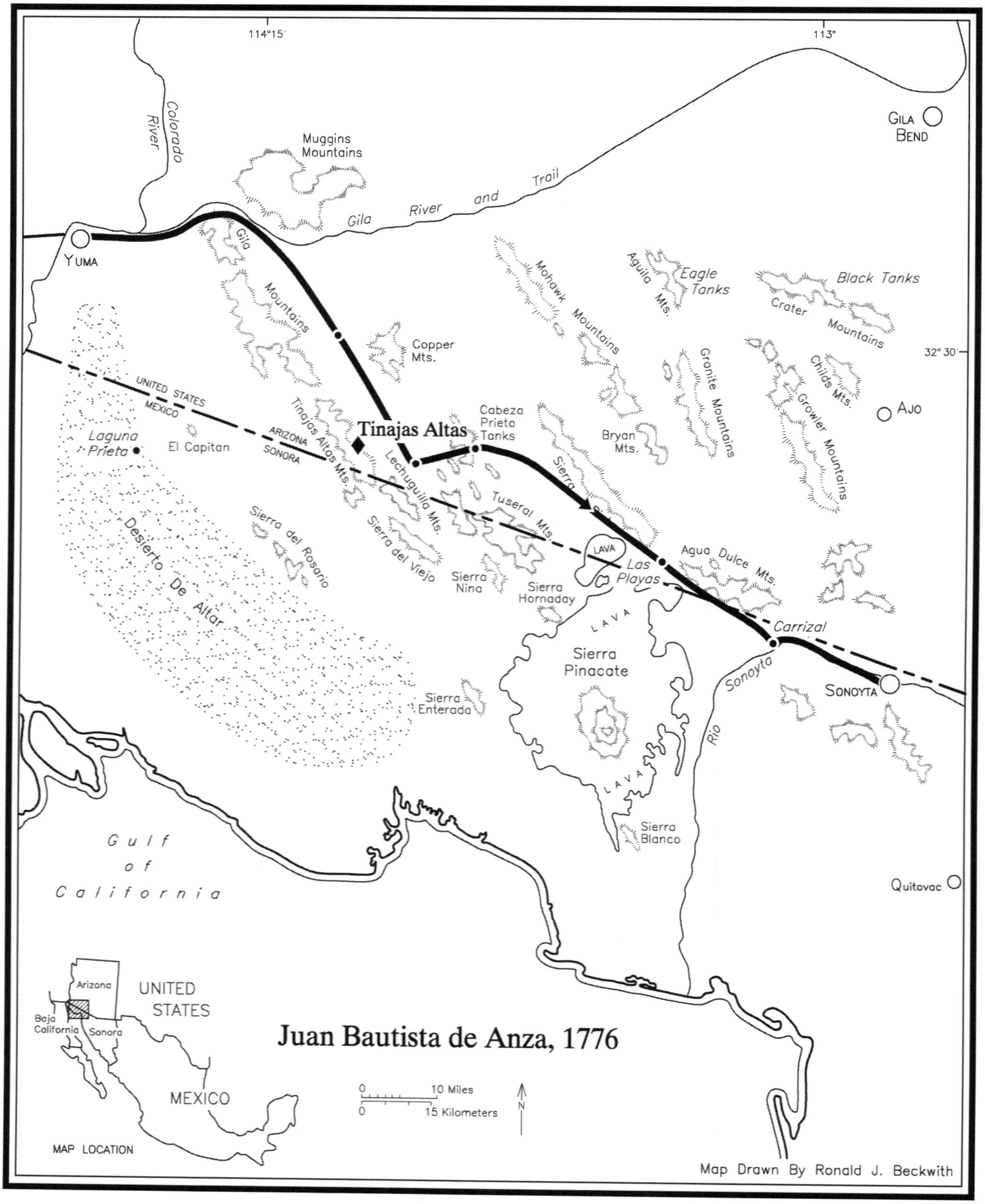

Juan Bautista de Anza's route, 1776. (By Ronald J. Beckwith)

Puerto Blanco. The distant "white" ridges of the Sierra Pinta beckon travelers to Heart Tank, which lies in a canyon downslope from the summit. (Photo by Bill Broyles, 2010)

in Anza's account amounts to nine leagues, but he does give the distance from Pozos de en Medio to Puerto Blanco as 17 leagues, which jibes with Font's measurement of 18 leagues. And, Bolton (1931:58) places Puerto Blanco as a "white pass" opening in the O'Neill Hills between Sierra Pinta and El Carrizal, likely, we believe, an easy passage at the southeast corner of Las Playas and near a site Hornaday (1908:181) calls La Represa Tank.

Today the *puerto blanco* is a smugglers' route through the western O'Neill Hills. Although it is not labeled on the topographic map, a person on foot can easily walk several miles of the original Camino here. The route curves slightly as it wends through some hills, hills lighter in color than the lava buttes to the south. But walking northward through the pass one is struck by how the whitish summit of Sierra Pinta looms on the horizon and marks the route to Heart Tank. From the low hills at Puerto Blanco, travelers could also see Cabeza Prieta Peak, which would have been their landmark after Heart Tank. The distances look formidable but straightforward.

> Sunday, May 19. Continuing our march from the last-named site without interruption during the whole night, going southeast, making a total of eighteen leagues in eighteen and a half hours [since 5 p.m. the preceding day], we arrived at half past eleven o'clock at the Arroyo de Sonóitac [Sonoyta River] and the site of El Carrizal, where we found water and fairly good pasturage in abundance and where we halted to remain the rest of this day. In this day's march and the two preceding we have lost six saddle animals, most of them being those which I left at the Colorado River, where they had improved very little in flesh." From the river named to Arroyo del Carrisal or de Sonóitac, 284 leagues. (Bolton 1930, 3:185–87)

Anza's group must have been relieved to reach Sonoyta. Knowing the distances they traveled and the difficulties of the terrain, modern readers find new appreciation of Anza's accomplishments.

Pedro Font's Diary

Pedro Font also left a diary of Anza's second expedition, including a description of its return from Yuma Crossing along the Camino. Font's is one of the first reports to make much mention of the region's vegetation or to acknowledge personal comfort and details like thirst. According to Font:

> Friday, May 17 [1776]. I said Mass. We remained here until afternoon in order from this river to take the road to Caborca. With the heat which we felt here today the burning of my mouth became worse, and with the salt water the animals became sick, for this river might well be called the Salado, so that today two mules and one horse remained on the road tired out and lost.
>
> We set out from the Laguna Salobre [on the Gila River] at four in the afternoon, and at a quarter to twelve at night halted at a patch of coarse grass which they call galleta [*Hilaria rigida*], a league after crossing the sand-filled arroyo which they call the Pozos de en Medio, having traveled eleven leagues, three to the south, six to the south-southeast till we came to the road followed by Señor Ansa [Anza] the first time, then two to the southeast. The road was level all the way, but over sandy country, dry and most difficult, with no trees except some scrubby mesquites near the sandy arroyo, a great deal of hediondilla [creosotebush], and some coarse galleta grass."
>
> Saturday, May 18. We set out from the Zacatal Duro ["rank pasture"] at five in the morning, and at a quarter to two at night halted near the Llano de Tuzal ["gopher plain"] before reaching the Puerto Blanco, having traveled eighteen leagues by a route much varied, three leagues east by north, two northeast, and one east-southeast, when we came to the Tinajas de la Candelaria, where we stopped to take a siesta from nine in the morning until a quarter to six in the afternoon. (Bolton 1930, 4:503)

The use or application of the name Llano de Tuzal may indicate that Font and his group crossed a large plain pocketed with gopher holes, or more likely, kangaroo rat dens, locally known as *tusa* (generally *tuza* means 'mole'). By the early 1900s when Lumholtz was writing about this region, the mountain range south and west of here was named Sierra Tuseral. Bolton's (1930) Anza map shows a Llano del Tuzal approximately east of Sierra del Tuseral and in the vicinity of Dos Playas, the Pinta Sands, and a tongue of the Pinacate Lava Field. (See also Fages's entry for October 26, later in this section.)

Font continues,

> This place consists of some wild and naked sierras or very high cliffs in which there are some concavities called tinajas [Cabeza Prieta Tanks], in which there is usually water. At this place there are nine tinajas formed in the rocks, one above the other, or one higher up than the other, and animals are watered by emptying those above with coritas. The animals were watered twice, there being more than a hundred of them; and in this labor almost all of the time was spent. Then, with the intention of going straight to El Carrizal in one march, without touching at the Tinaja del Aguaje Empinado [Heart Tank], which we left at our right [Bolton argues Font should have said left] because it is not possible for the animals to drink at it, we traveled three leagues to the northeast over malpais [necessary to leave Cabeza Prieta Tanks], six to the southeast, and three to the east-southeast. We now dismounted to rest and lie down a little while the pack-train arrived, without taking the bridles from the animals.
>
> Sunday, May 19. We slept for a while. The pack train arrived at four o'clock in the morning and went forward without halting.
>
> We breakfasted and some of us changed mounts to continue the journey. At five o'clock in the morning we set out from the Llano de Tuzal, and at half past eleven arrived at the Arroyo del Carrizal, having traveled ten leagues, three to the east-southeast to the Puerto Blanco, and the rest southeast by south. The road was level, but almost all the way over sandy land and all bad, with no other grass than a little galleta; and today three horses and one mule were left behind worn out and lost. The pack train traveled something more than nineteen hours without stopping and arrived at one in the afternoon. (Bolton 1930, 4:502–4)

William Hartmann (1989:70) wonders if Font's description of Tinajas de la Candelaria (nine-tiered tanks, 12 miles and four dawn hours from their camp) indicates Tinajas Altas, but the expedition was going east, not west, from Pozos de en Medio, ruling out Tinajas Altas. Further evidence that Tinajas de la Candelaria was indeed Cabeza Prieta Tanks comes from Lumholtz (1912:324), who wrote, "There are said to be as many as eight pools at Cabeza Prieta, each higher than the other," and from Broyles (1996b), who noted at least eight pools at Cabeza Prieta 16 inches deep or more, with a ninth pool likely occupied by a modern tank named Sheep Hollow. Lumholtz (1912:324) reminds us to appreciate the accessibility of modern tap water, when he writes about his visit to Cabeza Prieta Tanks, "The four lower ones were empty, and the water had to be scooped down from above from one to the other, so that we could fill our barrels and our animals might drink. It is tedious and rather primitive to have to depend on tinajas for one's existence, and these were even less accessible than usual, the entrance being through a narrow cañon."

To review, we now have a fairly clear picture of the network of waterholes used by Native Americans and early Spanish soldiers and priests. The key points here are that a usable land route had been established from Sonoyta to the Colorado River, one with alternate trails, water stops, and pastures. The travelers found no reasons to fear hostiles once they left the Quechan community on the Colorado River. They found no terrible terrain or great unknowns, no fearsome wild animals or insurmountable conditions. No one got lost or injured. The route was a long miserable trudge on foot or horseback, but it was do-able.

The Yuma Revolt of 1781 and Pedro Fages's Punitive Campaigns

In 1780, the Spanish government attempted to solidify its control over Anza's land route to California by establishing the mission of La Purísima Concepción and the village and mission of San Pedro y San Pablo de Bicuñer near the junction of the Gila and the Colorado rivers. Friendly relations between the Quechans and the Spaniards quickly evaporated as Spanish settlers seized Quechan fields and Spanish livestock ate Quechan crops. On July 17, 1781, the Quechans attacked the Spaniards while they were attending mass. Within two days, they killed more than 100 Spaniards, including Garcés and the three other Franciscan missionaries, and took 74 individuals captive, most of them women and children (Ives 1968; McCarty 1975; Officer 1987; Santiago 1998).

On September 9, 1781, Teodoro de Croix, *comandante general* of the *provincias internas*, ordered Lieutenant Colonel Pedro Fages to retaliate against the Quechan rebels and rescue or ransom their Spanish captives. Fages and his troops left the presidio of Pitic on September 16 and returned on April 25, 1782. The total number of soldiers and Indian auxiliaries is not known, but the entire party may have reached nearly 200. The punitive expedition marched from Pitic to Cúmaro west of Pozo Verde (southwest of Sasabe, Arizona, and south of the international border).

Then they headed north across the Papaguería, intersecting the Gila near modern Gila Bend. Descending the Gila, they reached the Colorado on October 18 and negotiated with Captain Palma, the erstwhile ally of the Spaniards, for the return of 48 captives. The Akimel O'odham (Gila Pimas), Jalchedune (Halchidomas), and Pee-Posh (Cocomaricopas) who accompanied them attacked the Quechan two days later, and the Spanish troops soon joined battle. Ives reports that Fages succeeded in rescuing a majority of the captives and recovering a few church relics. Nonetheless, the premature attack by the Indians "not only precluded further ransoming of [the remaining] captives, but made the situation of the Fages party, which was outnumbered by approximately twenty to one, untenable, particularly as they were burdened by rescued captives" (Ives 1989:230, n. 44).

After the fight, Fages and his officers decided to retreat to Sonoyta across the Camino del Diablo. On October 22, they reached the Laguna Salada (also known as Laguna del Agua Salada) on the Gila, where Anza's second expedition had turned southeast across the Camino five years before. According to Ives (1966a:64), the slough or basin holding the laguna is located 5.5 miles west of Wellton. The nature and exact location of this place is unclear to us. Possibly it was an oxbow along the meandering river, but modern irrigation projects have radically altered the landscape in this area. We can't explain why they favored this spot over others in the same reach, except to say they were looking for clear water, pasture, firewood, shade, a ford, security, and probably a breeze to drive away the flying insects.

Herbert Priestley (1913) translated and published Fages's entire diary of the campaign in 1913. Fifty-three years later, Ives published a slightly revised but thoroughly annotated version. Fages and his troops stopped at a "Tinaja Alta," which at first glance sounds like Tinajas Altas but raises perplexing questions. His brief account of the trek follows:

> October 24 [1781]. We broke camp [at Laguna Salada on the Gila] at midday, and traveling about one league toward the southeast and three eastward over very spongy ground full of gopher holes and very scant in pasture, we arrived at the place called Pozas de San Miguel, where there was a pool of excellent water in abundance. The pool must have been about twenty yards long,

two or three yards wide in places, and two yards deep. There was scarcely any pasture.

October 25. We broke camp at about ten in the morning, traveling eastward until eleven at night over very sterile, stony land with a few hills. After having traveled twelve leagues, we halted one league from La Tinaja Alta, as it is called, where there was very little pasture and no water, nor had there been any on this day's march.

October 26. We broke camp in the morning, and after traveling until midnight toward the east over land that was very sterile, spongy, and full of gopher holes [remember Llano de Tuzal], scant in pasture and without water, we arrived at the place called El Carrizal, where there was some pasture, though scant, and water, but it was salty. After having gone about one league, I ordered some soldiers to go with me to bring water to the camp from La Tinaja Alta, which is about one league from the road. Although they call it a watering place, the water is about a quarter of a league up a hill or cliff, which the people climb with difficulty. For this reason, it was impossible for the animals to drink. To a few of the very tired ones, water was given in small trays, because it was necessary to climb on all fours up the hill and go down to where the water was. Some of the animals were exhausted on account of the great scarcity of pasture and water. (Ives 1966a:68–69)

Agua Dulce Spring. A concrete catch-basin was built by US Fish and Wildlife Service in the 1940s. (Photo by Juel and Madeleine Rodack, November 1970. Courtesy Gayle H. Hartmann)

This account at first seems shuffled. The distance from Pozos de San Miguel to Tinajas Altas is about 20 miles, but Fages says they traveled 13 hours, which should have carried them nearly 30 miles. Further, there seems to be a disjunct in the narrative for October 26—if they camped one league from La Tinaja Alta, and then marched eastward from morning until midnight, they could not have sent soldiers back one league to La Tinaja Alta for water. Were there two places called La Tinaja Alta: one "eastward" twelve leagues from Pozos de San Miguel (October 25) and a second La Tinaja Alta another full day's travel "toward the east" from the first? At no time does he march south or west; rather he maintains the same vector "eastward" (actually southeast) and ends up at El Carrizal. A La Tinaja Alta 12 leagues eastward from Miguel and midway on their trek home indicates his first Tinaja Alta was either Heart or Cabeza Prieta tanks; the distances traveled better fit Heart Tank.

Further, does he imply two locales named El Carrizal, one a league from Tinajas Altas and one near Sonoyta? There is no carrizo (cane, reedgrass) at or near Tinajas Altas. One explanation is that their eastward line and time of travel are accurate and that they visited Heart or Cabeza Prieta tanks and then a day later visited an unspecified or unknown waterhole named La Tinaja Alta (singular) somewhere near El Carrizal. One candidate is Agua Dulce Spring, which is about two leagues from El Carrizal (they had already gone one league). The spring is "up a hill or cliff, which the people climb with difficulty" and where it would be difficult to take animals. The rocky arroyo bottom has ledges where it would be "necessary to climb on all fours up the hill and go down to where the water was." Or, perhaps it was a tinaja unknown to us in the mountains southeast of El Carrizal, mountains intriguingly named Sierra de los Tanques.

Vicinity of Pozos de San Miguel. (Photo by Bill Broyles)

It is perplexing that neither the Fages expedition nor other Spanish travelers through this particular area availed themselves of the bubbling springs at and near Quitobaquito Spring (these include Burro, Williams, and Aguajita springs). From El Carrizal, Quitobaquito lies only 5.5 miles across easy ground. We wonder if the Hia C'ed O'odham had a village there at that time (see Bell et al. 1980), and if the Indian guides steered the Spaniards away from it. Another explanation would be that the springs flowed less vigorously than in the twentieth century and could not accommodate a larger expedition.

According to Ives (1966a:69), "Pozas [*sic*] de San Miguel is in the channel of the Arroyo Coyote, east of the Wellton Hills. The pool described by Fages is a typical, short-lived scour channel. Usually, water is obtained here only by digging in the dry wash." Fages does not say why he followed a route that differed from either of the Anza expeditions. Fages's fast pace indicates that his entire entourage was mounted, as they were on his subsequent trips.

Fages and his troops rested their animals and gathered supplies in Sonoyta for nearly a month before setting out once again along the Camino to mount another attack on the Quechan and their allies. The first leg of the return trek was a grueling march of 30 leagues from El Carrizal to the Pozos de San Miguel. The expedition left on November 25 and traveled all that day and night and the following day, November 26. According to Ives (1966b:158), "The moon was gibbous and waxing at this time, so there was moonlight until long after midnight on the twenty-fifth." The party remained at Pozos de San Miguel on November 27 "so that the animals might rest, for they were much exhausted on account of the scarcity of pasture and water" (Ives 1966b:158).

Once Fages and his men reached the Colorado River, they tried to set an ambush for the Quechan but failed. Nonetheless, they negotiated the release of more captives and recovered the bones of the settlers and missionaries who had been killed. They kept the bones of the Franciscans intact

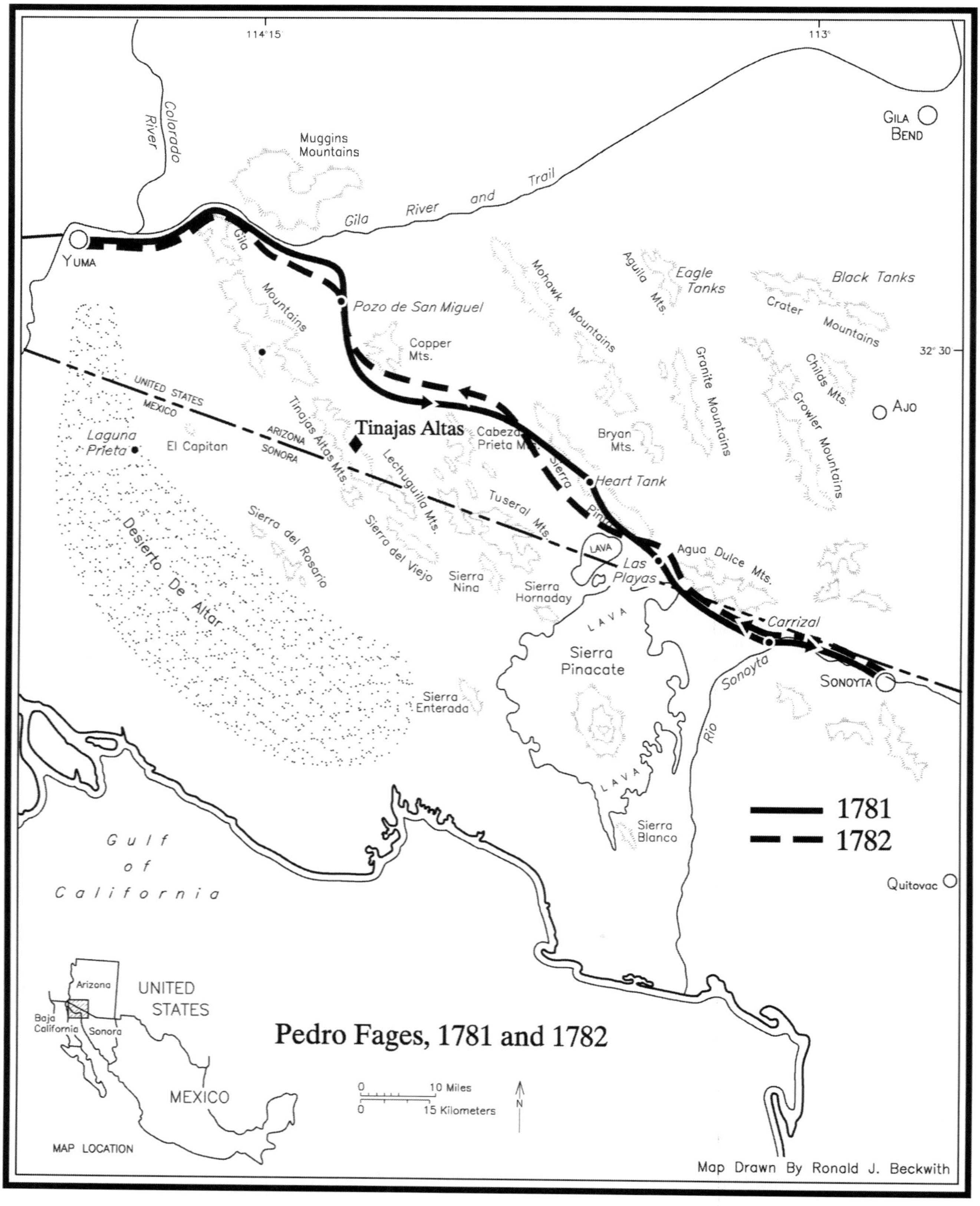

Pedro Fages's routes, 1781 and 1782. (By Ronald J. Beckwith)

but burned those of the settlers, keeping only their ashes. Then, on December 14, they followed the Camino back to Sonoyta once again, stopping only at Pozos de San Miguel before undertaking the journey to El Carrizal in a single march a second time. This time, however, Ives notes that there was no moon and the night was dark.

Fages reached Pitiquito on December 30, where he drilled his soldiers and awaited his next orders. They soon arrived, and on February 27, 1782, he and 39 of his troops rode off to reestablish the land route between Sonora and Alta California. He retraced his by-now familiar steps across the Camino, leaving El Carrizal on March 6. In Fages's terse words:

> March 6. We broke camp at this place at ten o'clock in the morning and [traveling northwestward] for the remainder of the day and part of the night, halted at a late hour after going twelve leagues. There was little grass and no water.
>
> March 7. We broke camp at this place at dawn, and after riding eighteen leagues . . . [omitted in Ives 1968] most of us came at sundown to the Pozas de San Miguel, at which place some of the men whose animals had become tired arrived some time after midnight.
>
> March 8. We remained at this place to let the animals rest, as they had been much exhausted in the preceding day's marches from long distances without pasture or water. Here we had to dig the wells out so that the animals might drink. The grass, called *galleta*, was very scant, and some distance from where we had placed the camp. (Ives 1968:227–28)

We, like others, have long read Fages's diaries as indicating that he stopped at Tinajas Altas (e.g., Ives 1966b, 1968), but that required reading La Tinaja Alta as plural, thinking his "eastward" direction on October 25 was really south, ignoring what would have been for him an exceptionally slow pace to Tinajas Altas from Pozos de San Miguel (13 hours to make 20 miles), and ignoring that on the return trip in March, 1782, he bolts straight from El Carrizal to Pozos de San Miguel and makes no mention of stopping by Tinajas Altas or any other tinaja with his men and stock that "had been much exhausted in the preceding day's marches from long distance without pasture and water."

If Fages's previous trip had included Tinajas Altas, surely he would have watered and rested there again. And the fact that he traveled so fast through open country and apparently did not stop at Cabeza, Heart, or Baker tanks or Pozo de en Medio may indicate that on his trips he traversed the Tule Desert from Las Playas to the northeastern corner of the Cabeza Prieta Mountains, passed south of the Copper Mountains, and continued northwest straightway for Pozos de Miguel midway across the Lechuguilla Valley and on a direct line toward the Gila River. That path follows the shortest, most direct rideable distance from Las Playas to Laguna Salada on the Gila. It may also indicate that he was traveling without an Indian guide or the latest reports on waterholes and pasturage. Mounted and racing to reach Yuma, Fages's troop didn't spare the horses or themselves. They didn't detour or deviate off course to find water or pasture. Fages was a professional soldier who pushed himself and his troops hard (Ives 1966b; Sheridan 1999).

In spite of his efforts, however, he failed to establish a land route to California. After his series of punitive expeditions, the route between Sonora and California was abandoned for nearly half a century.

Early Mexican Period Explorations

By 1815, as the wars for Mexican independence raged, more foreign than Spanish ships stopped at San Diego and Monterey. Argentine rebels under the command of Hippolite Bouchard even seized and occupied Monterey, California's capital, in 1818 (Rosenus 1995). Spanish and then Mexican officials realized that as long as they could get soldiers and supplies to California only by ships from the west coast port of San Blas, the province was in danger of being lost completely. During the chaotic years of the 1820s, the infant Mexican nation, which declared its independence in 1821, attempted to establish a new overland route or reopen the one that had been abandoned 40 years earlier after the series of punitive expeditions led by Fages (Bean and Mason 1962).

An event that precipitated these efforts was the arrival of a Cocomaricopa chief at Mission San Gabriel, California, in 1821. This leader, dubbed José Cocomaricopa, came from the ranchería of Avitucupaiba, six days northwest of Tucson, in order to trade for cotton and beads. He told Spanish officials that it would take 15 to 20 days to carry mail between San Gabriel and Tucson. Once again, Indian trading networks and trails conjured visions of communication and commerce between the Mexican frontier colonies (Bean and Mason 1962).

The Romero and Caballero Expeditions and the Reopening of the Sonora-California Land Route in the 1820s

José Romero, captain of the presidio of Tucson from 1819 to 1826, was ordered to travel to the lower Colorado River to lay out a possible mail route. Meanwhile, Dominican Felix Caballero, president of the missions in Baja California, set out from Mission Santa Catalina on the slopes of the Sierra Júarez southwest of Yuma to seek a route to Tucson (Bean and Mason 1962). They left Mission Santa Catalina on April 14, 1823.

Caballero and two Indian companions reached the Colorado River near its mouth on April 16. There they were greeted warmly by Quamaya (Cocopah) Indians, who may already have made contact with Caballero at Mission Santa Catalina. The Indians warned him of enemies to the north. The Cocopah were involved in a longstanding pattern of warfare along the Colorado and Gila that allied them with the Akimel O'odham (Gila Pima) and Pee-Posh (Maricopa) and pitted them against the Quechan and Mohave.

Accompanied by many Cocopah, the small party moved northeast through Cocopah territory. When they reached the northern limits of Cocopah settlement, Caballero left the Colorado and turned east with an escort of 11 Cocopah warriors on April 21 following the Cocopah trail. They ventured across sand dunes to a lagoon called Jailxquoac, which Bean and Mason (1962:95) identify as Laguna Prieta 15 miles southeast of San Luis Río Colorado and 36 miles west of Tinajas Altas. They left there at six that evening and continued on through more sand dunes until midnight, when they camped near a mountain range.

On April 22, they departed at 6 a.m. and reached the mountains an hour later, crossing over some low passes. Bean and Mason believe the mountains were the Tinajas Altas Mountains. At 8 a.m., they ran into the trail of the Quechan to the mission of Caborca, which must have run down the Lechuguilla Valley. They apparently followed the trail until two that afternoon, when they turned northward for 6 leagues until they reached Tinajas Cojuat. According to Bean and Mason (1962:95), "Sierra Cojuat may be the Copper Mountains, north of Coyote Peak. Cojuat is a Gila settlement mentioned by Manje and others. Its site is probably just west of Wellton and possibly indicates the Yuma frontier." Sierra Cojuat likely is the Copper Mountains, and Coyote Peak is their summit, not a peak north of them. But, Baker Tanks *is*, however, a logical place for Tinajas Cojuat.

The next morning, Caballero's party continued in a northeasterly direction, passing another mountain range, which we contend was Baker Peaks, and then the north end of the Mohawk Mountains and plain to reach the Gila River. They pushed on to Jalcehdume (Halchidoma) and then Pee-Posh (Cocomaricopa) territory before turning southward around Gila Bend. Caballero returned to Arizpe, Sonora, which continued to function as the capital of Sonora and Sinaloa.

What is interesting about the fragmentary account of Caballero's journey is its ethnic dimensions. His Indian guides clearly followed routes that minimized or avoided proximity to enemy territory and took them as quickly as possible—water and forage permitting—to the territory of their allies. It seems clear that Spanish travel across the region spanned by the Camino del Diablo was determined as much by the complex politics of the Colorado and Gila tribes as it was by the availability of water and pasture.

On June 8, 1823, Caballero, Captain Romero, and nine soldiers from the Tucson presidio set off for California, heading northwest to the Pima villages along the Gila and then descending the river through Pee-Posh (Cocomaricopa) territory. On June 26, four leagues southwest of Agua Caliente, west of the bend on the Gila River, their Pee-Posh (Cocomaricopa) guides halted, saying, "Now you can travel faster, for we will stay here to hunt hawks, and we will go back with our families, for if we go further on, the Yumas will meet us and will want to fight with us, and we will want to fight with them, but you do not wish to" (Bean and Mason 1962:19). The Romero expedition proceeded down the Gila nine more leagues past the Cerro de San Pascual and forded the Gila to its south bank. On June 27 at 3 p.m. they left the Gila and headed southwest toward the Sierra Cojuat (Copper Mountains). Bolton (1930:178) places Cerro de San Pascual at the northern Mohawk range, but we wonder if Cerro de San Pascual is Texas Hill, a prominent riverside landmark. "O'odham used to come all the way up from Mexico to take baths in the hot springs around the hill" (Broyles, Evans et al. 2007). Or, it may be Antelope Hill: Schneider (1993a) reports the Spanish names for Antelope Hill as Cerro del Metate or Cerro de Santa Cecilia.

According to Romero's diary:

> At 8:00 that night [June 27, 1823] we continued the march on the same route. At 12:00 we crossed the last sierras that there are east of the Colorado River through some easy passes and wide openings. According to conjecture these are the mountains called the Tinaja Alta. We looked at the road the Yumas have made to Caborca, that the padre had informed some soldiers about the route from Santa Catalina to Caborca during his trip. We did not take this route because the gentiles who guided us were afraid they might meet the Yumas, their enemies, and they did not wish to expose themselves or the padre to danger.
>
> June 28. At 2:00 in the morning we halted, having come out upon a plain. At 4:00 in the

> morning we left enroute to the west, crossing the sand dunes and sandy areas. At 11:00 we reach a lagoon called Yxlgqua [Laguna Prieta]. Distance: (Leagues 12).
>
> June 29. At 2:00 in the morning we left enroute to the south, inclining a little to the west. At 10:00 we halted, having arrived at the Colorado. We found it very high. Distance: (Leagues 14).
>
> Many Indians began to arrive—Cajuines and Quamayas. They brought pechita [flour made from mesquite pods], which was powdered, and balls of pechita and pinole, for we had used up all the food we had, and we were suffering from want. At 3:00 in the afternoon several captains came and led us to the first ranchería. Distance: (Leagues 1). (Bean and Mason 1962:20)

From Romero's account we can begin to reconstruct the ethnic geography of the Tinajas Altas region, or at least the Lechuguilla Valley. The Cocopah (Quamaya) guiding Caballero and Romero followed a trail that cut across the Yuma Desert from their territory along the lower Colorado to the territory of their allies, the Pee-Posh (Maricopa), along the Gila. That trail crossed the Tinajas Altas Mountains, perhaps at Tinajas Altas Pass or Cipriano Pass. The Lechuguilla Valley was a dangerous area, however, because the Quechan trail to the Río Sonoyta and the Altar Valley ran up the broad valley. The Tinajas Altas themselves may have been a dangerous and contested waterhole, more in the Quechan than the Maricopa or O'odham orbit. That may be one of the reasons why Indian, largely O'odham, guides did not often lead their Spanish companions to them.

Further, it is interesting that Romero describes a trail used by the Cocopah from the lower Colorado River and delta *northeast* to the Gila River, oblique to most of the traditional trails, which trend northwest. By traveling from the Colorado River to Laguna Prieta, then through one of several passes in the Tinajas Altas Mountains, they could safely and rapidly reach Baker Tanks and then the Gila River. Such a route covers about 100 miles and might be covered afoot in three to five days. This detour around Quechan territory near the Gila-Colorado confluence would reduce their chances of a hostile encounter, and this might indicate that Sedelmayr used Cocopah guides for the Colorado-River-to-Tinajas-Altas leg of his 1751 journey. An analogous situation occurred north of the Gila River where the Pee-Posh (Maricopa and Cocomaricopa) left a network of trails as they detoured across open desert to avoid their enemies who lived along the lower Gila and near its confluence with the Colorado (Becker and Altschul 2003).

Caballero and Romero were attempting to find an alternative route to California via Cocopah territory. On July 2, however, the Cocopah stole their horses and abandoned the Spaniards on a raft in the middle of the Colorado, where they almost drowned. "The road from Tucson is short . . . and it can be covered in ten or eleven days," Romero reported to the governor of California, "but there is the obstacle of the Colorado River, for from the month of May to that of August it remains in flood; and furthermore, the Indian nation that robbed us is there . . . according to my instruction, it only remains for me to see the road by which the Cocomaricopas travel"(Bean and Mason 1962:25). Romero's attempts to find that route, which crossed the Colorado considerably north of its junction with the Gila, were inconclusive. Yuma Crossing therefore remained the most common ford of the Colorado despite the sometimes hostile presence of the Quechans.

The Rising Importance of Tinajas Altas

It now seems clear that Tinajas Altas did not serve as an important way station and waterhole for Spanish travelers during the colonial period, although at least two expeditions stopped there. The most common route seems to have been to the east using water from Heart Tank, Cabeza Prieta Tanks, and Baker Tanks or shallow hand-dug wells along Coyote Wash. If this was the case, then why did Tinajas Altas become a major stopping place by the time of the California gold rush in the mid-1800s?

Tule Well apparently was not constructed until after 1860, so it would not have been a factor determining the route of the argonauts. However, nearby Tule Tank might have drawn some travelers to the west, away from Cabeza Prieta Tanks and then across the Lechuguilla Valley to Tinajas Altas. For whatever reasons, by the mid-1850s Tinajas Altas becomes a major and most reliable source of water along the Camino whereas Heart Tank and Cabeza Prieta Tanks apparently recede in importance in the popular imagination (see discussion of Kippen in chapter 4). This apparent change in routes may be an artifact of our lack of information on Mexican gold seekers and others crossing the Camino region. If we were to discover more accounts of Mexican sojourners along the Camino, we might find mention of alternate routes, including the more easterly Heart Tank-Cabeza Prieta Tanks-Baker Tanks trail. Until then, we are nudged to speculate from the fragmentary and ambiguous evidence at hand.

One potentially productive line of investigation would be to gather more information on the trails, water sources,

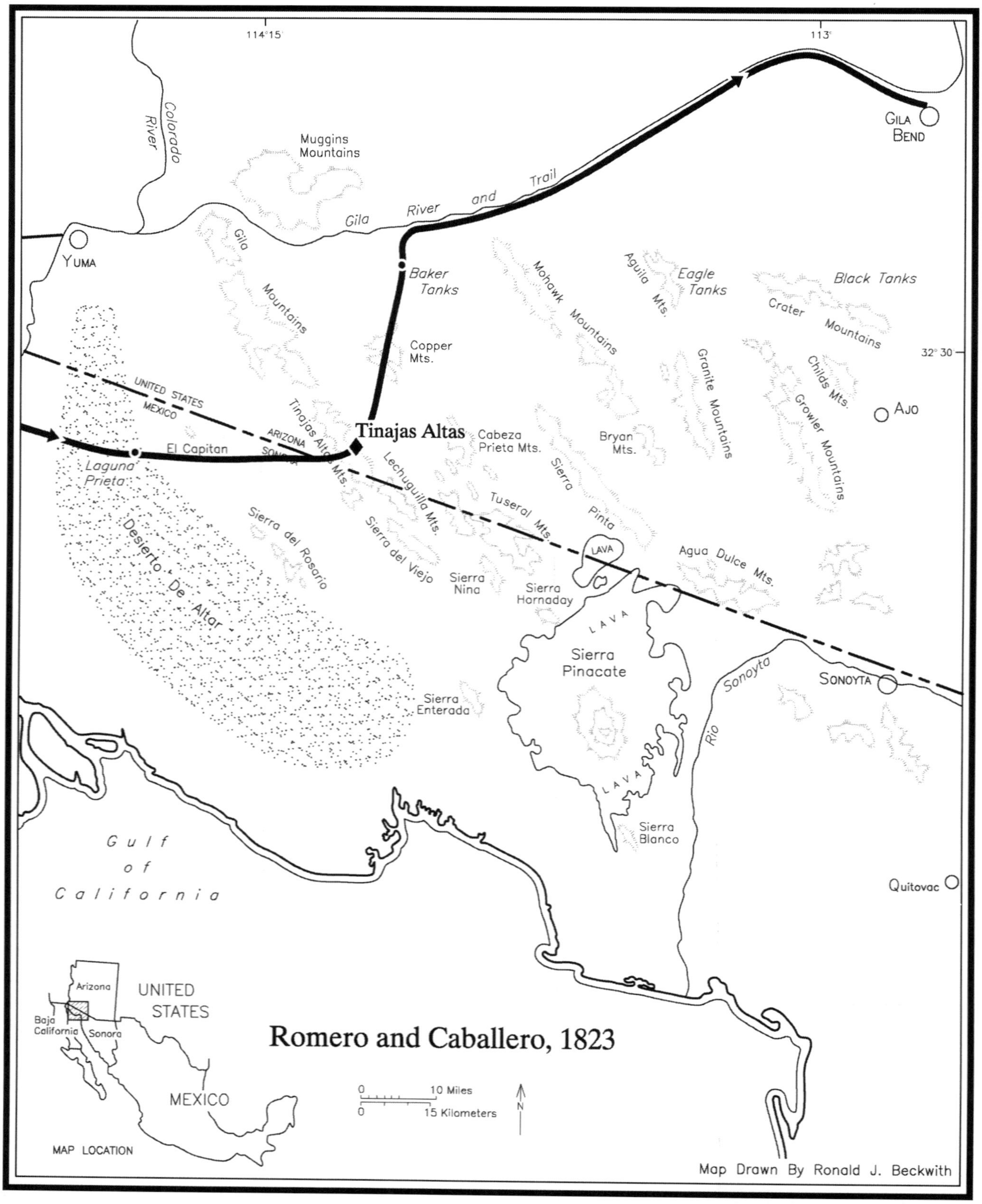

Romero and Caballero route, 1823. (By Ronald J. Beckwith)

and landmarks known and utilized by the different Indian groups that crossed the Camino region. Those groups would include the Cocopah, Quechan, and Pee-Posh (Maricopa) as well as the Hia C'ed and Tohono O'odham. Chapter 2 makes a start in this direction, but much remains to be done. Travelers in the late seventeenth and eighteenth centuries relied upon Indian trails and Indian guides, who would have had their own funds of knowledge about routes and water sources and their own motives for leading the missionaries and Spanish military officials in certain directions. Travelers following the Heart Tank-Cabeza Prieta Tanks-Baker Tanks route would have reached the Gila farther east than a Tinajas Altas-Dripping Springs route or Anza's cutoff west of the Gila Mountains. That might have brought them to Pee-Posh rather than Quechan territory. Did the Heart Tank-Cabeza Prieta Tanks-Baker Tanks trail represent a route controlled by the Pee-Posh, who were bitter enemies of the Quechan?

And were there other Quechan routes, or trails where Quechan may have been encountered? The Akimel O'odham (Gila Pimas) were staunch allies of the Pee-Posh (Maricopa) and enemies of the Quechan and their allies, who included the Mohave and Yavapai. The roles of the Tohono O'odham and Hia C'ed O'odham in the complex alliances and enmities of the River Yumans are less clear and may have shifted through time. The complex struggles of the River Yumans—and the relationships of the Hia C'ed O'odham with these different Yuman groups—undoubtedly played an important role in the routes followed by early European and Euroamerican pioneers.

Those pioneers may not have been aware of the reasons why certain guides led them over certain routes, however. Trying to reconstruct what routes Kino, Sedelmayr, Garcés, and Anza followed, and why they followed them, from the documentary record alone reveals the limitations of that record and the Eurocentric biases of the scholars who have struggled to interpret it. We falsely assume that the forbidding basin-and-range country was an empty wilderness of scorched rock and stunted vegetation. On the contrary, it was crisscrossed with trails made by many different peoples: Hia C'ed O'odham, Quechan, Pee-Posh (Maricopa), Cocopah, and other River Yuman groups whose identities we barely grasp. Those people came to those hidden tinajas to hunt bighorn sheep, gather food and medicine, hide from their enemies, and seek power. They followed those trails to trade, raid, visit, and explore. Kino and the Europeans who followed him came to know a few strands of a complex and ancient web, but they barely glimpsed the geographic networks of knowledge held by the people living in and around the Camino del Diablo region.

As more and more *gente de razón* ("people of reason," as the Spanish haughtily labeled non-Indians) settled in northwestern Sonora, and Indian populations declined, sojourners along the Camino may have grown less dependent upon Indian guides. Clearly, as time went on, horse-mounted prospectors, military officials, and even bandits developed their own knowledge about waterholes, pasture, and terrain across the basins and ranges north of the Altar Desert. They may, therefore, have chosen to shave miles off the journey to the Colorado River and beyond by heading straight to Tinajas Altas, where they either continued along the eastern slopes of the Tinajas Altas and Gila mountains to Dripping Springs, or crossed Cipriano Pass or Tinajas Altas Pass and sought Spook Tank before braving the bleak and sandy Yuma Desert. In a sense, their understanding of the geography of the Camino region overwhelmed the earlier O'odham-Quechan-Maricopa-Cocopah network of alliances and trails. And as it did, Tinajas Altas loomed larger and larger in the lore of the Camino del Diablo.

Gold Fever and the Camino del Diablo

The history and cultural significance of Tinajas Altas during the late eighteenth and nineteenth centuries cannot be fully comprehended without understanding what geographer Robert West (1993) calls the "Sonoran gold craze." Two significant events attracted non-Indian settlers to northwestern Sonora in the eighteenth century. The first was the establishment of the presidio of Altar on the Río Altar in the mid-1750s (Polzer and Sheridan 1997), which provided military protection in the western Pimería Alta. (Altar was the oft-transferred Sinaloa presidio, founded in 1595 in San Felipe y Santiago de Sinaloa far to the south.) The second event was the discovery of gold at Cieneguilla 12 leagues south of Altar in February 1771. The series of gold strikes that followed drew thousands of prospectors—as well as those who supplied or preyed off them—to one of the most arid regions in North America. According to West (1993:79), the discovery of gold at Cieneguilla south of the presidio of Altar was among the "first authentic gold rushes in North America." In his words, "Although characterized mainly by the formation of ephemeral camps, these gold rushes represented the initial development of western Sonora, which today, not through mining but through irrigated agriculture, has become the leading economic and demographic area of the state" (West 1993:83).

Cieneguilla was the first and biggest of these strikes. Several thousand prospectors extracted 1,000 *marcos* of gold worth 130,000 pesos during the first two months of the strike; by 1772, the *real* (royally chartered mining community) of

Spook Tank. (Photo by Bill Broyles, 2005)

San Ildefonso de Cieneguilla had swollen to 5,000 people. In January 1773, the Cieneguilla mining district contained an estimated 7,000 individuals, making it the largest settlement in Sonora (Del Río 1981; Escandon 1985; Stern and Jackson 1988). Pedro Corbalán, governor of Sinaloa and Sonora, described the region as "sown with gold" (Del Río 1981:165). Prospectors scattered across the landscape, throwing up crude shelters near their diggings. Even though the town of San Ildefonso de Cieneguilla grew up around wells drawing from shallow subsurface water found in local arroyos, the placers themselves extended for 12 to 14 leagues in the direction of "the four winds" (Del Río 1981:167).

Many of the miners were Indians, particularly Yoeme (Yaquis), who returned to their communities to participate in the fiestas of Corpus Christi (celebrated in late spring or early summer) and San Juan (June 24), as well as to plant and cultivate their summer crops from May to the end of October (Del Río 1981). Others, however, were members of a floating frontier population of *castas* (people of mixed race) including *mestizos* (Spanish-Indian), *mulatos* (Spanish-African), and *coyotes* (mestizo-mulato-Indian).

As Cieneguilla's surface deposits were exhausted, prospectors drifted to other strikes such as Santa Rosa de Buenavista, which was located at Palo Ensebado, eight leagues east from Cieneguilla. Cieneguilla's population itself declined to 775 in 1778, but additional strikes in 1802 led to the establishment of a new settlement called San Francisco de Asís to the northeast. By 1806, the Cieneguilla district's population had risen again to 5,000 (Stern and Jackson 1988). Both Indian and non-Indian prospectors alike migrated hundreds of miles between their home communities (if they had them) and gold strikes across central and northwestern Sonora (Del Río 1981; Escandon 1985). They followed a seasonal pattern—working the *placeres* (alluvial deposits) or *labores* (subsurface veins of gold) from the end of October until mid-May, and returning to their homes to tend their fields during the frost-free summer growing season from May through October (Del Río 1981).

Table 3.1. Population of the Altar District, 1834

	Females	Males	Total
Altar	595	632	1,227
San Francisco	321	285	606
Cieneguilla	298	289	587
Caborca	183	199	382
Pitiquito	138	177	315
Tubutama	179	197	376
Total	1,896	1,983	3,879

Source: Archivo Histórico del Estado de Sonora (AHES), carpetón 60, film 45, microfilm on file at the Arizona Historical Society/Tucson.

Most of the prospectors who searched for gold in northwestern Sonora and southwestern Arizona were not full-time miners, but rather jacks-of-all-trades, mining during the gold strikes while maintaining farms and livestock in their home communities. Many miners barely eked out a subsistence, bartering their gold for cloth and foodstuffs with itinerant traders employed by the seven merchants who controlled commerce in the Cieneguilla district (Del Rio 1981). In 1816, a census of the Cieneguilla district noted 838 permanent residents: 253 Spaniards, 288 castas, and 297 Yaquis and Níjoras (Stern and Jackson 1988). Níjora was a generic frontier term for Indian captive; it did not refer to a specific ethnic group (Dobyns et al. 1960). Captives arrived via a slave trade in which River Yumans and O'odham served as intermediaries. These Níjoras and their captors undoubtedly brought knowledge of lands and perhaps mineral deposits north of the Cieneguilla district. It is quite possible, then, that some casta and Níjora prospectors may have explored the dry mountain ranges north of Cieneguilla and the Altar Valley during the late eighteenth century. They undoubtedly knew of Tinajas Altas as the most reliable source of water between the Río Sonoyta and the Colorado.

The Presidio of Altar

The community of Altar, which developed around its military garrison, served as the center of supply and protection for this far-flung mining district on the Sonoran frontier. It lies in a relatively fertile river valley about 200 miles southeast of Tinajas Altas. In 1764, a decade after its establishment, there were 285 settlers in Altar, including the soldiers and their families. By 1822, the Altar district had 2,645 non-Indian settlers scattered across an immense dry landscape from Cieneguilla on the south to Quitovac on the north (Nye 1861; Stern and Jackson 1988). And as Indian populations in the mission communities of the Río Altar-Río de la Concepción river system dwindled because of epidemic Old World diseases and the loss of their farmlands to non-Indian settlers, more non-Indians moved into former mission pueblos like Caborca, Tubutama, Oquitoa, and Pitiquito (Jackson 1994; Radding 1997). By 1834, a census of the Altar district revealed a total population of 3,879, and went on to say, "There exist in the vicinity of this *villa* [Altar] three haciendas, sixteen ranchos, five *minerales* [mining camps], and two seaports that are not habitated. [A route] by land has also been discovered and is being traveled without risk to the territory of California" (Archivo Histórico del Estado de Sonora [AHES] 1834). This "route by land" statement indicates that the land route between the Altar Valley and California along the Camino del Diablo continued to be used on at least a semi-regular basis.

Historian William Mason (1963) called the route the Sonora Trail and shows Tinajas Altas on his map as a major stop along the way. He also lists a number of prominent early California individuals who traveled along the Camino, including Antonio Franco Coronel in 1834, who served as mayor of Los Angeles in 1853 and California state treasurer from 1867–1871, Luis Arenas (in 1839), a Sonoran native elected mayor of Los Angeles in 1838, Narciso Botello (in 1833), secretary of the Los Angeles town government in 1836, and Juan Forster (in 1833), an Englishman who owned the Santa Margarita y las Flores ranch.

The Altar Valley—Alta California Connection in the 1830s

During the 1830s, another series of gold strikes advanced the mining frontier north of the Altar Valley. According to Francisco Velasco (Nye 1861), the first was La Basura in 1835, eight leagues north of Caborca, where 2,000 to 3,000 prospectors converged until the shallow surface deposits played

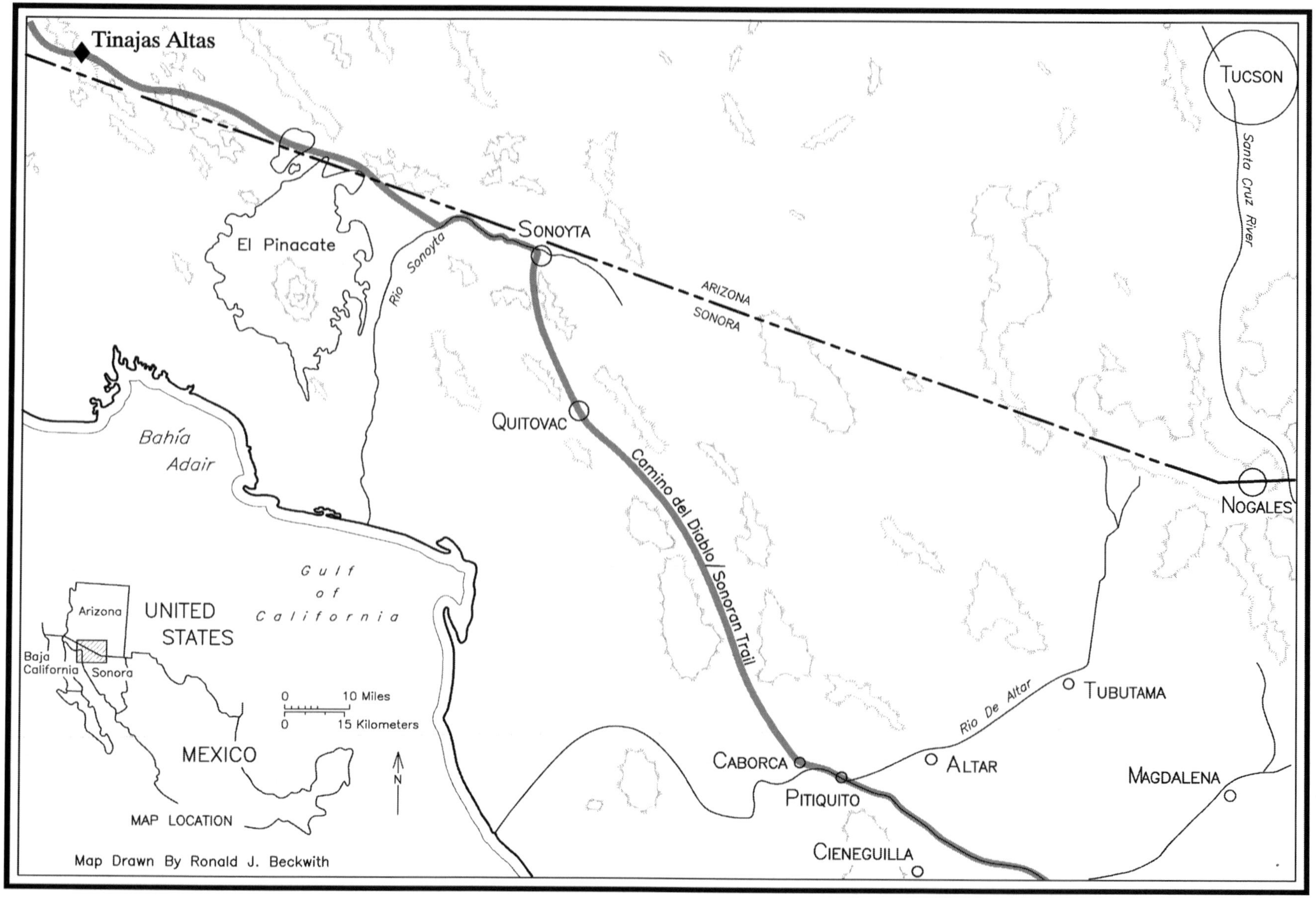

Region showing Caborca, Altar, Cieneguilla, Altar Valley, and the Sonoran Trail to Quitovac and Sonoyta. (By Ronald J. Beckwith)

out. The second was San Perfecto to the northwest and the third was Quitovac in 1835 near the famous oasis southeast of Sonoyta. Quitovac was the most substantial strike; after working the placers, miners dug shafts as deep as 10 to 15 varas (about 27 to 41 feet). Three leagues west of Quitovac, O'odham led Padre Faustino Gonzales to the strike of San Antonio, where Indian and non-Indian prospectors worked side by side until the O'odham rebellion of 1841 (Nye 1861). Strikes were also made at Alamo Muerto and at Las Palomas in 1835 west of Caborca. Those strikes yielded both gold and silver, although silver mining was limited by the shortage of quicksilver (mercury) necessary to reduce the silver ore (Nye 1861). At the end of the twentieth century a major mining company was still probing the Quitovac area.

The most common form of transportation along the frontier of northern Mexico was the mule train, and the demand for mules in an active mining district would have been particularly high. Mules were used not only to haul ore and support materials, but also were harnessed to a pole in the center of a circular open space (*patio*) and driven through sludge to mix quicksilver with crushed silver ore, a technique known as the patio process of silver reduction. Some of those mules were obtained in Alta California and driven back to northwestern Sonora over the Camino del Diablo. On August 11, 1838, for example, Lorenzo Martínez, justice of the peace at Altar, wrote to José Urrea, governor of Sonora, that Captain Juan José Tobar had transported 21 "beasts, both mules and horses" from California to Sonora that July, which is the hottest part of the year. Captain Pablo de la Portilla lent him six mules and gave him one horse; retired Sergeant Francisco Gastelum lent him six mules; the administrator of the mission of San Diego gave him two mules and one horse; and the administrator of the mission of San Luis Rey gave him two mules and three horses. The party left four mules along the Río Colorado, perhaps as a "gift" for the Quechan Indians who controlled the river crossings. They drove the rest to Quitovac, the most active mining district at the time (Martínez 1838). Those mules and horses probably watered at Tinajas Altas.

The late 1830s were also times of turmoil in northern Mexico as federalists (those who favored strong states) and centralists (those who favored a strong central government and weak states) fought for control over Alta California, Sonora, and New Mexico in the wake of the Texas revolt. In 1837, General José Urrea, who was born in Tucson in 1797 and was a national hero because of his distinguished service against the rebel Texans, led a federalist revolt in Sonora (Herring 1995; Voss 1982). In the spring of 1838, officials in the Altar district were pressured to make "voluntary contributions" to support a federalist expedition to Alta California, which left from Altar and may have traveled along the Camino del Diablo and watered at Tinajas Altas. The Camino served as an important route between Sonora and California during the chaotic power struggles that wracked Mexico in the decades following independence from Spain.

Quitovac and the O'odham Rebellion

Then the discovery of gold near the oasis of Quitovac strained relations between the O'odham and the non-Indian settlers of northwestern Sonora, eventually sparking a series of O'odham rebellions that affected travel along El Camino del Diablo. In late April 1838, Rafael Moraga led the militia company of Altar on an expedition to investigate and quiet the unrest. Sixty-four men assembled at the wells of Corona and Caborqueños; they were on foot because all available horses had been commandeered for the federalist expedition to Alta California. They were accompanied by Francisco Carreon, general of the O'odham in the Altar Valley. At the ranchería of Soñi, Moraga and his company met with the O'odham governors of Cubó, Carricito, Ayoma, and Tac. At Carricito, an O'odham settlement near the placers of San Perfecto, the governor complained that a miner named Diego Celaya had brought a crew of nine workers to dig out Carricito's waterhole in order to increase the water supply for the gold placers and the animals of the miners. The O'odham protested when Celaya built a diversion ditch to channel their water onto his diggings, so Celaya told the justice of the peace in San Perfecto that the O'odham were in revolt.

Moraga's expedition had to return to Altar after a few days on the trail because it ran out of provisions. Nonetheless, his diary of the brief campaign allows us to catch glimpses of the pressures experienced by the O'odham as Mexican settlers pushed them out of the Altar Valley riverine oases while prospectors attempted to seize the waterholes of their desert communities (Moraga 1838). Later that year, Santiago Redondo, subprefect of Altar, wrote the governor of Sonora that "the indigenous inhabitants of Quitovac, Sonoyta, Carrizalito, Soñi, and others scattered across Tribaipa and Cuboá found themselves very discontented because of the extortions and insults they had suffered from evil men, who had deprived them of their scarce water to supply those who inhabited the gold fields" (Redondo ca. 1838). The conflict erupted into open warfare in 1840 and 1841, when pitched battles between O'odham and Mexicans took place at Cóbota and the foot of Baboquivari Peak (Officer 1987). These pressures undoubtedly drove the Hia C'ed O'odham deeper into the northern and western reaches of the desert, including the Pinacate volcanic area *and* the region around Tinajas Altas.

El Camino del Diablo west of Tinajas Altas, in the vicinity of Arroyo de San Albino. (Photo by Bill Broyles, 2006)

The Mexican War and the California Gold Rush

The Camino del Diablo became an escape route for Mexicans exiting California after war broke out between Mexico and the United States. On September 1, 1846, Dionisio Gonzales, military commander and prefect of the district of Guadalupe [Altar], reported that José Castro, the commandant general of Alta California, had fled with 20 other officials and soldiers after US troops landed at Monterey and then invaded Los Angeles (Eisenhower 1989). Castro and his party crossed the Colorado and traveled the Camino to Quitovac, where he notified Gonzales of his arrival (Gonzales 1846).

Californio (Hispanic Californian) resistance to the US invaders continued after Castro departed. Captain José María Flores, who assumed command of the Californios after Castro left, even managed to defeat US garrisons in Los Angeles, Santa Barbara, and San Diego. But no evidence has yet been uncovered that Castro or any other Mexican commander led Sonoran reinforcements back to the fray. After US forces recaptured Los Angeles in January 1847, Flores escaped to Sonora as well, perhaps along the Camino, perhaps along the more well-traveled Gila Trail (Eisenhower 1989). US forces under General Stephen Watts Kearny and Lt. Colonel Philip St. George Cooke followed the Gila River into California. When Kearny's troops crossed the Colorado, Lt. William Emory cast his attention to the south, and wrote, "Departing from the ford . . . in the direction of Sonora, there is a fearful desert to encounter. . . . All accounts concur in representing the journey as one of extreme hardship, and even peril. The distance is not exactly known, but it is variously represented at from four to seven days' journey" (quoted in W. Hartmann 1989:81). The Camino remained a trail of dim legend for Anglo-American soldiers penetrating the Sonoran Desert during the Mexican War.

But all soon changed. The military movements into and out of California during the Mexican War paled by comparison with the California Gold Rush. News of gold at

Sutter's Mill on the Sacramento River drew argonauts from the United States, Mexico, and even Chile. According to historian Patricia Etter (1998), approximately 20,000 forty-niners followed the Southern Overland Trail to Yuma Crossing, most of them traveling north through the Santa Cruz Valley and west along the Gila River. Few US forty-niners, in contrast, dipped south through the Altar Valley to travel the Camino. Etter's exhaustive study of the Southern Route lists only three individuals who left records of braving the Camino on their way to California. The Archivo Histórico del Estado de Sonora contains mentions of armed parties of Anglo-Americans in the vicinity of Altar but no biographical details revealing who they were.

ANGLO ARGONAUTS ON THE CAMINO DEL DIABLO: DYE, HOBBS, AND HUBBARD

The earliest Anglo argonaut who left an account of travel in the region, and who may have traveled along the Camino, was Job Francis Dye. He first reached California as a trapper with mountain man Ewing Young in 1832 and embarked on a colorful career. During the next two decades, Dye hunted sea otters, raised mules, and operated a distillery. After the gold rush began, he made several voyages from Monterey to Mazatlán to purchase goods that could be sold at high prices to the prospectors. In February 1849, Dye bought 300 mules in Mazatlán and drove them back to California through Sonora and across the Colorado River.

Dye may have followed the Camino, but Dye's own account is unclear (Etter personal communication, 2007). "In Mazatlán I bought a band of three hundred mules and started via the Colorado for California, and cleared over $12,000 on the speculation, notwithstanding I lost 80 odd head in crossing the river Jila [*sic*] and the Colorado," Dye recalled. "While passing through Sonora the people warned me of the danger from these rivers and the savage Indians, stating that I never would get through the Apache country without losing all the stock; therefore, I took every possible precaution. I hired fifteen Mexicans to pack and herd the mules, and twenty-five New Yorkers, armed with rifles and shotguns to guard the train, and protect me through the Apache Nation" (Dye 1951:66–67). Whether Dye followed the Gila Trail or the Camino del Diablo, the story illustrates the perils of travel through the region.

Another equally vivid character named James Hobbs did travel the Camino to California and may have utilized Tinajas Altas, although his account also is unclear. Like Dye, Hobbs came west as a trapper and mountain man in the 1830s. After serving as a scout for Colonel Alexander Doniphan in northern Mexico during the Mexican War, Hobbs was working for an English mining company in Guanajuato when "the great California gold fever" broke out in Mexico and he "became infected with the disease" (Hobbs 1872:205). Hobbs made his way north through the northern Mexican cities of Zacatecas, Durango, Mazatlán, Culiacán, Alamos, Guaymas, and Hermosillo. There he met several Americans from California, including Captain Ankrim (perhaps William Ankrim, who settled at Yuma during the 1850s [North 1980]), David Brown, and Thomas Smith, who had come there to buy cattle because prices were so high in California. They hired Hobbs as an interpreter and convinced him to throw in with them. He purchased 450 head of his own at $10 per head and the party headed for Altar with a herd of 2,500 cattle.

In Hobbs' words:

> At a town called Altar, near the northern boundary line of Sonora, we engaged twenty vaqueros, or Mexican herdsmen, and bought jerked beef, and bread, and pinola [*pinole*, which is ground parched corn] to eat on the way when cooking would be inconvenient. We also purchased a lot of jackasses to use in place of pack mules; eight of which I required for my own use, and for my men.
>
> A number of Mexican men, women, and children, who were poor people going to California, joined our company for their own protection. We had for a guide, a Mexican who had come over the road from California to Hermosillo with the three other owners of the stock.
>
> A part of our route lay through a sandy country, destitute of rivers or springs. On the route is a great rock, with a cavity which is so large that in the rainy season it fills up, and supplies water for travelers the year round. It was a hundred and sixty or a hundred and seventy miles from Altar, and eight miles from our last watering place. We could drink at the rock ourselves, and bring enough water for our mules, horses, and jacks; but our cattle had to go dry, and before us was a stretch of sixty miles of sandy desert without a drop of water that we knew of, or a blade of grass. The desert was also subject to sand storms, which blew the fine sand up in clouds like fog, and swept it over the road, obliterating all tracks, and making traveling by compass necessary. We had no fears for our jacks, for they can forage for a living. A

bundle of rags, or a deck of cards is a fine lunch for one of them. The morning after leaving the watering place, there was great complaint on account of the scarcity of hats. The jacks had been around and eaten up ten of the Mexicans broad-brimmed straw hats, besides a lot of our bread and pinola.

While the Mexicans were lamenting the loss of their hats, I discovered a greater loss. I missed two of my jacks, and while trying to track them, I saw the moccasin tracks of two Indians. Taking a Mexican along with me, we rode about four miles, tracking the Indians easily, and then on seeing a smoke among the hills, we dismounted, and creeping around, saw one of the Apaches roasting a steak which he had cut from one of the jacks, which he had killed. The other Indian had started for a stream of water close by. I told the Mexican to attend to this one, while I crawled up near the Indian by the fire and shot him in the back. He fell forward on his face into the fire, with his jackass meat. The other one escaped from the Mexican, but we secured the stolen jack that remained, and returned to camp, glad indeed to find water for our stock. On reflection, I felt sorry that I had shot the Indian, as his theft had led me to discover water sufficient to save many of our animals, who were suffering. We sent back and picked up fifty or sixty cattle that had given out by the way, got them up and they joined the herd, when they were all safely driven to the stream of water. We remained there for two days resting and recruiting the stock, and then started for Fort Yuma, on the Colorado river. We arrived there the second day after leaving the stream that saved our stock. (Hobbs 1872:211–213)

Hobbs' account is fascinating yet elusive. The "great rock" with the large cavity could be Tinajas Altas, yet the distances Hobbs gives are somewhat troublesome. He describes it as "eight miles from our last watering place"—too short a distance if that last watering place was Tule Tank. Alternatively, it could have been Cabeza Prieta Tanks, which are closer to Tule Tank and about 60 miles from Yuma Crossing. Even more tantalizing is the "stream of water" Hobbs found while tracking his stolen jackasses. The stream must have been an ephemeral flow following a locally heavy rain. After rains, the impervious streambed between the tinajas and a short distance below Cabeza Prieta Tanks may flow for a few days. The place also was within a two days' ride of Fort Yuma, which at that time was actually Camp Calhoun (Walker and Bufkin 1979). Hobbs gives no dates for his trek along the Camino. Based on his later reference to the massacre of John Glanton's party of scalphunters "about a month before we arrived," however, Hobbs and his companions must have traveled it in May or June, because the Quechan killed Glanton's band on April 21, 1850. This would be the driest time of the year in the Sonoran Desert, which makes the "stream of water" all the more intriguing, if Hobbs meant it literally.

A third Anglo-American forty-niner who followed the Camino was Cal Hubbard. Unfortunately, the only available account of his ordeal is an article in *True West* entitled the "The Devil's Highway" by Tom Bailey (1959). The account is written like a pulp Western, with treacherous guides and steely-eyed showdowns. Nonetheless, it mentions Hubbard and others talking to reporters from the *Alta Californian,* an early California newspaper, so there must be articles about the Hubbard party in that periodical. According to Bailey, Hubbard was leading a wagon train from Memphis bound for the gold fields in 1849. When the party reached Arizona, presumably somewhere along the Santa Cruz River, Hubbard told them they could choose between the Gila Trail, where they would have to face the Apaches, or the Camino del Diablo, where they would have to face their thirst. "Fellow travelers, it is your right to say which way you will go to California—north or south. But let me tell you this: These damned Apaches here in Arizona are the worst savages on the entire continent. If we were a caravan of men I would say take the northern route and fight them. But I have my wife and little one to think of. I vote for the Camino del Diablo" (Bailey 1959:7).

Hubbard put the matter to a vote. Fifty adults selected the Gila Trail and 36 chose to follow Hubbard. In Sonoyta, three "ruffians" rode out of the cottonwoods and offered to guide Hubbard's eight wagons. "Is it really necessary that we have guides? I hear the route is plainly marked by the wagon ruts," Hubbard asked. "One of them explained that there were long stretches of shifting sands that completely obliterated any rut marks. If they got through safely they would need good guides" (Bailey 1959:7).

The guides identified themselves as Jake Fleagle, Charlie Free, and Francisco Morales. They claimed they had served as guides for the US dragoons during the Mexican war. As the wagon train pulled out of Sonoyta, Hubbard's adopted son, Piny Shoot, fell in with the guides, and was soon to follow his own devil's highway. Bailey does not say what time of year they were traveling, but he mentions the dryness and the intense heat, so it may have been June or early July.

"The heat became more oppressive as the caravan moved deeper into Mexico. It was so hot that grease ran out of the axle hubs. In the distance little heat devils danced and not a living creature stirred" (Bailey 1959:8).

Three days out of Sonoyta, the wagon train was crawling across the desert at the rate of seven or eight miles a day. "I can't see what God made so much land for!" Hubbard wailed. On the fourth day, the other members of the party who had chosen the Gila Trail rejoined them, scared off by rumors of Apache attack. By the eighth day, the group's water had run out. Hubbard's wife gave birth and the baby died within an hour. Hubbard's wife died the next morning. Hubbard wanted to remain at her grave but his friend Price convinced him to go on for the sake of his little daughter.

Hubbard accosted the guides and asked why there were no wagon ruts along the route they were following. The guides pointed to the sun and said they were on the right trail for Yuma. They said there was a spring four miles ahead. The spring was dry. Hubbard broke out a barrel of whiskey and filled everyone's tin cups. "The whisky was wet and it helped, but there was no helping the poor horses" (Bailey 1959:8). Apparently, neither Hubbard nor his chronicler understood the correlation between alcohol and dehydration.

As the party trudged onward, tongues thickened and blackened except for those of the guides. The guides convinced Hubbard to gather up all the canteens so they could fill them up at a waterhole five miles ahead. They also talked him into unhitching the horses and oxen and letting them go to water. Hubbard staggered into the desert after them, on foot and delirious. He passed out. "When he opened his eyes again day was breaking and he was wringing wet. It was raining. It wasn't just an ordinary drizzle but a real desert downpour. Water at last! Sent from the heavens" (Bailey 1959:42).

When Hubbard walked back to the wagon train, the survivors were filling buckets and barrels with water. But 12, including Hostetter, Asa Savage, Doris Ruark, Mrs. Zahner and Mrs. Whiteside, had died. Another woman died a day later. Hubbard sent some men after the guides, including his adopted son, Piny, and others to gather the stock. They found Charlie Free dying from a gunshot wound to the chest. Free told Hubbard the guides had planned to lose the wagon train in the desert so everyone would die of thirst. Then they intended to return and loot the victims. Piny had been an accomplice from the beginning, hiding water for them. Free and Fleagle quarreled and Fleagle shot him and left him for dead.

Bailey's account is long on drama and short on facts, although it does state the wagon train was 12 miles west of the Camino, which if true may indicate they were out in the dunes somewhere west of the Tinajas Altas Mountains. Instead, it concentrates on Hubbard's quest for revenge, which he finally slakes when he confronts Piny outside a Placerville saloon and shoots him dead. "This is the kind of thing we do to traitors," Hubbard said. "What have you to say for yourself?" (Bailey 1959:45). But Piny the traitorous was beyond words.

SONORAN MINERS ALONG THE CAMINO DEL DIABLO

Most forty-niners who followed the Camino del Diablo to California were Sonorans or Mexicans from other states in the Mexican Republic who passed through Sonora and wanted to avoid the dangers of Apache attack along the Gila Trail. According to historian J. M. Guinn (1911:31–32), "The first foreigners from beyond the limits of California to reach the mines were Sonorans or Sonorense [*sic*], as they were sometimes called. While they were credited to Sonora there were among them bands from Sinaloa, Chihuahua, and Durango. Many of these had had some experience in placer mining in their own country, and the report of rich placers in California, where gold could be had for the picking up, aroused them from their lazy self content and stimulated them to go in search for it." Guinn (1911:32) goes on the say that the "first gold placers discovered in California—those located in the San Fernando mountains—discovered by Francisco Lopez in 1841, had been worked mainly by Sonorans, most of whom left the country at the time of its conquest by the Americans. These returned with their numerous relatives."

Guinn assumed that most of the Sonorans traveled the Anza Trail, starting at Tubac and following the Gila River to Yuma. From the evidence presented here, however, it is clear that many did indeed take the Camino and stop at Tinajas Altas. Guinn noted, "They traveled in squads of from fifty to one hundred their meager belongings packed on mules or burros. They came in the early spring and returned to their native country in autumn. From this fact came the term, the Sonoran Migration. Some of them brought their women and children with them," he added (Guinn 1911:32). Later, he states, "No record was kept of the number of Sonorans coming into California." In a memorial to Congress in 1850 asking for the establishment of a Custom House in San Pedro, the memorialists say, "At least ten thousand Sonorans pass through Los Angeles on their way to the mines each spring, generally returning to Mexico in the autumn." These migratory birds of passage carried to their homes a considerable amount of gold. A thousand dollars was sufficient to elevate a peon to a plutocrat. This migration began in 1848, reached its maximum in 1850—from that time it gradually decreased until in 1854 it had almost entirely ceased. Their persecution by the Americans, the foreign miner's tax, and the increased amount of labor required to get the gold, convinced these migratory nomads that there was no place like home, so they went home and stayed there" (Guinn 1911:33).

Apparently most of the thousands of Mexicans came up through Sonora, and many must have traveled on the Camino past Tinajas Altas. When Benjamin Hayes, a young lawyer from Missouri en route to California, was waiting to cross the Colorado River in late December 1849, he struck up a conversation with the son of Agustín de Iturbide, the first emperor of Mexico, and a Colonel Carrasco. Both were members of the Mexican party surveying the new international border (demarking lands that changed hands following the Mexican War, not the Gadsden Purchase). According to Hayes' diary, "The Col. estimates the number who have passed here this year at 12,000, of whom about one-half are Sonorians" (Wolcott 1929:46). Doris Marion Wright (1940–1941) concluded that 8,000 Mexicans, most of whom were from Sonora and Sinaloa, journeyed to California in 1849. Even more passed through in 1850, the height of the rush. In 1850, Francisco Velasco (Nye 1861) estimated that there were 4,000 to 5,000 people in the Altar district, most of them drawn by the gold placers in Sonora or the promise of gold in California. Many of those settlers possessed a wide variety of frontier skills including prospecting and placer mining.

As Guinn observed, the number of Mexican miners soon plummeted after the California legislature passed the Foreign Miners Tax Law, which levied a $20 per month fee on "foreign," particularly Hispanic, prospectors. Another humiliation inflicted upon the Mexicans was the abuse they suffered when they attempted to cross the Colorado River. Unsavory ferry operators, including military men like Lt. Cave Johnson Couts, charged exorbitant fees and even robbed miners of weapons and gold. In November 1850, a party of returning miners from Ures reached Altar and complained to Antonio de Urrea, prefect of the district, that Americans had attacked them after they had forded the river and entered Mexican territory (Officer 1987). That party obviously returned to Sonora via the Camino del Diablo.

While they were in California, however, the Sonoran miners bequeathed two techniques to the other prospectors: dry washing, and the *batea,* which Guinn claims was an Indian basket but more likely was a wooden bowl made of mesquite or some other hardwood. Both techniques had been employed in the gold placers of northwestern Sonora, which contributed many miners to the California digs.

Even after the California gold rush was over, hundreds of Sonoran prospectors traveled the Camino to reach the gold strikes along the lower Gila and Colorado rivers. Those strikes began in 1857, when Jacob Snively discovered the gold that led to the boom town of Gila City (Sheridan 1995). As the Report of the International Boundary Commission, which resurveyed the border between 1891 and 1896, noted:

> During the early 1860s there was a large influx of Mexicans from Sonora to the gold diggings on the Colorado River, and an enterprising Mexican dug two wells near the road, in the Tule Mountains, built a small adobe house, and occupied it with his family for two years, for the purpose of selling water to travelers. But the deaths from thirst along this route became so frequent that the road was soon abandoned, and for over twenty years had remained unused. Having accidentally learned of the former existence of these old wells, which are not shown on any maps, arrangements were made to have them cleaned out as soon as possible. This was done by a party of Mexicans and Papagos from Sonoyta, who found them almost filled up.
>
> These important wells are situated on the road about 5 miles north of the point where the line crosses the summit of the Sierra del Tule, and when thoroughly cleaned out yield a supply of about 500 gallons a day. The water is beautifully clear, but owing to the presence of minerals has a vile taste and is very unwholesome. Near the wells stand the ruins of the old adobe house, the only building between Quitobaquita [*sic*] and the Colorado River. (International Boundary Commission 1898:24–25)

Between 1848 and the Civil War, then, there were at least two waves of Mexican migration to gold fields north of the new international border. This was the heyday of the Camino—the era when it became the infamous Devil's Highway. In the dramatic words of the International Boundary Commission (International Boundary Commission 1898, 2:26):

> The road between Agua Dulce and Yuma is for the most part very heavy, and, until Tule Wells were reopened, no certain supply of water could be counted on between Agua Dulce and the Colorado River. This road is appropriately called by the Mexicans El Camino del Diablo. When traveling it for the first time, alone or with but few companions, it is hard to imagine a more desolate or depressing ride. Mile after mile the

Tule Well in 1925. (Photo by Godfrey Sykes [Sykes 1927: fig. 3])

> journey stretches through this land of 'silence, solitude, and sunshine,' with little to distract the eye from the awful surrounding dreariness and desolation except the bleaching skeletons of horses and the painfully frequent crosses which mark the graves of those who perished of thirst—grim and suggestive reminders when the traveler's supply of water is running low. In a single day's ride sixty-five of these graves were counted by the roadside, one containing an entire family, whose horses gave out and who, unable to cross the scorching desert on foot, all perished together of thirst. Their bodies were found by some travelers during the following rainy season, and were all buried in one grave, which is covered with a cross of stones and surrounded with a large circle of stones, inside of which not a bush nor a blade of grass grows. Near by lie the skeletons of their horses and the broken fragments of their water bottles. During the few years that this road was much traveled, over 400 persons were said to have perished of thirst between Sonoyta and Yuma, a record probably without a parallel in North America.

Four hundred was probably a conservative estimate. In a series of articles, writer and hiker John Annerino (1996b, 2009) traced the history of deaths in Arizona's southwestern deserts. Concerning the 1850s to 1900, he stated, "During this period, historians estimate between 400 and 2,000 people died of thirst, making the Camino the deadliest immigrant trail in North America." This is the era when graves dotted the Camino and clustered at the base of Tinajas Altas.

"Various estimates have been made as to the deaths from thirst and other mischance between Sonoyta and the Colorado during these few years of wild excitement; some placing the numbers at hundreds, others at thousands," wrote Godfrey Sykes (1927:64) of the Desert Laboratory in Tucson, who traveled the Camino in 1925. "The facts can never be known, but it is significant that over fifty rude graves can be counted still in one small area at the foot of the rocky chine which contains the Tinajas Altas. Numerous others can be identified along the route, in spite of the obliteration due to time, wind, and occasional torrential rains, and of the probability that the actual line of travel was several miles in width except at certain focal points such as the tinajas."

Unfortunately, no firsthand account of travel by these mid-nineteenth century Mexican sojourners has yet been unearthed. At present, the only tale of the Camino by a Sonoran in this period is the second-hand account collected by James Barney published in *Arizona Highways* in 1943. Barney heard the story about 1920 from Francisco Salazar, an 88–year-old merchant from Hermosillo. According to Salazar:

> Sí, señor, when I was a young man—full of the spirit of adventure—I traveled over the terrible Camino del Diablo, the old trail or road from Sonora to California. It was called the Camino del Diablo because thirst, hunger and Apache Indians took a fearful toll of those who chanced to pass that way. The victims who fell by the wayside along that dreaded desert highway, died unprepared and unshrived, without the comfort of priestly admonition, and so, perhaps, did not reach the gates of heaven. 'Quien puede decir?' Anyway, this deadly trail became known as 'El Camino del Diablo' (The Devil's Highway), a fitting and appropriate title.
>
> As I said before, señor, I was born in the state of Sonora, Mexico, on September 10, 1832. When 17 years of age, I joined a party of about twenty persons—most of them from Sonora—who were preparing to start for the gold fields of California. I was the youngest member of the party and I induced half a dozen other young fellows to join the expedition, one of them—Miguel Costello—becoming my partner in our adventure. There were eight or nine married men and our guide was Don Alejandro Cordova, who had already made a journey to California over the Camino del Diablo and had returned to Mexico by water, landing at the port of Guaymas. From there, he had journeyed overland to Sonora, where he agreed to become our guide for the sunny, golden land of California.
>
> We started on our journey in the spring of 1850, soon after the news of the gold discovery reached Sonora. We took the regular road to California by Sonoyta, San Domingo, Quito va-quita [*sic*], to the south of the Sierra Salada, to Tinajas de Tule and then over the Tule Desert to the Tinajas Altas (High Tanks), an important stopping place on the road to California; from the Tinajas, we followed the real Camino del Diablo for miles and miles across the Sonora Desert to the Colorado River. This stretch of desert is the most dreary and desolate that can possibly be imagined, a sandy, waterless waste—50 miles across—into which hardly a living thing ever ventures.
>
> Some twenty miles to the west of the Tinajas was a permanent landmark called Las Cuevitas (Little Caves), in some rough, volcanic knolls to the south of the road, which at this point went through a pass between Las Cuevitas and the Sierra del Gila. [Las Cuevitas are tafoni caves in the Butler Mountains.]
>
> We were pretty well equipped for the long and dangerous trek, having burros and small Mexican mules for pack animals and being well-armed to resist (if necessary) Indian attacks. For food, we carried large quantities of carne seca (sun-dried beef), pinole (parched corn or wheat ground to a very fine meal) and panocha (a sweet product made from sugar cane).
>
> Our water bags—absolute necessities on such a trip—were made of "vaqueta" (leather) well smeared on the outside with tallow and then covered with a woven mat of grass. They were slung across the back of the pack animals, one half hanging on each side. These water-bags were then (and yet are) made by the Indians of Mexico.
>
> Not all of us had horses to ride, but those who did, had very good animals. My own horse was an excellent steed, large, strong, and dependable. Each member of the party carried "dos pistolas" (two pistols), a "carabina" (shotgun), and a "bota de vaqueta" (a sort of leather canteen for carrying water for individual use).
>
> We had to travel at night for fear of the Apache and concealed ourselves and stock during the day. We had selected the spring time to make our journey, [for] during the fall and winter months . . . was the most dangerous period to

travel over the Camino del Diablo. At that time of the year, the Apache were the most active in their raids and attacks. By then the farmers in the rancherías and pueblitos of northern Sonora had gathered their crops of grain and corn and stock in general was fat and in good condition. That was the raiding time for the Apache and they swarmed along the trails and roads and in the vicinity of waterholes and small communities, and we selected the nighttime in which to move along the dangerous trail, as Apache seldom attacked after dark.

Many people—especially "los Americanos"—thought that the winter months were the best for traveling over the Sonora Desert, but the danger from blood-thirsty savages was then very great. In the summer-time, the Apache were inclined to remain in their mountain haunts and but infrequently visited the hot desert country, which could furnish but little comfort or sustenance.

Our journey as far as Sonoyta was without incident, except that we always had to be on the lookout for Apache. Between Sonoyta and Tinajas Altas—a distance of some 80 miles—there was only one place Tinajas del Tule at which water could be obtained, and between the Tinajas and the Colorado, there was absolutely no water of any kind.

When nearing Tinajas Altas, we camped in a secluded spot some distance from the tanks, fearing an Indian ambush. We then drew lots for the purpose of selecting three men who were to go after water; it fell to my lot to be one of that group. Water was not plentiful at the Tinajas and we had a great deal of difficulty in getting sufficient for our needs.

We would first bring water for the stock and then for our own use. All these activities had to be performed after night-fall under the supervision of our guide.

Travelers had to be careful in approaching the few waterholes along the road, as they were favorite places of ambush for hostile Indians. At Tinajas Altas, in the old days, Apache were always skulking in the nearby hills and arroyos, ready to pounce upon weary or weakened parties or some careless and unwary traveler. The Indians sneaking from place to place would signal to each other by imitating the cry or call of various wild birds and animals, such as the owl, crow, coyote, etc. At times around the tanks these calls could be heard continually, giving to the surroundings a weird and unearthly setting, as if the evil spirits of countless generations were hovering about the place.

The immediate area around the Tinajas was a vast graveyard of unknown dead and the road from there to the Colorado was marked the whole way by the dried carcasses of mules, horses and cattle and the scattered bones of human beings, slowly turning into dust. In such a region but little time can be given to conventional things and the dead were left where they fell to be sepulchred (if at all) by the fearful sand storms that sweep at times over the desolate waste.

We started from the Tinajas with as full a supply of water as we could carry and managed to reach the Colorado River without serious mishap. We hurried over this part of the desert as fast as we could and did not dare make a fire while crossing for fear of the Indians and so had to eat our meals cold. We struck the Colorado some miles below its junction and camped on the California side for a short time to recuperate. We were then among the Yuma Indians but they caused us no trouble. As a rule the Indian tribes of the Southwest were more unfriendly to the whites than they were to the Mexicans. There was much in common between the Indians and the Mexicans (the latter all having some Indian strain) but nothing at all between the whites and the Indians. We continued our way to California and in the month of September, 1850, reached Placerville (first called Hangtown), where much placer gold was still being taken out. (Barney 1943:18–19)

Salazar's account, if true, is interesting for a number of reasons. First, he mentions no alternate route such as up the Mohawk Valley. Nor does he note any sources of water other than Tinaja del Tule and Tinajas Altas. Finally, he presents the danger of Apache attack, not the availability of water, as the major determinant of the preferred season to travel the Camino. According to Salazar, summer was the safest season to make the journey because the Apaches tended to remain in their mountain homelands.

Salazar was an old man remembering a journey 70 years earlier. Even though he returned to Sonora via the Camino and had to hide from Apaches two-thirds of the way between the Colorado River and Tinajas Altas, he apparently never made the trek again. As he told Barney (1943:19), "Eventually, we reached our homes in Sonora, where I bought a sugar plantation, married, and settled down. Gracias a Diós!" During the gold rush, then, the Sonoyta-Tule-Tinajas Altas route appears to have been the most popular for travelers from Sonora to California.

It is also interesting to speculate about the presence of Apaches. Beginning with the discovery of Cieneguilla in 1770, Apache raiders preyed upon the placer mines and missions of northwestern Sonora. Mateo Sastre, governor of Sonora, communicated frequently with the viceroy about Apache hostilities at Cieneguilla and the missions of the Altar-Concepción drainage. Sastre reported that a militia had been formed in Cieneguilla, and Papagos [O'odham] reported that they had killed 10 Apache males and 21 females and had captured 6 Apache girls (Sastre to Bucareli y Ursúa 1772–1773). By the time the gold rush occurred, Apaches—probably Western Apaches from the Aravaipa and Pinal bands (see Goodwin 1971)—had 80 years of experience in the western deserts. By then, they undoubtedly knew routes, settlements, and waterholes throughout the area. Lumholtz (1912:291) mentions Apache raids on Hia C'ed O'odham at Sierra Blanca north of Puerto Peñasco and George Kippen mentions them near Ajo (see chapter 4).

Banditry along the Camino del Diablo

It is also possible that Apaches occasionally received the blame for raids by O'odham, Anglo bandits, and perhaps even Quechan attackers. As noted earlier, O'odham unrest intensified in the late 1830s and early 1840s as miners attempted to monopolize O'odham water sources. And banditry by both Mexican and Anglo-American outlaws flared along both the Camino and the Gila Trail and at Yuma Crossing during the height of the gold rush. Thousands of travelers bearing livestock and gold must have presented inviting targets to Indian and non-Indian raiders alike. Trouble continued during the early 1850s. On September 19, 1851, Antonio Urrea, prefect of Altar, complained to the governor of Sonora about widespread "demoralization" in his district because of Apache hostilities, Papago [O'odham] unrest, and the presence of bands of "foreigners" who had no respect for Mexican authority or Mexican laws (Urrea 1851).

But Anglos were not the only outlaws on the Camino. One such desperado was Joaquín Murrieta, a disillusioned Mexican forty-niner who struck back at the Anglo invaders. Murrieta became a folk hero in California, Sonora, and Chile, who all claimed him as their native son. After reviewing often conflicting evidence, anthropologist James Officer (1981) rejected the notion that Murrieta was a Chilean, many of whom did migrate to California during the gold rush, and affirmed that he was a Sonoran, probably from the *municipio* (roughly analogous to a US county) of La Colorada, an important mining district.

As a Sonoran, Murrieta became intimately familiar with the Camino. According to Frank Latta (1980), Murrieta specialized in rounding up wild horses in California and driving them back to Sonora. To do so, he organized four gangs that followed a defined trail between 1851 and 1853, when Murrieta was supposedly shot and killed by California Rangers. The trail—La Vereda Caballo (The Horse Trail)—ran from Mount Diablo east of Oakland in California to Rancho La Berruga (now Rancho El Bámori) southwest of Pitiquito in northwestern Sonora. Murrieta apparently first traveled to California over the Camino in 1849. Three years later, he took his widowed sister and her two-year-old son back to California with him, again over the Camino. Manuel Rojas (1986) argues that there was a second trail as well—La Vereda Mesteña (The Mustang Trail)—that dipped south into Baja California, crossing the Colorado near its mouth and following the Sonoran coast past Puerto Peñasco to Rancho Berruga. Presumably they watered at the coastal pozos (see Lumholtz 1912; Ezcurra et al. 1988).

Latta (1980:423) indicates that Murrieta's horse gangs made Tinajas Altas a regular stop on their drives, along with Tinaja Escondida [Cabeza Prieta Tanks] and Tinaja Corazón (presumably Heart Tank). The gangs never drove more than 300 to 350 head of horses at a time because of feed and water conditions along the trail. "No drives were made during the summer months, except the final droves in July of '53 when all Murrieta Gangs planned to return to Mexico to stay," Latta (1980:403) notes. "The season for starting the rodeo at Mount Diablo began on the 16th of September just after they had celebrated Mexican Independence Day, and did not extend beyond the first of May. The amount of water in the natural tanks and springs along the desert portion of the trail determined the size of the bands and the frequency with which they could be started from the Cantúa." The gangs generally made eight drives a year. "It took more than six weeks of hurried driving to take a band of 300 horses from Arroyo Cantúa to Rancho La Berruga in Mexico and return to the Cantúa, 1,200 miles round-trip, more than 30 miles a day," Latta (1980:403) states. One member of the gangs, Teodoro Valenzuela, claimed to have made at least 20 drives over the trail, including the Camino. "Lupe Valenzuela, son of Teodoro, told me that often his father jokingly stated that he knew by name every individual cactus and clump of brush on the trail between Cucamonga and the Horse Gang Rancho in Sonora," Latta (1980:16) reports. "Surely no person ever knew El Camino del Diablo as well as did Teodoro Valenzuela."

Joaquin Murrieta died in 1853 from loss of blood after being shot, and the horse drives also ended in 1853. Whether he or his gangs robbed emigrants along the Camino itself is not known, but driving herds of 300 horses across that desert terrain would have left them little time for anything else. Nonetheless, others apparently did prey on travelers along the route. We have Hubbard's account of nefarious "guides," which is echoed by Raphael Pumpelly's description of his own trek along the Camino in 1861. Cipriano Ortega, the founder of Hacienda Santo Domingo, reputedly robbed travelers along the Camino del Diablo with his brother Bartolo (Hoy 1990, 1995). Thirst was not the only danger along the unforgiving trail. The Camino was a high-risk gamble, whatever the stakes.

Filling the Water Bag. Murray Chappel, Raven Butte Tank, 1913. (Photo by E. A. Goldman. Courtesy Neil B. Carmony. Cropped from original)

4 Surveyors to Campers
1854 to the Present
Bill Broyles ansd Gayle Harrison Hartmann

In the last century and a half Tinajas Altas has reflected the quintessential frontier image: a point in the middle of nowhere, a trailside rest area on the very edge of the earth, a key milestone on a journey to a new beginning, an exam where failure means death. It ranks with other American pioneer landmarks; it rivals the Oregon Trail's Chimney Rock, Independence Rock, and Devil's Gate in importance. But, it is a way station and no one stays. Occasional, transient visitors—not inhabitants—have wearily plodded through. While here, they rested, watered, played, prospected, botanized, mapped, studied, hid out, and worked. But, with the exception of those who are buried here or whose ashes are scattered here, few lingered.

Outside of some locations protected within national parks and monuments, probably few locales in Arizona have been the focus of so many modern articles or photographs. The place's fearsome reputation has grown to the point that it has become a place people love and hate at the same time. Anyone who has traveled El Camino del Diablo remembers Tinajas Altas. For some, its reputation preceded it. Noted Arizona cartographer and historian Don Bufkin first heard of Tinajas Altas as a boy in California but, because of its remoteness, it was one of the last places in Arizona he visited (Don Bufkin, personal communication 1991).

Zoologist and explorer William T. Hornaday (1908:156) never reached Tinajas Altas, but he noted his trepidation from afar: "north-westward along the Boundary, beyond the Tule Desert floor and its cheerful stand of creosote bushes, rose the scowling mass of the awful Tinajas Altas Mountains, where the water is little and bad, and many a poor traveler has died of thirst and heat." But those who have been there and lived to tell their tale were moved to speak freely and often, as we'll see in this chapter.

Just Passing Through: The Travelers

As we saw in chapter 3, the Tinajas Altas route, between the towns of Caborca in Sonora and Yuma in Arizona, was in common usage when the United States purchased the land south of the Gila River. The Gadsden Purchase bargain was made in December of 1853, but the treaty was not ratified by the US Senate until April 25, 1854. The acquired territory included both Tinajas Altas and part of what was then known as the Caborca-Yuma Trail, including the dreaded stretch along the border from Sonoyta to Yuma, called the Camino del Diablo. Travelers on this route went afoot, by horseback, or by wagon. They carried few amenities and survived on basic, limited staples; today's campers and backpackers are incomparably better prepared and supplied than those who ventured across this land before the advent of autos, ice chests, and duct tape. A canteen and a gun were basic equipment. Hunger and thirst were common, but few complained. Discomfort and peril were expected. To talk of heat, cold, or thirst was to talk about the obvious and to dwell on the unchangeable. This doesn't mean that early travelers fell in love with the countryside or enjoyed the trip; in contrast to modern travelers who go into the outback seeking beauty, adventure, and personal satisfaction, they labeled the scenery "forbidding," "harsh," and "bleak." The trip was genuinely dangerous, and they reached the dingy outposts of civilization with great relief and a sense of salvation.

The Tinajas Altas route was only one of three main trails between Sonoyta and Yuma. One headed north from Sonoyta, passed through Ajo, and then went to the Gila River at Gila Bend or veered northwest from Black Tanks to the river, but it had proven dangerous during the 1849 Gold Rush due to hostile Apaches, and it was the longest of the three. A second branch from Sonoyta followed a portion of the Caborca-Yuma trail, and at a dry lake called Las Playas followed Padre Kino's old trail to Heart Tank, Cabeza Prieta Tanks, Baker Tanks, and then north to the Gila River. The third, the Tinajas Altas route, was the most challenging of the three. It turned *westward* from Las Playas and went to Tinajas Altas and then either northward to the Gila River near Wellton or northwestward to the confluence of the Gila and the Colorado at Yuma.

Most travelers who took the Tinajas Altas route used the eastern fork of the Camino along the east side of the Gila Mountains, since it was safer and easier (Irwin ca. 1970). Other groups took the western fork through Tinajas Altas Pass, wending their ways along the west flank of the Gila range, or heading straight toward Pilot Knob on the banks of the Colorado River, fighting soft sand and scant shade on the way. These routes are confirmed by geologist Kirk Bryan (1925).

In 1861 geologist, world traveler, and mining engineer Raphael Pumpelly made a harrowing trip across the Camino del Diablo on horseback, dodging bandits, heat, and thirst.

His trip is reported in his fascinating book, *Across America and Asia* (Pumpelly 1870), which includes his adventures in Japan and China. Pumpelly's companion was explorer Charles Debrille Poston, often referred to as the father both of Yuma and the state of Arizona. They decided on the Camino del Diablo route even though, "owing to two successive rainless seasons, many of the usual watering-places . . . were dry" and noted that "the thermometer ranged by day between 118 and 126 degrees in the shade, rising to 160 degrees in the sun" (Pumpelly 1870:52; 57, 58). As they struggled through the Tule Desert, an unexpected rainstorm saved them from a long, hot ride of nearly thirty hours before they would find reliable water at the Gila River.

Pumpelly (1870:58–59) notes that

> during the night the sky was overcast with black clouds, and there came the first rain that had fallen on this desert for more than two years. Never was storm more welcome; both we and our animals enjoyed heartily its drenching torrent. Before day-break the sky had cleared, and with the rising sun began the heat of another day. A broad sheet of water, only a few inches deep, covered the *pleya* [*sic*] for miles before us, and banished from our minds all fear of suffering.

Two days travel after the rain he remarks,

> Our next camp was made at the Tinaje Alta [*sic*] or high tanks. Here at the head of a long ravine in the mountains, there is a series of five or six large holes, one above the other, worked in the granite bed of the gorge. . . . After a succession of dry seasons it sometimes happens that travelers arrive here already dying from thirst. Finding no water in the lower holes, they climb in vain to the higher ones, where, perhaps, losing strength with the death of hope, they fall from the narrow ledge, and the tanks, in which they seek for life, become their graves.

Yuma resident Rosalia Coz-Villa passed by Tinajas Altas in 1874 (Irwin ca. 1970) when her family made a pilgrimage from Yuma to Magdalena, Sonora, to the shrine of San Francisco Xavier (Father Kino's patron saint). Because of marauding Papagos (O'odham) at that time, the group went well armed and arrived safely at Magdalena on Mexican Independence Day, September 16. This group surely stopped at Tinajas Altas on both legs of the trip. The record of this trip suggests that the route was a common road for families commuting and visiting in Mexico, as has been noted by later scholars such as Forbes (1952) and Ives (1984). A few Mexican families used the trail as late as 1875 to commute between California and Mexico on the Camino del Diablo, which was then generally called the Yuma Trail or Sonora Trail (Mason 1963). But the depletion of gold in the California fields and the coming of the railroad along the Gila River in 1877–1880 ended the incentives to travel this route, so few people thereafter passed Tinajas Altas on their way between Mexico and California.

Raphael Pumpelly, 1900. (Pumpelly 1918: frontispiece)

One early, well-known Anglo figure in the area was Tom Childs Sr., who came to Arizona Territory from Mississippi in about 1850; he used the trail during his travels from California to Sonora. In 1853 he explored for minerals in the Ajo area and in 1862 traveled across southern Arizona as a volunteer with Confederate Army officer Sherod Hunter (Finch 1996). On that trip, they followed the Gila River route and watered at Baker Tanks, but did not use Tinajas Altas, farther to the south. Childs and his son, Tom Jr., who was born in Arizona City (later named Yuma) in 1870, later

Tom Childs Jr. and Marta Childs, his wife. (Photo courtesy Oscar and Fillman Bell)

traveled throughout southwestern Arizona and likely visited Tinajas Altas on various occasions (Childs 1954; Hoy 1999; Van Valkenburgh 1945). Tom Childs Jr., ranched and prospected widely in the area and married a Hia C'ed O'odham woman, Marta García, whose family knew the Tinajas Altas region (Bell et al. 1980, Hoy 1999). Childs Valley and Childs Mountain, northwest of Ajo, were named for the family.

MERCHANTS AND WAGONEERS

One of the first English-language accounts of traveling the trail is left to us in a diary by George Kippen, who regularly traveled the trail from Yuma to Sonoyta and Ajo on business for the Ajo mine from 1854 to 1858. In an article in the *Journal of Arizona History*, Bill Hoy (1995) recounts and annotates Kippen's diary. For example, on July 29, 1854, Kippen writes that from Yuma he "dispatched a Mexican to the Tinajas Altas to see if there was water at that point." Evidently there wasn't sufficient water, because Kippen took "the Cabeza route," heading south from the Gila River 42 miles east of Yuma (the vicinity of modern Tacna). As was discussed in chapters 1 and 3, this route, perhaps taking advantage of Baker Tanks, was preferred by Kino and other early Spaniards. Kippen returned to Yuma by way of Gila Tanks (likely what are today called Black Tanks in the Crater Range north of Ajo) and the Gila River, so he again avoided Tinajas Altas.

On another foray, Kippen left Yuma for Tinajas Altas on January 18, 1856. "Tues. 19. Arr. Tinajas Altas. No water below 5th tank." So on Wednesday, January 20, he and his party carried water down from one tank to another most of the day so their animals could reach it at the bottom tank. The next day they rode on to Las Playas, which at that time had good grass. Not all such trips were successful. Kippen reports that on June 18, 1856, "Yancy starts for the river [Yuma] via the 'Cabeza Prieta' route. Fri 19 Yancy returns not being able to get through."

Kippen went from Ajo to Yuma in December 1856 for supplies by either the Tinajas Altas or the Cabeza Prieta route,

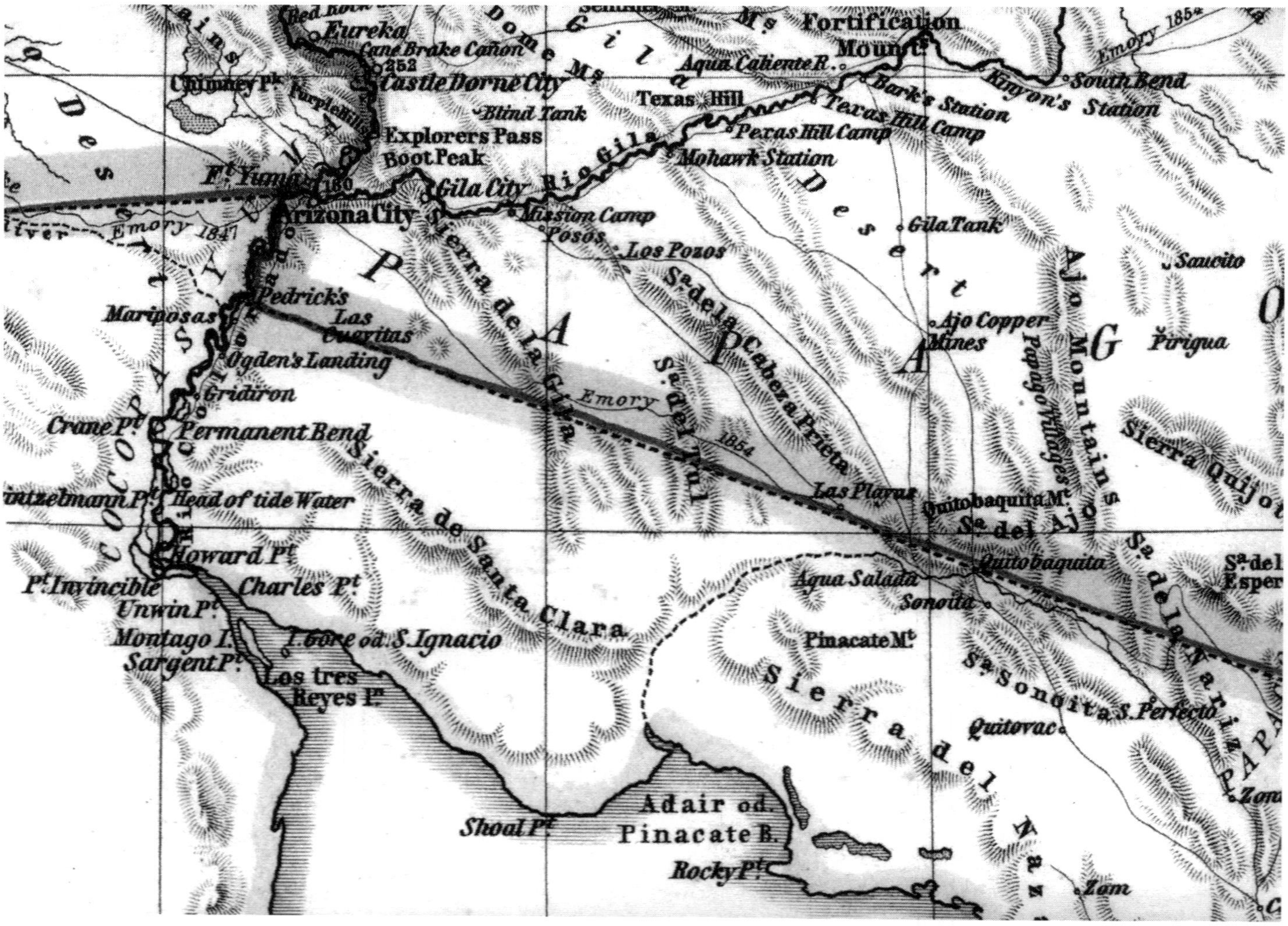

From Stieler's map of Arizona, 1875, including the routes used by Kippen 20 years earlier. Adolf Stieler was a noted German cartographer whose maps went through many editions long after he died in 1836. (Stieler 1875)

but the supplies had not arrived from San Francisco. A day later others from Sonoyta arrived looking for the supplies. They left for Ajo in early January with the supplies, though co-workers named Dunbar and Yancy had returned to Ajo and then back to Yuma in the meanwhile by unspecified routes. This sort of long-distance commuting was apparently common, though not without peril. While still in Yuma, Kippen reports:

> Wed. 14 [1856]. A Frenchman comes in afoot from the Tinajas Altas route to Sonora. Most horribly cut by 2 Mexicans who had murdered and robbed his 2 companions, but failed in killing him. He succeeded in saving his money, 800.00, but [they] robbed his companions at 1,000.00 in gold coin.
>
> Thurs 15. Yancy and myself went in pursuit of the two Mexicans who murdered the Frenchmen. Did not succeed in overtaking them, but found the bodies of the murdered men about 23 miles from Ft. Yuma. The most horribly cut men that I ever saw! Buried them as well as circumstances would permit and returned to Ft. Yuma. (Hoy 1995:244)

Due more to the financial condition of the company than to the difficulties of transportation, chaos erupted at the Ajo mine while Kippen was in Yuma waiting for supplies: the food ran out and the hungry workers quit. Kippen hastened back from Yuma via Antelope Peak to Las Playas (and presumably past Cabeza Prieta Tanks) to reorganize the operation as hastily as possible and to try to retain workers. He reported marauding Indians:

> Sunday March 22 [1857]. Dario reports the death of Marcelino by Apaches in the valley be-

> tween the mines and Sonoita. Miscl: There has been all this month many Apaches in the valley and vicinity of the mines committing robberies and killing one of our peons. Also stealing 3 mules from the mines. (Hoy 1995)

Kippen mentions perhaps a dozen trips he or his associates made from Ajo to Yuma by any of three primary routes in use at this time: (1) west to Tinajas Altas and then north to the Gila River just east of Yuma; (2) west to Cabeza Prieta Tanks and then north through the Tule Desert to the Gila River; and (3) past Gila Tanks north of Ajo, then west along the Gila River. These trips ranged in size from a single horseback rider to trains of 12 wagons hauling ore to Yuma and returning with supplies to Ajo. Kippen mentions that these trains were operated by a man named Tomlinson and by Louis Jaeger, but it is unclear if Tomlinson worked for Jaeger (McClintock 1916; *Ajo Copper News* 1917). Kippen quit the mine to tend Louis Jaeger's store in Yuma.

Jaeger's full name was Louis John Frederick Jaeger (1824–1892), and he was a wagoneer and merchant who came west from Pennsylvania with the forty-niners to California. He didn't find gold, but he did join a group that re-established a commercial ferry across the Colorado River at Yuma in 1850 (Hargett 1967; Patzman 1963). Jaeger branched out from being just a ferry owner to also running freight wagons throughout southwestern Arizona. He occasionally used the Tinajas Altas route to transport supplies to the mine at Ajo and to return with copper ore to be shipped from Yuma by river steamer, then by ship to Wales for processing. He traveled at night to avoid the scorching heat (Irwin ca. 1970), but shortages of water were the real threat. Undoubtedly Jaeger stopped at Tinajas Altas coming and going.

Louis J. F. Jaeger, date unknown. (Courtesy of the Arizona Historical Society/Tucson; image no. 23459)

Apparently, his trip was not an infrequent event, however, especially during Fort Yuma and Yuma Crossing's heyday in the 1850s and early 1860s. A fragmentary diary, the "Diary of a Ferryman and Trader at Fort Yuma, 1855–1857," probably written by Louis Jaeger, recorded daily life at Fort Yuma between December 11, 1855 and July 2, 1857 (Beattie 1928, 1929). The diary was recovered from an old trunk near the adobe house that Jaeger once occupied southwest of the town of Colton, California. "The main business at the junction was furnishing beef and perishable provisions to the garrison and to the river steamers," Beattie (1928:91) noted in his introduction to the diary.

> The writer of the journal was connected with one of the two business firms thus engaged and had the beef contract at the Fort. . . . When rain filled the *playas* or dry lakes, the *pozas* [*sic*] or waterholes, the *tinajas* or natural tanks of Sonora, he made personal trips with wagons over the dreaded *Camino del Diablo* and visited the towns of Sonoita, Altar and Caborca in search of supplies. Two such trips are described in the fragment we have. In one he obtained water from the *Tinaja Alta* [*sic*] on the mountain side, carrying it down to his train. When the waterholes were found empty, his party and his teams suffered thirst. (Beattie 1928:91–92)

One such Jaeger trip took place from July 11 to July 21, 1856. The brief entries reveal the salvation of summer rains. On the journey from Fort Yuma to Altar, the ferryman wrote, "July 24. We had a cloudy day & arrived at 5 1/2 o'clock at Sonoita made 20 miles this drive in 8 hours & plenty of water at the posa & caught up with B. . . & left at 5 o'clock & went on 3 miles & camped. . . & we had a shower of rain—a heavy rain—made 3 miles this drive" (Beattie 1928:121). On the return trip, he wrote more detail:

> Aug. 20. We had a fine day & John and I left at 9 1/2 o'clock & arrived at Agua Dulce at 1 1/2 o'clock made in 4 hours drive. 20 miles & Slaton took on the wagon 12 miles & brought back the mules at 3 1/2 o'clock.
>
> Aug. 21. We had a fine day & left at 6 o'clock & stopped 1 1/2 hours at the wagon & took breakfast & then pushed on to the playa & went ahead to hunt for water & found some plenty right in the road & 2 miles off the road also at the laguna & arrived at 1 1/2 o'clock at the playas & at 3 1/2 o'clock the wagon also & camped. Made 25 miles & with the wagon 15 miles in 5 1/2 hours drive.
>
> Aug. 22. We had a fine day & left at 12 1/2 o'clock & found a heavy road & had to turn back again & take a new start & they got up with the team at 4 o'clock & I left them at 20 minutes after 4 o'clock & arrived at Cabesa Prieta [*sic*] at 9 3/4 o'clock and camped. Made in 5 1/4 hours. Made this ride in 7 hours in all from the playa 30 miles.
>
> Aug. 23. We had a fine day & laid over all day & found water up at the Cabesa Prieta. I went up & Jose Ortez & took us 1 1/2 hours to go up & stopped 1/2 hour & took us 2 hours to get back—a very bad road made 10 miles & I sent back a man to the wagon & they got up at 4 o'clock & they took up their mules to water also.
>
> Sunday, 24. We had a fine day & laid over all day & I left in the evening at 8 3/4 o'clock for the pozas & the team at 12 o'clock in the night & I went on 18 miles drive in 4 hours. Arrived at 12 3/4 o'clock.
>
> Aug. 25. We had a fine day & we left at 25 minutes after 5 o'clock also & arrived at 10 minutes before 10 o'clock. Made 18 miles in 4 1/4 hours, and in the evening pack train got in from Altar for the river loaded with provisions & poor Francisco died in Sonoita in Sonora also. Thank God that he has taken him to his home; he is better off than we are.
>
> Aug. 26. We had a fine day & the wagon got in at 7 1/2 o'clock & I am going to leave with the pack mules this evening at 5 1/2 o'clock & went on 9 miles & camped & wagon going to leave tomorrow.
>
> Aug. 27. We had a fine day & I left at 5 o'clock & went on 12 miles at the foot of the mountain & arrived at 9 o'clock & left again at 3 1/4 o'clock & got home at 8 o'clock. Made drive in 4 3/4 hours. Made 46 mile drive in all 8 3/4 hours. Found things all right. (Beattie 1928:124–125)

On that trip, the ferryman made no direct mention of Tinajas Altas, but used Cabeza Prieta Tanks. On a second trip to Sonora in September 1856, the ferryman's party *did* water at Tinajas Altas:

> Sept. 13. We had a fine day & I was up at Fort also & busy all day getting ready to leave & got off at 5 o'clock in the afternoon at Hinton's & loaded the wagon—got on about 2500 lbs & made camp at 12 o'clock at night—made 15 miles & mules very tired—heavy pulling also & Jones in company with us & I took along $300 dollars also & bought Sapatro's flour & pinole.
>
> Sept. 14 (Sunday). We had a fine day & wagon aint got up yet & left at 12 1/2 o'clock & made Tinajas Altas at 8 o'clock—made drive in 7 1/2 hours—made 40 miles & Jones got in at 4 o'clock & his mules & horses very tired out also & very little water in the Tinajas Altas also.
>
> Sept. 15. We had a fine day & I & Jones hard (at work) passing down water & Slaton & John got in at 10 o'clock & had a hard time to water the mules also & I and Jones left in the evening at 6 o'clock for the playas also & Slaton & John going back after the wagon in the night.
>
> Sept. 16. We had a fine morning & arrived at playa at 8 o'clock & I went in after the water 2 miles off the road but did not find any water, then stopped 2 hours & pushed on to Agua Salada & 3 miles this side met Walker & he gave me some water & I was very hard up for water & I was caught in a storm & heavy rain 2 miles this side of Agua Salada also—made camp at 4 o'clock made 75 miles this drive in 18 hours & Walking took in . . . & cheese in & stuff also.
>
> Sept. 17. We had a fine day & Jones & Thomas got in in the morning & it rained behind them also & Rayas [Reyes?] got in off the Cabesa Prieta road also & it rained heavy on that

Las Playas hold rainwater for weeks following showers, and even in modern times Las Playas' mud bedevils travelers. (Photo by Bill Broyles, March 2010)

> road also & I sent after Slator with water also Jose Ortez [José Ortiz?]. (Beattie 1928:127)

This section of the diary reveals why the Camino was so treacherous. Until the heavy rain fell, the ferryman found no water along the stretch near the playas [Las Playas?] where water had been plentiful in July. His team could not continue. However, if the mud at Las Playas was too wet, it turned to impassible goo, and Jaeger would have to take the alternate route past Baker Tanks instead of Tinajas Altas. Luckily for Jaeger, a rain storm delivered water to the tank and playas.

In his September 20 entry, he mentions two mules dying at Tinajas Altas from eating some sort of "poison weed." He also notes that travelers reported "no water at the Tule Tinaja & Tinajas Altas" (Beattie 1928:128). Depending upon the rains and the impact of other travelers, a tank holding water on one day might be dry within a few days later. Louis Jaeger became known affectionately as Don Diego, and a set of tinajas in the Aguila Mountains bears his name (Lockwood 1968).

AN AMERICAN ATTACK ON MEXICO IN 1857

A bizarre episode of border history arose in 1857, when Henry Alexander Crabb and a group of California filibusters invaded Sonora via the Camino del Diablo. Crabb was a forty-niner and a lawyer who rose through California state politics until he lost a race for the US Senate. After that defeat, he turned his attention southward to Sonora, where his marriage to a member of the prominent Ainsa family gave him access to Sonoran political leaders. One of those leaders was Ignacio Pesqueira, who was engaged in a bitter struggle for power with Manuel María Gándara. Crabb visited Sonora in the spring of 1856, where he supposedly allied himself with Pesqueira. In return for mining concessions and large grants of land along the border, Crabb promised to organize a military expedition of Anglo adventurers who would help Pesqueira defeat Gándara. Unfortunately, there is no written record of these agreements, and several Sonoran historians have questioned the Pesqueira-Crabb connection (Officer 1987).

After Crabb returned to San Francisco to recruit volunteers, his brother-in-law Agustín Ainsa was arrested and charged with high treason by the Sonoran state government, which was controlled by Gándara at the time. Crabb proceeded to enlist 90 men into the Arizona Colonization Company, a quasi-military organization. They were to march overland to Altar and join a much larger party of 1,000 men supposedly sailing from San Francisco to Puerto Lobos on the Sonoran coast. Then the combined army was to mass "with their allies in Sonora [to] make their issue open and in strong force" (*New York Times*, April 15, 1857, quoted in Forbes 1952:8). Crabb fulfilled his part of the mission. In January 1857, Crabb's overland force sailed from San Francisco to San Pedro, California. They followed the forty-niner's trail via Warner's ranch southeast across the Borrego Badlands of California and reached the Colorado River on March 1. When asked what their horses would eat in the dry desert ahead, members of the group joked, "Oh, we'll ride them into the stores and eat calico" (Forbes 1952:8).

After resupplying at Yuma, the Crabb expedition marched east along the Gila River about 45 miles to a place that came to be called Filibuster Camp, east of the modern community of Tacna. Apparently, members of the expedition carved their names into the bark of trees there. Then the remaining 89 members of the group turned south down the Lechuguilla Valley. Forbes (1952) claims their first stop for water was Tinajas Altas and from there he says they traveled southeast to Tule Tank. But Forbes's conclusion does not fit the fact that N. S. Bunker, who was on the trip, says they watered at Cabeza Prieta Tanks, which would be their logical route if they had just marched east to Filibuster Camp near the head of the Cabeza Prieta route. To reach Tinajas Altas they would have had to backtrack, and no one who had just come from Tinajas Altas and Tule Tank would bother to march at least 15 miles out of their way to Cabeza Prieta Tanks. At any rate, they left 20 of the weaker members behind while Crabb and the rest continued on to Sonoyta. It was a typically harrowing trip along the Camino. According to N. S. Bunker, who sent a letter from Dunbar's store near Sonoyta before he was killed:

> When we left the second team [not far out of Fort Yuma], we packed the animals with the baggage to a watering place called Cabeza Priete [*sic*](midway of the Camino del Diablo). The team was left on the desert with Charlie [Parker, who had been accidentally shot in the knee on board the steamer to San Pedro] and two others for 6 days on account of our not being able to get the wagon on to the water which was three leagues. There we took water to them every day, Charlie not being able to move with crutches. Crabb and the main body started on to Sonoita, leaving Capt. McKinney in charge of us all, 16 in number, to come on without delay. But on starting we found that the animals would not travel over two miles per hour, and finally when within 25 miles of water, they all gave out. Here we left Mr. Long and Charlie, while we drove the animals to water. Four mules and two horses died on the road. Our allowance of water was less than a quart to the man for the trip. Again we packed water to Mr. Long and Charlie until Capt. McKinney left us to get some animals to pull us to Sonoita. The morning he left us, John Edmonson said he would pack water for the last time for his horse had nearly given out. He took a horse and mule, carried them some water, and to our great astonishment he came to camp with Charlie and the wagon, having thrown out all the blankets, provision, and ammunition, too. We received four jacks and three yoke of cattle the next day, and sent the jacks out to pack in all our things and what provisions that were not stolen by a band of Mexicans, 100 in number, going to Cal. for fear of a revolution in their country.
>
> From this place we went to Sonoita. Here we remained two days when Capt. McKinney left us and took all of the men but Parker, Harrison, Mr. Long and myself. Mr. Long was our commissary, and he remained to take care of us. We were the only invalids in the company. When McK. left us it was to meet Gen. Crabb who was some days in advance. (*Sacramento Daily News*, May 26, 1857, quoted in Forbes 1952:12)

By the time Crabb reached Caborca, his supposed alliance with Pesqueira had evaporated. Pesqueira had defeated Gándara and was in control of the Sonoran government. Crabb's party was attacked outside town and then the local defenders took refuge in Caborca's mission church while Crabb and his men occupied a series of houses across the plaza. They attempted to rush the mission and failed. Indian allies of the Sonorans shot flaming arrows and set the thatched roofs of their buildings on fire. Bullet holes seen today on the exterior walls of the church are purported to be from this battle.

Tule Tank. (Photo by Bill Broyles)

The famous church at Caborca, Sonora, La Purísima Concepción de Nuestra Señora de Caborca, built between 1797 and 1809. (Photo by Carl Lumholtz 1912:146)

Finally, on the night of April 6, Crabb convinced his men to surrender. All but a sixteen-year-old boy were shot by firing squads the next morning. Crabb's head was cut off and pickled in vinegar so it could be carried in triumph to the state capital of Ures. Apparently, the pickling was unsatisfactory, however, and the rotting head was thrown away on the trip. Mexican reinforcements also captured and executed McKinney's rear guard of 16 as well as the 4 men who remained behind at Dunbar's trading post near Sonoyta. In the eyes of Sonora and the Mexican republic, the defeat and execution of Crabb and his filibusterers was fitting retribution for the Mexican War and other abuses and invasions Anglo-Americans had perpetrated during the past decade.

Smugglers and Covert Immigration

Because the Tinajas Altas region is unpopulated and frontier-like, it attracts people trying to enter the country illegally or with illegal goods. Tinajas Altas itself has long been on a smuggling corridor; indeed, Tinajas Altas Pass just north of Tinajas Altas is popularly referred to as Smugglers Pass. Thus, Tinajas Altas has seen its share of criminal and law enforcement activities.

When writing of his 1871 journey across El Camino del Diablo, Raphael Pumpelly (1870) noted that he was trailed by cutthroat bandits when he traveled past Tinajas Altas. He eventually outwitted them, but not without grave worry and extreme vigilance.

In an article written for *Arizona Highways*, Barney (1943:17) mentions that "by the same route [Yuma-Caborca Trail] also went the notorious Bell Gang, which, for a time, terrorized California soon after the discovery of gold. This band of ruffians entered Sonora and proceeded to commit even more heinous crimes than in California, until driven out of the state by the angered Sonorans." Tinajas Altas has always been on the far edge of civilization. It is in a *despoblado*, an uninhabited region, that people pass through on their way to elsewhere. With no permanent settlement, it required no military presence or permanent lawmen. Because it was not strategically located, except as a watering hole for a few nameless travelers, neither Mexico nor the United States saw any need to defend it with troops. Kirk Bryan (1925) reported that at the time he was writing, no one lived between Ajo and Yuma except the watchman at the Fortuna Mine.

Tinajas Altas was a stopover and waterhole for an assortment of *—istas* who traveled by horseback and on foot between Sonoyta and Mexicali during the political upheavals in Mexico in the early twentieth century (Bill Hoy, personal communication 1996). Charles Sheldon (1993), who visited Tinajas Altas in November of 1913 to hunt desert bighorn sheep, encountered one such group. When he and his party arrived at Tinajas Altas, they found a party of 30 Mexican revolutionaries camped there. Their commander's surname was Gallego and their "secretary" was Francisco Mendoza. They were hospitable and permitted Sheldon's party to

Charles Sheldon's party at Tinajas Altas in 1913. (Photo by E. A. Goldman. Courtesy Neil B. Carmony)

Charles Sheldon and party camped near Tinajas Altas, 1913. (Photo by E. A. Goldman. Courtesy Neil B. Carmony)

obtain water, with Mendoza even writing a letter permitting Sheldon to cross the border.

In Sheldon's journal (1993:23–24), published as *The Wilderness of the Southwest,* he wrote:

> I remained and had a most interesting conversation about the revolution. They are retreating from, but not running away from, a fight near Calexico, [Baja California]. They are also bringing a load of ammunition which is hidden with eight or ten mules. It will be brought in later. They could not stay here, and left to get feed for their horses, six miles away. It was a highly picturesque crowd, all armed with Winchesters (I saw two or three Krags), with belts of cartridges around their waists and over their shoulders (all softnosed bullets). Some of the horses were good ones, some with terrible backs, and some nearly exhausted. I noticed a pair of field glasses in a saddlebag. They had a blanket apiece, several smooth boards for making tortillas, which seemed to be their exclusive food, with coffee. Some were middle class and able, and some sad-looking peons. All had canteens. When they rode off, about ten walked. They told me they had lost horses from lack of water, etc. They will go on to Tule Tanks and then to where they can get cattle.

Sheldon found additional tracks of 8 to 12 Mexicans at Tinajas Altas when he returned there on December 14, 1913, but learned no more about their identity.

Sheldon camped at Tinajas Altas again in 1916 on his way to hunt bighorn sheep in the Sierra del Rosario. When Sheldon and his party arrived on March 4, an unidentified man was camping there. On March 14, a Ford motorcar arrived with "a special messenger to notify me of the condition on the border. U.S. troops were about to cross over and punish [Pancho] Villa for his raid on Columbus [New Mexico]" (Sheldon 1993:111).

Another American who visited Tinajas Altas during this unsettled period was Kermit Roosevelt, the eldest son of President Theodore Roosevelt. He writes of his visit, which took place in 1911, that "[a] couple of months previously a band of insurrectos had been camped by these tanks, and two newly made graves marked their contribution. The men had been killed in a brawl" (Roosevelt 1920:73–74).

Roosevelt (1920:74) also tells a tale of murder at Tinajas Altas that was reported to him, although it may be a campfire yarn.

> Forty-six unfortunates perished here at one time of thirst. They were making their way across the deserts to the United States, and were in the last stages of exhaustion for lack of water when they reached these tanks. But a Mexican outlaw named Blanco reached the tanks ahead of them and bailed out the water, after carefully laying in a store for himself not far away. By this cache he waited until he felt sure that his victims were dead; he then returned to the tanks, gathered the possessions of the dead, and safely made his escape.

One of the few early lawmen in the area was Jefferson Davis "Jeff" Milton, who worked for what was then called the Customs Patrol. For a while in the 1890s, he was assigned to work the borderlands between Sonoyta and Yuma. Milton rode horseback along the border, stopping for water at Quitobaquito, Tule Well, and Tinajas Altas. He observed that signs lasted seemingly forever: wagon tracks, human foot prints, hoof prints. Milton himself "was apt to say, 'I passed that point twenty years ago,' and ride by to show his horse tracks in proof" (Haley 1948:160).

Milton worked in all seasons, riding mainly at night and carrying three one-gallon canteens. His route was westward from Tucson to Sonoyta to Tinajas Altas and up to Wellton. Sometimes, he'd retrace his route; other times he'd ride back to Sonoyta by way of Cabeza Prieta Tanks and Heart Tank to Ajo and then back to Tucson. After a few years Milton was reassigned to another post, but he returned in 1904 with the mission to enforce the ban on Chinese immigration. He confided, "[I] required four horses or mules in order to transport from point to point myself and outfit, consisting of not only feed and water for myself and horses, but also bedding and cooking utensils" (Haley 1948:347). He even made one trip on horseback from Quitobaquito to Yuma in summer. "By the time he reached Yuma, his horses could hardly walk, and they stood for days cooling their feet in a ciénega, or marsh, along the Colorado River" (Haley 1948:350). Milton spent the next few weeks working immigration cases around Yuma before heading back to Sonoyta.

About the only thing that bothered Milton were cone-nosed kissing bugs, which Milton called Hualapai tigers in deference to their blood-sucking bites (Haley 1948:384). Unlike many outdoors men of his time, on several occasions he refused to kill rattlesnakes that had not bothered him. In 1907, Milton helped guide the Carnegie Laboratory expedition to the Pinacate volcanic region of northwestern Sonora (Hornaday 1908), but by 1909 even the tough Milton was tired of the arduous desert patrols and got himself reas-

Jeff Milton near the Camino del Diablo in 1907. (Photo by John M. Philips, courtesy of the Arizona Historical Society/ Tucson; image no. 71879. Compare Hornaday 1908:93)

Jeff Milton camped near Tinajas Altas with a group of border patrolmen. Information with the photo is ambiguous, indicating that the date was perhaps 1939 or 1944, which would have made him 78 or 83 at the time. Milton was born November 7, 1861. (Photo by Earl Fallis, courtesy of the Arizona Historical Society/Tucson; image no. 52703)

signed to the grasslands and cooler mountains of Hereford, Arizona.

Milton returned to the borderlands in 1930, this time in a stripped-down Ford. He and a partner, Nick Collaer, drove from Tucson to Sonoyta and then to Tinajas Altas, looking for tracks of illegal immigrants all the way, or "cutting sign" in Border Patrol parlance. In the dunes west of Tinajas Altas they managed to get their car stuck and nearly had to walk out; Milton was 69 years old at the time (Haley 1948). Soon after, the US Border Patro l took over immigration duties and Milton retired, though he did make one last trip there about 1939 or 1944 with a group of border patrolmen led by Earl Fallis.

On August 21, 1967, a US Border Patrol station was opened at Tacna (Corder 1987), although agents had worked in that area for decades before then. Since that time, the Border Patrol has patrolled the Tinajas Altas area daily by vehicle or by aircraft looking for illegal aliens or smugglers (see Broyles and Haynes 2010). The Tacna station was replaced in 1989 by a new station at Wellton 11 miles (17 km) to the west, but patrols of Tinajas Altas continue around the clock. The list of dedicated agents—men and women—who have served at Tacna and Wellton is long and distinguished. A new Wellton station opened in 2012. A border fence, or in some places a border wall, was completed in 2009. Camp Grip, a semi-permanent Border Patrol camp west of Papago Well, was established in 2002, so that agents can work closer to the border.

Surveyors and Cartographers

By the later nineteenth century more visitors came to Tinajas Altas to study the region and its features. They included curious surveyors and cartographers. Little was known of the geography, botany, mammalogy, geology, or archaeology of the Tinajas Altas region when it became part of the United States. Indeed, these sciences themselves were in their infancy in 1854 as the United States government prepared to define its southern boundary and to learn more about the resources and potential of the region (Bufkin 1983).

The first systematic survey of the lands acquired in the Gadsden Purchase of 1854 was the Andrew Belcher Gray

Border Patrol agent Mike Ringler along El Camino del Diablo, 2005. (Photo by Bill Broyles)

party's survey of a southern railroad route from Texas to California, including a branch into Mexico, to the Gulf of California. The group roughly followed the 32nd parallel and prepared a report for the Texas Western Railroad Company, taking notes on the terrain. Charles Shuchard, an artist who accompanied the expedition, made sketches of many places along the route; unfortunately, the original drawings were destroyed in a fire at the Smithsonian Institution in 1865. Shuchard's drawing "Tinajas Altas-High Tanks" (Bailey 1963:86) is the first known likeness of Tinajas Altas.

Of Tinajas Altas itself, Gray reports nothing, mentioning the place only in passing: "East of the Tinaja Alta or high tank range, lie the famous Sierra del Ajo, now in United States territory" (Bailey 1963:90). Gray saw no difficulties in traveling through southwestern Arizona except the absence of water. He foresaw rich markets of copper, cotton, and cattle, making the rail line both profitable and strategically important as a short, easy route between the East and West coasts. This proposed line shows up on some old maps.

Peter Rainsford Brady, a soldier and engineer who accompanied Gray, had his own, more personal account of the trip to Tinajas Altas.

> From the Agua Dulce to Yuma was 120 miles, water half way at some tanks in the mountains called Tinajas Altas (high tanks), and high they were, too. We left camp about one p.m. and followed a trail leading northwest, rested about three hours in the middle of the night and then up and forward. About two p.m. we reached the water, and a fearful job it was, too, to get up to it and pack it down in canteens to our thirsty animals. It was nearly dark when we got through watering our animals, then we took some pinole, jerky, and sleep. We were up the next morning at dawn. Then another three hours job of watering the animals and off again for Yuma over a dry and sandy plain. (Bailey 1963:221)

Along the trail "there were lots of animals scattered along the road where they had died from exhaustion and then dried up like mummies. On this stretch of desert there had been neither ravens nor coyotes to devour them" (Bailey 1963:221). The Brady party was ecstatic to reach the Gila River.

In 1854, Congress directed William H. Emory to conduct a survey of the Mexican border from Texas to California. This survey laid the international boundary, erected border marking monuments (later replaced by the ones there now), and measured distances to a number of points along the way. The men faced many hardships, and a focus of their misery and fascination was Tinajas Altas. Lt. Nathaniel Michler (1857–59:115–16) records some of the earliest and most vivid images of the Tinajas Altas region. The land was scorching hot and blindingly bright, and travel here was a "torment." His writings provide a primary source of information that has been used and reused in virtually every essay on Tinajas Altas to the present day. Michler's accounts appear in the three-volume work *Report on the United States and Mexican Boundary Survey* by William H. Emory (1857–59:115–16).

Charles Shuchard's drawing of Tinajas Altas, in the A. B. Gray report of 1856. (Bailey 1963:86)

A photographic view similar to Shuchard's drawing, showing Tinajas Altas from Mesa of the Dead. (Photo by Tad Nichols, 1933)

> Well do I recollect the ride from Sonoyta to Fort Yuma and back, in the middle of August, 1855. It was the most dreary and tiresome I have ever experienced. Imagination cannot picture a more dreary, sterile country, and we named it the "Mal Pais." The burnt lime-like appearance of the soil is ever before you; the very stones look like the scoriae of a furnace; there is no grass, and but a sickly vegetation, more unpleasant to the sight than the barren earth itself; scarce an animal to be seen—not even the wolf [coyote] or the hare to attract the attention, and, save the lizard and the horned frog, naught to give life and animation to this region. The eye may watch in vain for the flight of a bird; to add to all is the knowledge that there is not one drop of water to be depended upon from Sonoyta to the Colorado or Gila. All traces of the road are sometimes erased by the high winds sweeping the unstable soil before them, but death has strewn a continuous line of bleached bones and withered carcasses of horses and cattle, as monuments to mark the way.
>
> Although I traveled over it with only four men in the most favorable time, during the rainy season of Sonora, our animals well rested and in good condition, still it was a difficult undertaking. On our way to the post from Sonoyta we met many emigrants returning from California, men and animals suffering from scarcity of water. Some men had died from thirst, and others were nearly exhausted. Among those we passed between the Colorado and the "Tinajas Altas," was a party composed of one woman and three men, on foot, a pack-horse in wretched condition carrying their all. The men had given up from pure exhaustion and laid down to die; but the woman, animated by love and sympathy, had plodded on over the long road until she reached water, then clambering up the side of the mountain to the highest tinaja, she filled her bota (a sort of leather flask), and scarcely stopping to take rest, started back to resuscitate her dying companions. When we met them, she was striding along in advance of the men, animating them by her example.
>
> On our return we had to ride to the "Tinajas Altas," forty-five miles, the first night to reach water; and the second one over sixty-three to "Agua Dulce," where we managed to obtain some by digging. During this time our poor mules plodded through the heavy sand without rest or food. It was over this country one portion of the new boundary line was to be traced; the road of which I have just spoken runs immediately along the line, and is the only practicable one connecting California and Sonora.
>
> Before completing the work immediately on the river, a party had been sent out to make a reconnaissance in the direction of the line, and principally to examine the country for water. Anxious to have no delay, as the hot weather was fast coming on, and the river-bottom having become so infested with mosquitoes as to make life unendurable and labor of any kind impracticable, we commenced prolonging the line without awaiting the return of the reconnoitering party. Mr. Phillips, in the performance of this duty, reached the Tinajas, but there found very little water, and what there was, difficult of access; and, although directed to some new waterholes by a Papago Indian, still he only found sufficient to last a short time for a small number of men and animals. Mr. Aleman, of the Mexican commission, also endeavored to travel the road; but meeting the party first sent out, and hearing their report of the entire absence of water from the river to Sonoyta, was compelled to turn back.
>
> The escort and provision train likewise made an attempt, but it was found almost impossible to advance more than a few miles with the heavily-loaded wagons. My own success was little better; starting from our camp with a light spring-wagon and six good mules, I managed to make twenty-five miles in twenty-eight hours' constant travel; an express then reached me from Mr. Aleman, informing me of the unfavorable account of the search for water.

Michler (1857–59:124) glowingly wrote that a member of the survey, Arthur Schott, made "a large and interesting collection of botanical plants and of natural history, besides making careful examinations of the geology of the country; he has also taken the views [sketches] of the scenery along the line, which accompany this report." Two of Arthur Schott's drawings (labeled 59 and 60) are of the mountains south of Tinajas Altas at boundary marker IV. One view looks eastward and the other, west. Both depict craggy terrain.

View from Monument No. IV looking west towards Monument No. III. (From Emory 1987:pl.60 fol. p. 120, v. 1)

If the written record is any indication, few people passed Tinajas Altas between this first border survey in 1854 and the second one in 1891–1896. David Du Bose Gaillard of the US Army Corps of Engineers headed the resurvey of the international boundary. The team examined the stretch between Sonoyta and Yuma in June of 1893. By this time the southern Gila Mountains had come to be called the Sierra de las Tinajas Altas. Gaillard (International Boundary Commission 1898:25; see Hayden and Friends 1993) reports:

> Many years ago the Papagos [O'odham] were accustomed to camp at . . . Tinajas Altas for the purpose of hunting "big horns" or mountain sheep, which then, as now, constituted the principal inhabitants of these desolate sierras. In the vicinity of the tanks are still seen the remains of their old camps, around which are strewn the horns of mountain sheep—as many as twenty or thirty pairs have been counted at a single camp.

Those camps have since nearly disappeared, obliterated by time and visitors. Regarding the importance of Tinajas Altas, Gaillard wrote:

> That this water supply was a very uncertain one when the road from Sonoyta to Yuma was much traveled (as was the case during the rush to California in "the fifties" and to the Colorado River diggings in the "sixties") becomes painfully apparent from the number of graves, about fifty in all, on the bluff just east of the tanks. These graves, covered by stones laid on the ground in the form of a cross, mark the resting place of thirsty travelers who had pushed on to the tanks, hoping to find water there; but disappointed in this, and too weak to journey on, had perished miserably of thirst; their sufferings aggravated in many cases by the knowledge that the water which they craved could be obtained in one of the tanks but a few yards above them, had they but the strength to climb it. (International Boundary Commission 1898:25)

Gaillard (International Boundary Commission 1898:26) believed that Kino had visited here and that the "Tinaxas" name on Kino's maps referred to Tinajas Altas, though we now know that Tinaxas refers to Baker Tanks, a place Gaillard did not visit. He noted:

Boundary marker 190 about 1896, looking west. (International Boundary Commission. 1898. Photo Album, plate 190)

Boundary marker 190 in 1998. Note the matching skyline in the preceding photo. (Photo by Gayle H. Hartmann)

> Two other tanks, of which no previous knowledge had been obtained, were discovered a short distance above the heads of two valleys which indent the east side of the mountains between 1 and 2 miles north of the boundary. They each had a capacity of about 600 or 800 gallons.

One of these tanks may be known today as Surveyors Tank.

Gaillard's survey team erected large, concrete markers at intervals along the border. These are the ones that dot the border today. The two boundary markers nearest to Tinajas Altas are 190, southeast of Tinajas Altas, and 191, south of Tinajas Altas. The survey report contains a log of boundary markers; due largely to the precipitous terrain, the greatest difference between the US measurements and the Mexican measurements along the boundary occurred at monument 191, with a discrepancy of 2.04 meters (International Boundary Commission 1898:193).

The boundary survey markers give good reference points for photographs. Robert Humphrey, botanist and grasslands expert at the University of Arizona, revisited most of these markers in the 1980s to do a repeat-photography analysis of vegetation changes at the sites as compared with the International Boundary Commission 1890s photographs. His book *90 Years and 535 Miles* (Humphrey 1987) records and photographs most monuments. But because of the rugged terrain, he was unable to reach monument 191, so like the original photographer he set his tripod and camera

"Graves of the victims of thirst, below Tinajas Altas." (Photograph from Gaillard 1896:596)

Today the graves have been obliterated, as this approximately matching photo shows. (Photo by Bill Broyles, 2010)

on the bajada at the foot of the mountain range. In 2010 David Taylor, a professor from New Mexico State University, worked to re-photograph all of the boundary monuments. His results likely will be published in a book; his photographs of the borderlands, aliens, and Border Patrol appear in the book *Working the Line* (Taylor 2010).

In addition to writing the official government report of the boundary survey, Gaillard (1896) also authored a popular article in *Cosmopolitan.* In our opinion it is the best single piece on Tinajas Altas and this remote part of the Sonoran Desert. He combines the trepidations and the fascinations of the desert in a firsthand, reliable account of the region. He describes Tinajas Altas as a

> more reliable water supply . . . fifty-six miles from Yuma, where the waters of occasional rains were collected in a high mountain cañon, and, passing through a narrow rocky gorge, hundreds of feet above the ground, came tumbling down the rocks in a succession of cascades. At the foot of each of these, during the course of countless ages, a deep, circular basin, or tank, has been worn in the solid rock. Of these tanks there are seven large ones and a number of smaller ones,holding in all fifteen to twenty thousand gallons when full. The lower tank alone can be reached on horseback, and to reach the next three requires a stiff climb and a cool head. The upper ones can only be reached by ascending, to a height of several hundred feet, the steep gorge on the right. Even then, except in the case of the highest tank, it is necessary to lower oneself down a rope, fastened to the rocks above. There is nothing in external appearances to indicate that water can be found here, and the fact of their existence is almost unknown to Americans. (Gaillard 1896:602)

Gaillard (1896:602) also records how some travelers perished before reaching the tanks and at the tanks themselves: "Too feeble to climb to the next . . . [they] perished miserably, their horrors aggravated by the thought that the water . . . was but a few yards off." He made note of "fifty graves, near the foot of the tanks, marked by rough stones piled in the form of a cross" and "sixty-five graves in a single day's ride of a little over thirty miles" east of Tinajas Altas (Gaillard 1896:603).

Approaching Las Tinajas Altas. Grave with stones in form of a cross. (Photograph from Gaillard 1896:596)

Graves on Mesa of the Dead, November 23, 1913. (Photo by E. A. Goldman. Courtesy Neil B. Carmony)

Mesa of the Dead a century later from roughly the same view as the photo at left. (Photo by Bill Broyles, 2010)

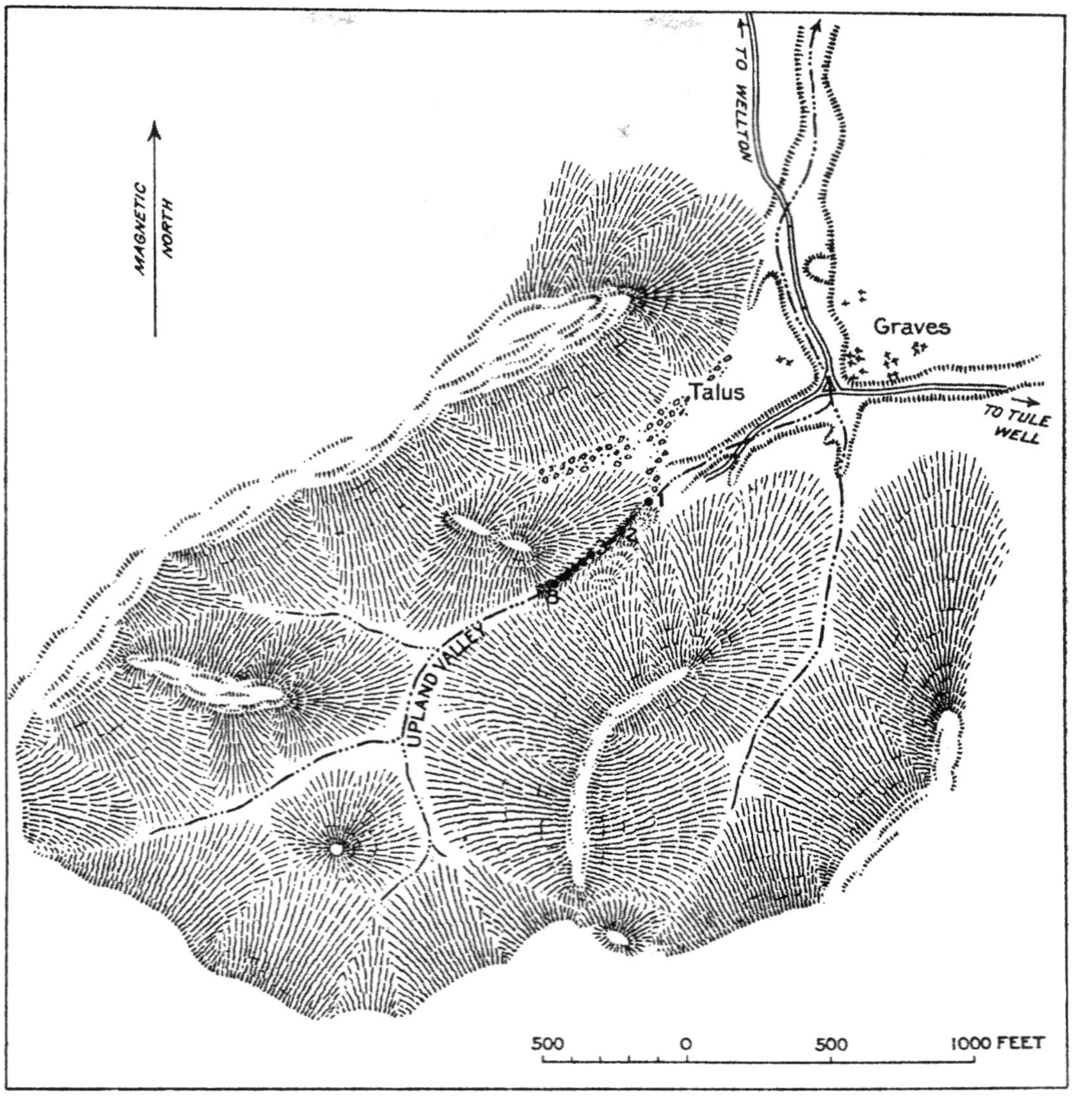

"Sketch map of Tinajas Altas. From plane-table sheet by Kirk Bryan and sketch by C. G. Puffer." (Bryan 1925:132)

Gaillard (1896:603) also recounted Michler's 1855 story of the woman who returned with water to her fallen comrades, and the story of a party of 17 men and one woman who ran out of water 25 miles east of Tinajas Altas and were found by Don Pedro Aguirre of Buenos Ayres [Aires], who was en route from Yuma to Altar, Sonora. Michler rescued them. Gaillard (1896:603) notes,

> Nor are these horrible experiences confined to the past alone. The desert still claims its victims, and not a month passes but that some inexperienced prospector yields up his life in the search for the fabled mines of the desert-sirens which have lured scores of victims to their deaths. On two different occasions prospectors were rescued by our own [survey] parties when death seemed to them to be inevitable.

But disregarding the perils of the desert, Gaillard (1896:605) fell in love with it.

> in spite of all that has been said, there is an unexplainable fascination about the desert—a charm which, after a while, every one feels, and which causes one to look with longing for the time when once more his eyes may rest upon the well-remembered scenes, and once more he may stand amid that weird, appalling—but now to him soothing-silence.

Surveying was arduous in this terrain and climate. The land lines, which define ranges, townships, and sections on topographic maps, have never been completely surveyed and marked for the Tinajas Altas area as of 2010. Surveys of adjacent areas and partial borders with Range 17 West occurred in 1940 and 1941. The southern boundary of Township 13 South, Range 16 West was surveyed by Benjamin J. Kinsey (Cadastral Engineer) and his crew between November 22, 1940, and February 13, 1941. The western and northern boundaries of Township 13 South, Range 16 West were surveyed between December 28, 1940, and February 1, 1941, and between January 15, 1942, and February 3, 1942 (US Bureau of Land Management 1940–1942). Preliminary projections of range and township lines have been made for Township 13 South, Range 17 West, and Range 18 West.

Modern surveyors use satellite photographs, global positioning system units (GPS), and all-terrain vehicles to determine positions and elevations, a far cry from the pioneers' method of lighting fires atop peaks at night in order to triangulate with transits.

Cartographers

With our modern maps and hand-held navigation equipment, we sometimes forget what it must have been like to migrate along El Camino del Diablo, relying on what someone told us in a smoky cantina deep inside Mexico or on our ability to follow months-old footprints after rain and wind storms have swept aside most tracks. We can only speculate how people without maps were able to locate themselves. How was navigation of this region accomplished when waterholes were so cryptic and camps so sparse? Some assume that native peoples used mnemonic devices such as those Bruce Chatwin (1987) describes in *Songlines*. The locations of places and the routes to them were incorporated within songs and chants.

Some petroglyphs may provide water information, just as early European travelers sometimes aligned rocks to point toward a waterhole. In some instances, rock drawings may have functioned as navigational aids, and there is considerable rock art in the vicinity of Tinajas Altas. As we know, however, the "meaning" of rock art (if any) is notoriously uncertain. In a comprehensive survey of the rock art of the Picacho Mountains, Wallace and Holmlund (1986:147–48) discuss a large abstract glyph that appears to represent the ridge upon which it is situated. On his work along the lower Colorado River, Albert Schroeder (1952:44) noted that "[a] few of the linear drawings observed appear to be maps," and in central Arizona a petroglyph showing water pockets has been interpreted as a map (Dockal and Smith 2005). A petroglyph at Antelope Hill may be a map (Boma Johnson, personal communication, and Johnson 1986). Because Tinajas Altas was such a critical landmark to native peoples, it may have been recorded on a rock map, but none has yet been identified.

The earliest paper map of the Tinajas Altas area that has survived was made by Father Eusebio Kino and is dated 1696–1697 (Burrus 1965:Plate 9). It shows no tinajas or trails in the vicinity, indicating that when native guides took him on his first sojourn into the region it was, for him, terra incognita. Kino's 1701 map, printed in Paris in 1705, is the first to depict places he called "Agua Escondida," "La Tinaxa," and "Aguaje de la Luna" (Burrus 1965:Plate 11; also see the map at the beginning of chapter 3).

Maps immediately following Kino's were imaginative but unreliable. For example, Nicolas de Fer pirated one of Kino's map, reprinted it in 1720 and sold copies. Unfortunately, he plagiarized Kino's large-scale regional map dated 1695–1696, which was drastically unreliable and already outdated by Kino's own newer maps (Burrus 1965). In a series of maps that followed, the cartographers struggled to get the geography straight and the features in correct proportion. New information trickling in from travelers was added to new editions of the maps, but few mappers actually saw the region and none visited Tinajas Altas. Some travelers, for example the Franciscan Pedro Font, a member of the 1776 Anza expedition, sketchily mapped the terrain, but it wasn't until the A. B. Gray expedition of 1854 that the region was portrayed with some semblance of accuracy for major features or routes (A. B. Gray 1857).

Each cartographer tried to fine-tune the trails, rivers, mountains, and place names. Because of its remote location, Tinajas Altas remained vaguely situated and phonetically spelled. The name was variously Tinajealta, Tinajaalta, Tinachas Altas, Tinajos Altos, and Tinajualto. Some maps put it east of longitude 114° and others, west. Actually, the lowest tinaja is at 32°18'42"N, 114°03'00"W (NAD 27).

The first map with the modern name Tinajas Altas and location was the US Bureau of Topographical Engineers map of 1859 (corrected in 1867). This map is sufficiently more accurate than its predecessors that we must wonder if the Bureau conducted its own survey instead of relying on the information developed by the Emory Boundary Survey of 1854.

Other important maps include the 1894 General Land Office map, the Army Corps of Engineers map of 1918, Bryan's map of 1922, and of course, the USGS maps of 1942, 1962–1965, and 1990. These all brought significant improvements to the accuracy and completeness of regional cartography, in some cases feeding off locally made and sometimes hand-drawn maps that supplemented place names and routes.

Perhaps the most interesting map was that of Carl Lumholtz (1912). It incorporated the latest surveys of the time, added Lumholtz's own peripatetic observations, and

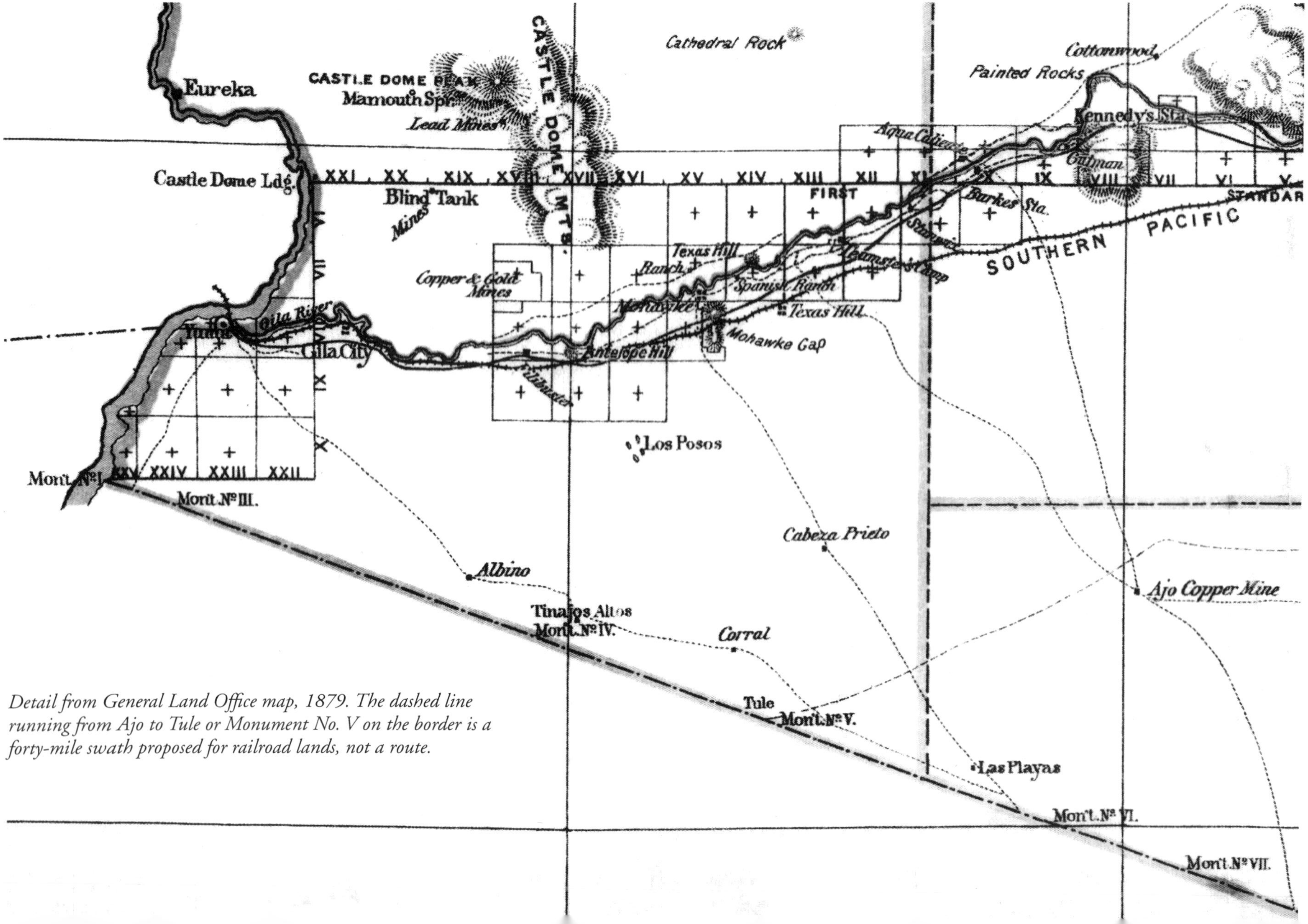

Detail from General Land Office map, 1879. The dashed line running from Ajo to Tule or Monument No. V on the border is a forty-mile swath proposed for railroad lands, not a route.

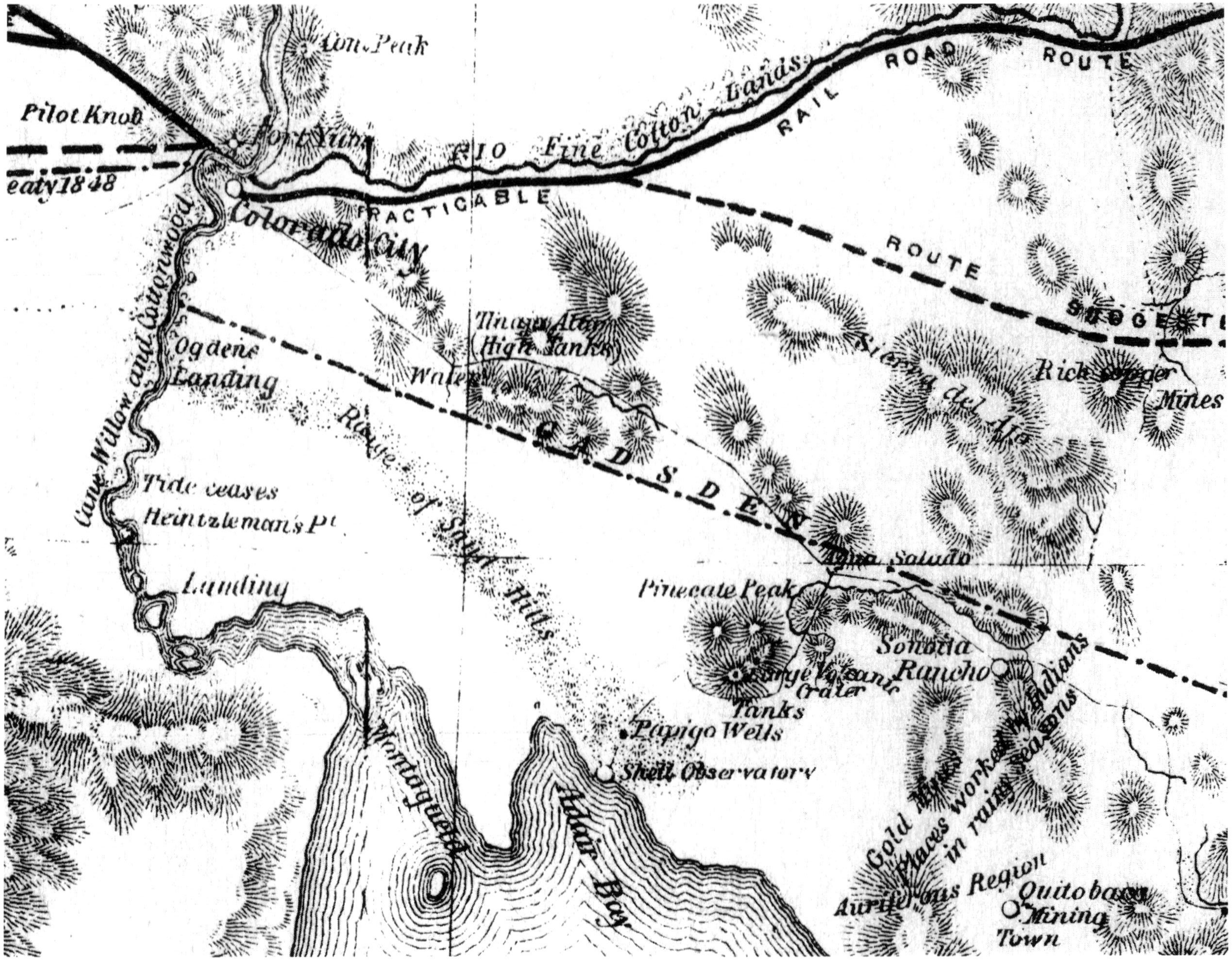

Detail from the A. B. Gray map, 1857.

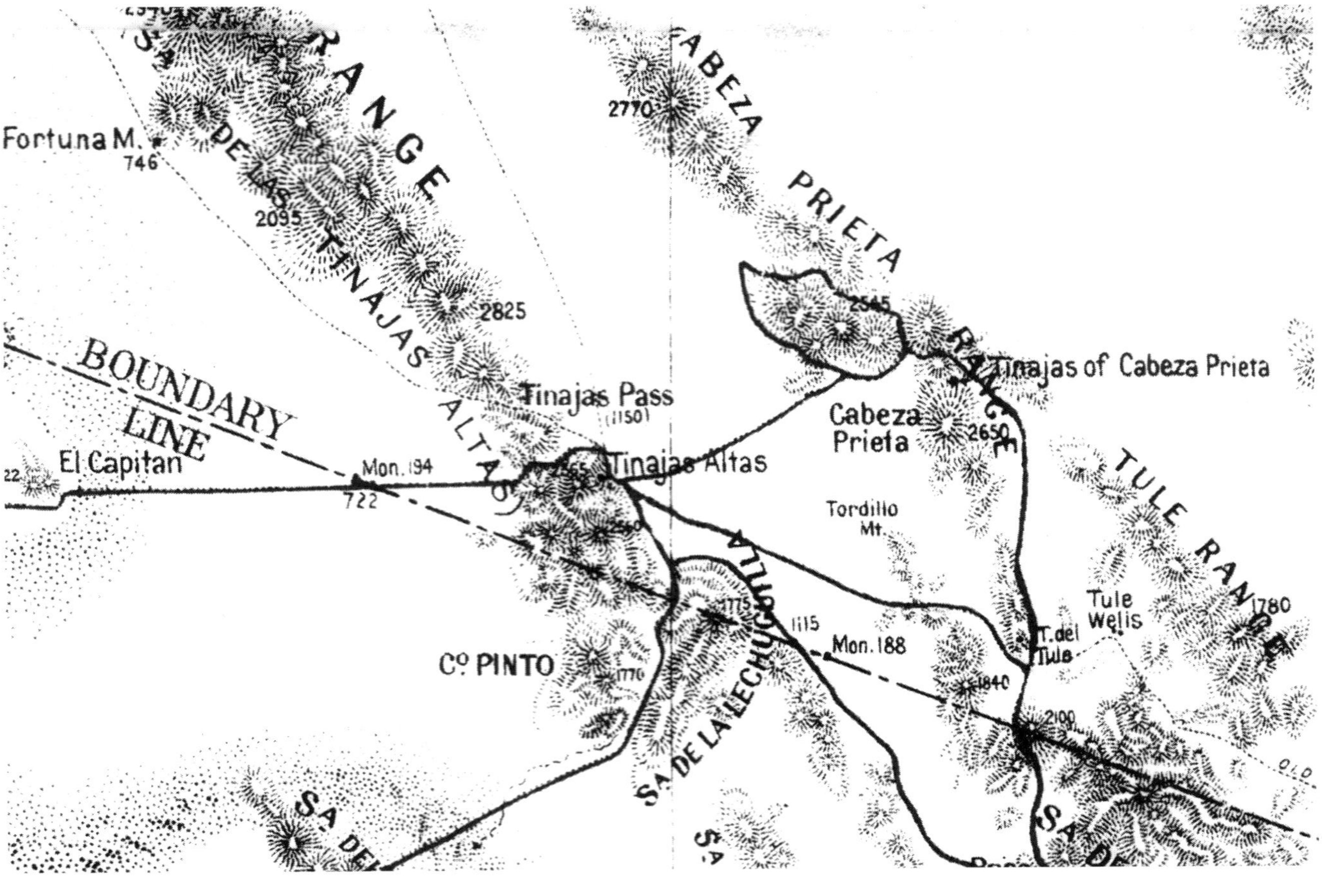

Detail from "Map of Southwestern Arizona and Northwestern Sonora Comprising the Region Formerly Called Papaguería. From Original Surveys Made by Carl Lumholtz. Drawn by A. Briesemeister." (Lumholtz 1912: insert) The darkest lines show Lumholtz's cross-country routes afoot and horseback, and do not represent trails or roads.

was well-proportioned on a regional scale so that the reader was able to gain an understanding of the places' relationships to each other. This map is still reproduced and used as an underlay for new information. Its sister map shows the traditional villages, ranches, and camps of the Hia C'ed O'odham and Tohono O'odham.

A general geologic map of the area was first produced in 1924 by Kirk Bryan. In 1933 Eldred D. Wilson produced a more detailed map, *Geologic Map of Southern Yuma County, Arizona*, which was updated and published in color by the Arizona Bureau of Mines and University of Arizona in 1960. Its scale is 1:375,000, and it shows Tinajas Altas in Mesozoic granite and related crystalline intrusive rocks, with Quaternary sand, silt, and gravel forming the valley floor. The Tinajas Altas Mountains are shown as Laramide granite.

We know of at least a hundred maps depicting Tinajas Altas or its region, and likely there are scores more. Many of these maps are located at the Arizona State Capitol Archives (ASCA) or are in the University of Arizona Map Collection (UAMC) in the Main Library (see, for example, Charlene Baldwin and Jack Mount's *Maps of the Pimeria* [1991]). Other universities, including Tulane, host substantial collections of historical Spanish Colonial and American maps. The private David Rumsey Map Collection, now associated with Stanford University, holds many if not most maps that include the Tinajas Altas region; some of these may be viewed on the internet. Over the years, changes in the spelling have become standardized, and the location of Tinajas Altas, as well as the locations of relevant geographic features, important trails, and other major waterholes, has become clearer and more precise.

Father Eusebio Kino's maps can be found in *Kino and the Cartography of Northwestern New Spain* by Ernest J. Burrus, S.J., published in 1965. We are still looking for maps of the American portion of the region made during the period 1777 to 1849. It is hard to believe, especially during the 1849 Gold Rush, that many of the rank novices who attempted the arduous journey to California could ferret out a safe route without some sort of map. Kino's maps were not sufficiently detailed to allow someone to locate routes or waterholes or campsites with forage for livestock. Perhaps by the mid-nineteenth century the routes were etched into the ground, so that any pilgrim might simply read the "map" with his feet.

The route of the Camino del Diablo, the main trail in the region, is not always straightforward. It is complex, especially when it passes through the Cabeza Prieta Mountains. The maze of canyons and indistinguishable landmarks still confuses those who venture there without maps. Each year US Border Patrol agents find people wandering the region, hopelessly lost (Hank Hays, personal communication 1984; Joe McCraw, personal communication 1997).

Cabeza Prieta Peak, though not the high point of its namesake range or of the region, is sufficiently distinctive to act as a compass for people traveling in the region—it is plainly visible from Tinajas Altas and it has a set of important tinajas near its northeast base (Broyles 1996b; Lumholtz 1912).

For several reasons, we lack a comprehensive Native American map of the region. Modern scholars have been slow to learn the indigenous names that still exist (many have been lost in the shadows of the past), and local tribes have been reluctant to reveal those names still known, some of which are associated with religious ceremonies and sacred places. Further, some groups, such as the O'odham, name sites but not entire mountain ranges. Indigenous people did not define their territory in modern, platted terms (Broyles, Evans et al. 2007). Ethnographer Robert K. Thomas (1991:393) commented on the difficulty of getting Hia C'ed O'odham to generalize their geography:

> Papagos do not, for instance, name ranges of mountains. They name peaks. It is difficult to pin Papagos down as to what a range of mountains is. Also, they will not put a boundary between their country and the country of other tribes. They will say, 'We traveled here, and we gathered here, and the country of the Yuma is farther down the river.' But they will not say that the Gila Mountains were the boundary between the Papago and the Yuma. It is easy to get them to say this is the original country of the Aliquippa people, but if you ask them where their country is, they will invariably give you their fields.

Maps lure the human mind to places unknown and to places remembered. They reveal settlement patterns and our relationship with the land—from where we put our camps and drill our wells to where we set aside wild, natural lands confirming our heritage and respect for ground so that generations to come may enjoy and understand its wonder. They are especially important in difficult and dangerous places such as those found around Tinajas Altas.

Scientists

The scientific mysteries of Tinajas Altas have drawn a number of researchers, and their reports suggest they valued the place as much for its tranquility and beauty as for its data. Researchers were included in the first boundary survey team; in 1855 Arthur Schott, of that team, sketched and collected specimens for taxonomic classification in distant labs and archives.

In 1900, the American anthropologist, W. J. "Don" McGee, who always wrote his name as WJ McGee without the periods or a space, was studying the people of the Papaguería, and he traveled from Caborca to Sonoyta to Yuma along a route he knew as the Old Yuma Trail. In his 1901 paper, McGee gives us important information on the history of the famous name, Camino del Diablo. He acknowledges that some before him knew one stretch of the Old Yuma Trail as El Camino del Diablo, although he himself used the English name. He attributes the name El Camino del Diablo to the Mexican migration period before 1840 (McGee 1901:105), and notes that it again lived up to its name during the Gold Rush of 1849. The first use of the term in print may be by the International Boundary Commission (1898, 2:26) when they note, "This road is appropriately called by the Mexicans 'El Camino del Diablo.'"

Where did the name come from? Its origin is a puzzle. Gaillard (1896), the International Boundary Commission (1898), and McGee (1901) mention that the road between Agua Dulce and Yuma was called El Camino del Diablo, but McGee in 1901 still refers to it primarily as the Old Yuma Trail, and Lumholtz (1912:197) notes, "Many recent authorities speak of this part of the trail as the *Camino del Diablo*, but nobody in Sonoita [sic] knows it under that name." Other names in the history books include The Devil's Highway, Sonora Trail, Sonoyta-Yuma Trail, Yuma-Caborca Trail, Camino del Muerto, and Old Yuma Trail. As late as 1908, James M. Barney's *Map of Yuma County, Arizona* calls it "Old Camino Real," and a 1942 USGS map (1/250,000 scale) shows it as a dirt road and labels it "Santa Ana—Tijuana Highway" (Santa Ana being in Sonora on the road to Hermosillo).

Then Lumholtz (1912:197) adds a new wrinkle to the story: "possibly there may be some confusion arising from the name of a small cañon between the low ranges which the trail passes near Tule well, called Cajon del Diablo."

McGee's (1901) article on the Old Yuma Trail included a map of the trail route from the Baboquivari Mountains to the Colorado River indicating nearby geographic features.

W J McGee. Professionally William John McGee went by W J McGee and used no periods for his initials, but his family and close friends called him Don, because when they were children his younger brother had difficulty pronouncing John and said what sounded like Don. (E. McGee 1915: 38, frontispiece)

The old cemetery of lost travelers was still visible, and he noted "three-score cross-marked graves" (McGee 1901:138) on what has become known as the Mesa of the Dead, a term he coined when he referred to "The Mesita de los Muertos with its two-score cross-marked graves" (McGee 1906b:283). Although Kermit Roosevelt (1920:73–74) notes that there were approximately 150 graves, other early reports such as those of Gaillard (1896:603), Lumholtz (1912:242), and Bryan (1925:422) estimate the visible number as between 50 and 65 in that early period of 1896 to 1925. We may never know whether Roosevelt exaggerated, whether he simply counted too many clusters of rocks, or whether he may have encountered another 60 to 85 graves along the Camino del Diablo. The total number of graves continues to intrigue us.

McGee returned in 1905, ostensibly to study meteorology, but really to regain his flagging health in the salubrious dry heat. His visit there from May 20 through August 28, 1905, is one of the longest recorded stays at Tinajas Altas. This stay led to several scholarly publications. One was a meteorological paper, "Climatology of Tinajas Altas, Arizona" (McGee 1906a). He reported a high temperature of 117.4°F and only 0.12 inches of precipitation from May 21 to August 28, 1905. A second paper from his 1905 visit to Tinajas Altas was titled "In the Desert," and was published by his sister, Emma, after his death (E. McGee 1915). In it he waxes enthusiastic about the virtues of desert camping, even here in the wildest, hottest, driest, most remote spots in North America. "The effect of desert life on my health has been all that I anticipated, and I return to civilization entirely well and with a store of experiences well worth the trouble of securing" (E. McGee 1915:86). Although he attributed his malaise to the aftereffects of typhoid fever, more likely he was in the initial stages of prostate cancer; he died of it in 1912. He chronicled his last days in "Symptomatic Development of Cancer," an article published posthumously in the journal *Science* (McGee 1912:348–350).

A third paper, "Desert Thirst as Disease," included the classic desert tale of prospector Pablo Valencia, lost without water in this enormous desert in August 1905 (McGee 1906b). After eight days and nights and 120 miles (192 km) of travel, all on one day's water, Valencia struggled back to within earshot of McGee's camp at Tinajas Altas and let out an "inhuman bellow," awakening McGee and his Tohono O'odham camp-manager, José. As Broyles (1982:376) noted in his retelling of the story, "With McGee's kind and educated help, Pablo survived. And, . . . within two months he rode again." "Desert Thirst: The Ordeal of Pablo Valencia" (Broyles 1982) recounted the author's own ordeal when he ran out of water in an attempt to retrace Valencia's route. Because of coincidences and discrepancies in similar McGee stories, Valencia's ordeal may have been fewer days (Broyles 2006).

McGee's campsite for his summer stay was somewhere on the bench immediately east of the lowest tank. He had a wagon, but otherwise his equipment was minimal—he slept on the ground and did not erect a tent or permanent shelter. He reports having an hour's climb to fetch water, implying that the lowest tank was dry.

The next prominent scientist to stay at Tinajas Altas was Edgar Alexander Mearns (1907:19), who visited the tanks from February 14 to 23, 1894, with his assistant, Frank Xavier Holzner. Mearns' trip along the border from El Paso to San Diego was part of the International Boundary Survey

From the top, Tinajas Altas looking east, circa 1933. The Camino del Diablo runs from lower center to upper right of the photo. (Photo by Tad Nichols)

to improve the mapping of the frontier line between the two countries. They made camps at Tule Wells (now Tule Well), Tinajas Altas, and then at Adonde, a small settlement along the railroad line 25 miles (40 km) east of Yuma. The camp at Tinajas Altas, Station Number 63 on his trip, was near the lowest tank, though Mearns does not specify its exact location.

Mearns (1907:122) reported, "This important station is at the east base of the Gila Mountains, beside the lowest of a chain of natural rock tanks, in a steep ravine, containing an unfailing and almost inexhaustible supply of good water." However, he says, "The upper tanks are easily overlooked and difficult of access, which facts afford the most plausible explanation of the loss of the lives of many persons whose bones and graves were thickly scattered about our camp."

Mearns and Holzner made the first botanical collections at Tinajas Altas. They gathered botanical, mammalian, and herpetological specimens and listed them in the report. The list may be of value in illuminating what plants and animals might have been available for indigenous people utilizing the area at the turn of the nineteenth century. Mearns (1907:122) writes:

> Around the base of these mountains the last of the giant cereus were found on the Boundary Line. One species of paloverde, the small-leaf horse bean, and the ironwood, indigo, thorn, acacia, mesquite, and nolina complete the list of trees which are mostly confined to arroyos around the base of the mountains. Species of lotus, *Agave*, *Cotyledon*, *Fouquieria*, *Covillea*, *Ephedra*, *Datura*, *Nicotiana*, *Terebinthus*, *Cucurbita*, *Euphorbia*, and *Physalis* were common. Of cacti, a species of *Cylindropuntia*, called "tasajo," and another resembling the choya, one small species of *Platopuntia* that we only saw on the mountains of the Western Desert, a small, long-spined *Cereus*, the handsome, red-spined

> *Echinocactus lecontei*, and several species of the genus Cactus were the common sorts. *Cereus greggii* and one or two other cacti were found more sparingly. [See appendix 5 for a current plant list.]

Other botanists who followed included G. L. Harrison and G. O. Belden on March 5, 1927. Thomas H. Kearney, Robert H. Peebles, and G. L. Harrison collected plants at Tinajas Altas on March 29, 1930, and Kearney returned in 1935. Kearney and Peebles (1960) later published *Arizona Flora*, a standard reference book on Arizona plants. Numerous botanists have conducted surveys and made plant collections at Tinajas Altas in recent years, including Thomas and Rebecca Van Devender, Annita Harlan, and Richard S. Felger. Generally, specimens from these collections are deposited at the University of Arizona's Herbarium in Tucson.

Amazingly, Mearns noted no signs of archaeology; it would be interesting to see if his field notes do. One man did come looking for signs of ancient people as well as living Native Americans. Carl Lumholtz was a Norwegian ethnologist, linguist, and all-around scientist who wrote travelogue books with detailed studies of the lands he visited. His books include *Among Cannibals* (1889), *Unknown Mexico* (1902), and his classic of the Sonoran Desert, *New Trails in Mexico* (1912). (The 1971 edition of *New Trails in Mexico* contains a Lumholtz bibliography compiled by field historian Bernard Fontana.) This was an interesting era when explorer-travelers could set out into a region, discover, draw, and describe what they saw, then publish book-length accounts avidly consumed by readers from Europe to America. Lumholtz first visited Tinajas Altas on January 23, 1910 (Bowden 2007; Eek 2007). He reported,

> Tinajas Altas is a landmark in the local and recent history of that desolate region. It is a dismal looking place and, after having been accustomed to the absolute cleanliness of nature, the experience of again meeting with pieces of cast-off clothing, rusty tin cans, and other cheerless marks of human occupancy which were apparent here was not a pleasing one. There is a large space for camping at the lower end of the little valley which begins at the foot of the mountains where the principal pool is. An embankment of gravelly detritus shuts off the view toward the east, while high granite rocks, bare and hard, rise forbiddingly on the other sides. Large boulders which were once detached from the rocks above are lying at the base of the steep mountain sides. (Lumholtz 1912:239)

Edgar A. Mearns, February 5, 1909. (Photo courtesy of the Arizona Historical Society/Tucson; image no. 49511)

Lumholtz's book shows one photograph of Tinajas Altas (Lumholtz 1912:facing page 240), a general view from the east. Two trails leading toward the tinajas are apparent in the photo.

Lumholtz (1912:322), met two prospectors at Tinajas Altas who showed where Lumholtz's friend, "W. J. McGee, 'walking barefoot' he said for the sake of his health, had camped in a hut among the rocks near the lower tank." McGee himself wrote that he slept there on the ground.

Lumholtz (1912:240) noted that

> [a] prospector in Tucson, who ascended to the top, described to me a large sand basin above the tinajas, a few hundred feet south of which are a number of caves where pictographs are seen as well as pounding-holes in the rocks. Not having been there myself, I give the information with reserve. The Sand Papagos had an important camp at Tinajas Altas.

Carl Lumholtz, 1921. (Chicago Daily News *photo, courtesy Chicago History Museum. Image DN-0073047 from a glass negative and here cropped by permission)*

Lumholtz does not mention the historical inscriptions, pictographs, petroglyphs, grinding ovals, or mortars in the lower valley or near the tanks themselves. Of the caves with pictographs in the upper valley, Rodack (1968:27) notes in her article in *Desert Magazine*, "We failed to find them, but the upper canyon is well worth further exploration." The Tinajas Altas Archaeological Survey, conducted between January and April 1998, thoroughly surveyed the canyon and found no evidence of rock art (pictographs or petroglyphs) in the valley above the tanks (Hartmann and Thurtle 2000).

Lumholtz (1912:241) continues, "The locality harbors many unpleasant memories; in the days of the California gold excitement, as well as later, many people passed by here, and the lowest and most accessible pool did not last very long." He then recounts Gaillard's story about three exhausted prospectors who perished at the foot of the cliffs, their fingers worn to the bone from clawing at the rock. He adds that others were killed in fights over the meager water, and still others were murdered by robbers who lurked waiting for travelers who might be carrying valuables or gold.

> The dead were buried on the top of the embankment that shuts off the view toward the east. A small cross of stones embedded in the ground indicates where a person is buried underneath. We counted fifty-four of such crosses here; time has destroyed as many more and there were many, no doubt, who were buried without any sign to mark the spot. Holes had been dug in the ground at three or four places in hopes of finding riches, for tradition has it that travelers, their animals having given out or for other reasons, buried their treasures here, and even the camping place had been searched for these. (Lumholtz 1912:242)

Lumholtz and his companions left Tinajas Altas on January 24 via Tinajas Altas Pass, a little to the north of Tinajas Altas, on their way westward to the Colorado River. He commented on wildlife, including bighorn sheep and a mosquito that bit him! In March of the same year, 1910, he returned to Tinajas Altas on his way to the Sierra del Rosario, which he had seen from a distance in January but had not had time to visit. This time he encountered other people at Tinajas Altas.

> Four unwelcome strangers made their appearance, driving up their wagon outside of the little valley. They were evidently American citizens from somewhere on the Gila River, most of them Mexicans, and the owner of the wagon and fat horses was an unpleasant old Italian. They were not prospectors, and they were a queer lot—very different from the ordinary Mexicans—and they gave me an uneasy feeling because of their underhanded and evasive ways. Probably they were nothing worse than treasure hunters, but it was a relief to have them continue their journey next morning, while we recuperated here another day. (Lumholtz 1912:320)

Some researchers occupied themselves with collecting animals and plants for museums and specimens. Alfred Brazier Howell, a mammalogist with the US Bureau of Biological Survey (now known as the US Fish and Wildlife Service) collected mammals at Tinajas Altas from June 1 to 7, 1918. His field notes are in the Smithsonian Institution (Howell 1918).

A number of floral and faunal specimens were collected by other scientists from the Tinajas Altas area. In his compilation and encyclopedia of Arizona mammals, museum director and mammalogy professor Donald F. Hoffmeister (1986) lists many of the mammals taken earlier at Tinajas Altas for scientific study. For example, a bobcat (*Lynx rufus*)

Research biologist Dwight Mallery (left), noted botanists Ira Wiggins and Forrest Shreve, and photographer Tad Nichols at Tinajas Altas, 1933. Tad was a university student at the time. (Photo by Tad Nichols; see Nichols 2007)

specimen now at the University of Kansas was probably collected by Edward Alphonso Goldman at Tinajas Altas in 1913 (see also Sheldon 1993). Tinajas Altas was also the type locality for four species of rodents: *Neotoma albigua mearnsi* Goldman, *Neotoma lepida flava* Benson, *Spermophilus harrissii saxicolus* Mearns, and *Perognathus intermedius phasma* Goldman (Hoffmeister 1986:577), although the latter three subspecies have been supplanted by more recent research.

A number of eminent scientists associated with the Carnegie Desert Laboratory have done work in the Tinajas Altas area. The desert laboratory, located in Tucson on Tumamoc Hill, was established in 1903 under the direction of Frederick W. Coville and Daniel T. MacDougal as a botanical research station of the Carnegie Institution of Washington, D.C. (Coville and MacDougal 1903). During the 1930s its illustrious staff included Forrest Shreve, T. Dwight Mallery, Godfrey Sykes, and Ira L. Wiggins (Bowers 1988; McGinnies 1981). Forrest Shreve would author over 60 studies and articles on botany and biogeography. Wiggins wrote extensively on plant classification and biogeography. Together they produced the seminal and nonpareil *Vegetation and Flora of the Sonoran Desert*, a treasure that defined the range and subdivisions of the Sonoran Desert (Shreve and Wiggins 1964).

From 1925 to about 1936, Godfrey Sykes (1931) and Dwight Mallery (1936) conducted a precipitation survey that included Tinajas Altas. Their survey confirmed the obvious aridity of Tinajas Altas and gave a regional view of precipitation patterns. An overview of this precipitation study is provided in McGinnies (1981:101–16); part of the study is discussed in Turnage and Mallery (1941). Mallery studied weather and climate and how they affected plants. He published several important articles about plant physiology (McGinnies 1981) and was president of the Tucson Natural History Society, which pushed for the founding of Organ Pipe Cactus National Monument.

Mallery had a string of rain gauges stretching from Tucson nearly to Yuma. Tad Nichols (2007), a noted Southwest photographer, recalls traveling with Mallery and Shreve along the Camino del Diablo. Even in the 1930s, the road was very primitive and each foray by automobile was an adventure

Paleoecologist Thomas R. Van Devender (left) and botanist Richard S. Felger, 2006, Sonoran Desert National Monument. (Photo by Benjamin T. Wilder)

that tested everyone's courage, savvy, and perseverance. Nichols (2007:88) writes that Mallery had "several gauges, two north and south of Tinajas Altas, and one on the east side of the Lechuguilla Valley. We'd leave Tinajas Altas and drive north. Then he had one at Raven Butte stuck up in the lava somewhere. The whole trip was about five or six nights, with two nights at Tinajas Altas."

The irrepressible Godfrey Sykes (1951) also conducted a rain-gauge survey across El Camino del Diablo. He recounts his trip of July 1925 to take readings on precipitation in the gauges and to replenish them with oil, which forms a surface layer retarding evaporation of any rainfall before the next semi-annual reading. The trail was quite sandy in spots, and several times Sykes was forced to single-handedly dig out his car when it became stuck. The daytime temperature reached 112° to 115°F and Sykes, on the verge of heat exhaustion, was forced to rest in the shade under his car, recording with dry irony the performance of his own body throughout the perilous situation. By the time he reached Tinajas Altas, he had lost about 14 pounds from dehydration, but still had sufficient energy to climb to the top of the tanks. He noted that the tinajas were almost dry with "only two of the nine holes containing any water at all and these only some few gallons" (Sykes 1927:72).

Eldred D. Wilson (1931), a geologist with the Arizona Bureau of Mines at the University of Arizona, conducted a geological survey in the Tinajas Altas area in 1931 as part of his statewide assessment of mineral potentials. He reported no valuable minerals in the area, the nearest ore (gold) being found at Fortuna Mine on the northwest side of the Gila Mountains, some 24 miles (38 km) from Tinajas Altas.

Thomas Van Devender, paleoecologist, conducted a pioneering packrat midden survey at and near Tinajas Altas in the 1970s. Pollens and plant parts found in ancient midden deposits reveal climatic changes at Tinajas Altas in the Holocene and Pleistocene (Van Devender 1990). He and his colleagues were able to identify and date the area's plant-life to 40,000 years ago (see appendix 5). He concluded that during the Pleistocene epoch the Tinajas Altas area was more like the upper Mojave Desert is today, with a plant community featuring Joshua trees, single-leaf piñon, Mojave sage, California juniper, and Bigelow beargrass, whereas at present we find an association of paloverde, ironwood, bursage, and catclaw acacia (Van Devender 1990, 2007).

Archaeological Studies and Documentation of Changes at Tinajas Altas

The prehistory and history of Tinajas Altas have been of particular interest to many visitors. Ethnologist W. J. McGee (McGee 1901:138) examined

> the Indian mortars ground in the ledges and great boulders about every pool—mortars numbering hundreds if not thousands, some but a few decades old, but most so ancient that the polished bowls [differ from] the unpolished rocks around them—mortars recording the visits of uncounted generations of devotees, to whom each laboriously-wrought basin was at once symbol of and invocation for precious

> food and life giving water. One boulder bore 40 pits in its upper face, another 28; and up to the highest tinaja reached they are found in corresponding profusion. Most of them must have antedated padre Kino.

The eminent desert archaeologist from San Diego, Malcolm Rogers, first visited Tinajas Altas in 1929. In his system, Tinajas Altas became site A-56. He saw it as having one acre in the open and half an acre of rockshelters and caves representing what he called Amargosan II, Yuman I, II, and III, and Papago cultures. He recorded boulder windbreaks, cobble hearths, rock pile supports for roof poles, a cremation, broken and worked seashell, little flaked stone, at least 100 bedrock mortars, and "a great pictograph site, there being many shelters, and small pockets in the white granite for painting surfaces. . . . More pictographs are reported in a canyon 1 mile west of here, which was not visited in 1929" (Rogers 1929:n.p.). Rogers also noted "nine crosses constructed of boulders set into the soil (4' to 6' in length). There have been more, and these nine have since been destroyed." Two pages of drawings accompany Rogers' site report.

Over the decades, much of the historical and prehistoric artifactual material has been removed, presumably by subsequent visitors, although shell, chipped stone, and potsherds are still common. In addition, Tinajas Altas provides an interesting case study of the survival time of a desert cemetery. Many reports indicate the decline in the number of visible crosses on the Mesa of the Dead, from possibly as many as 150 after the 1849 gold rush to 9 in Rogers' 1929 report, to what seems to have been the final cross, photographed during a visit by Gayle and William Hartmann in 1971. The decline in numbers of visible crosses is also discussed in W. Hartmann (1989:189). It is clear that over the years the crosses and the mounds have been disassembled, and graves were intentionally disturbed.

During Carl Lumholtz's (1912:294) visit to Tinajas Altas in March of 1910, he reported that American prospectors at Papago Mine told him about "some suspicious looking Mexicans at Tinajas Altas who seemed destitute of everything except arms and full cartridge belts, and were very non-committal in regard to their errand." When Lumholtz and his two companions neared Tinajas Altas, they found unexplained fresh tracks in the sand near Sierra Lechuguilla. At Tinajas Altas Lumholtz (1912:308) noted:

The last historical grave on the mesa at Tinajas Altas, October 17, 1971. (Photo by William K. Hartmann 1989:189)

> Since I had been at this place last some bushes had been burned in the arroyo, and there had been dug two long, deep holes in the ground which served immediately to solve the mystery surrounding those Mexicans we had heard of, who were so strangely behaving and so well armed. They had been engaged in an occupation dearly loved by many of their countrymen—that of digging for hidden treasures. To bury money was the usual method of guarding it during the period of unsettled conditions in Mexico, and rumor has it that a great deal of plata (silver) was once deposited in this place for safe-keeping. People who returned from California in the [eighteen-] fifties, or people who robbed those who did not return, have also been credited with the burying of treasures here.

Photographer Tad Nichols (2007:90) noted excavation of graves in the 1930s. He remembers that "at the base of Tinajas Altas there were the remains of many graves. Most have been dug up, with many holes there. People had dug the graves out. There were also some marked with rocks."

This report is confirmed Randall Henderson, editor of *Desert Magazine*, who described significant damage before February 1940 when he visited the area with a group of friends. He comments that "vandals have excavated most of these burial places, and nothing remains to mark the old cemetery today except pits in the sand" (Henderson 1940:10).

Another nod comes from Yuma historian Catherine Irwin's (ca. 1970) comment that even before her visits beginning in the 1950s, many of the grave sites at Tinajas Altas had been excavated and looted. Irwin (ca. 1970:31) noted, "Long before 1947, when an inspection station for controlling hoof-and-mouth disease was established on the little mesa at the foot of Tinajas Altas, the graves there had been dug up in hopes of discovering treasure. Then, there was some reason in believing gold might be buried in the area; now there is no excuse for such desecration."

Additional damage was caused by the construction of the inspection station (Gale Monson, retired US Fish and Wildlife Service, personal communication 1997; Hal Coss, retired U.S National Park Service, personal communication 1997). Conversations with employees of the Bureau of Animal Industry camp indicate that significant damage was caused by the construction of that camp in 1947 (Thomas 1989). The remains of two concrete pads can still be seen today, marking the site of the inspection station cabin, and it is in the middle of the terrace where many of the graves were located. The area is now used as a parking lot at the end of the Tinajas Altas access road. Once the prominent graves had been altered or destroyed, visitors unknowingly drove their vehicles across the low mounds and rock alignments, and campers probably took the rocks for their own campfire rings. In addition, over time the hooves of livestock and wild animals, the work of digging rodents, wind, and rainfall runoff eroded other graves.

The post–World War II period, with its easy mobility offering access to out-of-the-way places, has witnessed an increase in damage to locations of historical interest. Motives range from ill-advised good will to pure vandalism. For example, perhaps in the name of historical research and education, a bedrock slab of granite with bedrock mortars was removed from Tinajas Altas in April 1956 and put on display at the Arizona Historical Society in Tucson. These mortars can be seen on the south side of the society building at 949 East Second Street in Tucson. Tom Peterson, former chief curator at the Arizona Historical Society, reports that they were chiseled out at the behest of an Arizona Historical Society board member or friend, lifted by crane, and trucked to Tucson by Tucson Warehouse and Transfer. Drill holes are still visible from this activity about 300 yards east of the lowest tank near the south wall of the canyon. A similar slab with mortar holes was taken to a private residence near Wellton. It would be fitting if this slab were returned to Tinajas Altas, or at least to public view at the Arizona Historical Society in Yuma.

The area has also been picked over by metal detector buffs. Wayne Winters (1963) was a treasure hunter, and photographs in his book *Campfires Along the Treasure Trail* show him using his metal detector on the flat below Tinajas Altas and then using a shovel to dig into a spot where his detector indicated metal. The captions read, "Treasure buried at Tinajas Altas? Maybe, for the metal detector indicated the presence of a metallic nature beneath this pile of rocks at the age-old camping spot" and "No treasure, but instead a horde of old tin cans buried by some camper who even though he was far from civilization, was no litterbug. Many ancient graves are in the area" (Winters 1963:86). He also recounts,

> With a companion, I left Wellton . . . [and] some hours later we arrived at Tinajas Altas and found evidence of many camps. There were parts of old wagons, the running gear of a Model 'T' Ford, a bunch of tin cans liberally sprinkled with beer bottles. Hoping to locate some of the 65–odd graves reportedly at this place, we were disappointed. We did find one weather-beaten headboard in the area. It was so old that the carved words and date had faded into the grain of the wood itself, making deciphering an impossibility. (Winters 1963:87)

Arizona Republic newspaper reporter Bob Thomas (1989:T3) described a more devious incident: "In the late 1960s, a band of motorcyclists from California drove across the Camino del Diablo, digging up what graves they could find and taking the skulls, law officers said."

Other graves may (or may not) be located in the area. Madeleine Rodack (1968:26), for example, introduces the mystery of the "Number 14 site" several miles east of Tinajas Altas near the Camino del Diablo:

> Lumholtz relates the story of a family of 14 who were either killed by Indians or died of thirst buried near this [ironwood] tree [Lumholtz 1912:238], but we were unable to locate any sign of their graves. Seven miles further, however, we came upon an immense Figure 14 laid out in large stones. Could this have been where the family died? If so, how could Lumholtz have confused it with the area of the ironwood tree? Another mystery of desert and legend. Near the

O'Neill's Grave in 2010 and in 1927 (inset). (2010 photo by Bill Broyles; 1927 photo from Sykes 1927:67)

> 14 are a couple of piles of stones, one in the shape of a cross, the other a small circle.

Is the "14" really a marker of death along the trail? It is generally presumed, but unproven, that the 14 dates from a land survey of the area, not from a funeral. Another pile of stones topped with a cross (made from what appears to be two pieces of agave stalk wired together) is located in the canyon itself, just to the west of the terrace where the cemetery was located. It was photographed by Gayle Hartmann in 1989 and was recorded in 1998 as part of the Tinajas Altas Archaeological Survey. Its location under a creosote bush and the consolidated appearance of the rock pile suggest it has some antiquity.

While looking at remaining grave sites along El Camino del Diablo, one wonders how many of the grave sites reported by early visitors were really graves in the modern sense of full-body interment. Burying would be a difficult, impractical chore. Soil in many places is quite shallow, even for diggers with shovels, which few historic travelers had. Survivors in a traveling party were frequently on the verge of death themselves and unlikely to dig holes for the deceased. It likely was common for survivors to leave the corpse to the elements, to cover it with rocks, or to bury the body in the nearest sandy arroyo bottom, where storms would eventually erode the bones. It is possible that survivors honored the dead by forming rock crosses at a later, more convenient time or when they reached a camp; this would be more in the spirit of modern roadside memorial crosses, *cruces de memoria* (Griffith 2005).

Further, in a land where it may be years between visitors, some graves and crosses may have been fashioned by passers-by who came upon skeletal remains years after the victim died. Considering that some "graves" are surficial cross-shaped rock-alignments and some (like O'Neill's grave near Papago Well) are low, linear mounds with vertical crosses made of iron or wood, it is possible that some of the former sites are shrines in support of the safe journey of travelers, while more of the latter are actual burial sites. Both the rock alignments and the mounds are memorials and deserve to be preserved as examples of our heritage and respected as tributes to our dead.

Treasure hunters and "rag pickers" would do well to remember that travelers through this region brought few if any valuable encumbrances with them. For example, despite popular notion, it is quite inaccurate that the Spanish

The Camino del Diablo tribute to historical travelers was erected about 1988 by the US Air Force. (Photo by Bill Broyles)

expeditions were legions of well-provisioned troops, or that they casually stashed treasures in obscure caves, or that they carelessly abandoned suits of now-priceless armor. Not only did Native Americans comprise two-thirds to three-fourths of Coronado's large expedition that passed through eastern Arizona in 1540, but

> [b]oth indigenous and European members of the expedition were outfitted almost entirely with native arms and armor. In general there were only stray pieces of European metal armor, an occasional coat of mail, gauntlet, or neckpiece here and there. Only a few of the captains possessed anything approaching a suit of metal armor. Instead, of the 287 European men listed on the muster roll, 261 (91 percent) declared that they were taking native arms and armor with them, usually exclusively, but sometimes supplemented by an item or two of European armament. . . . Meanwhile, only 45 men (16 percent) carried European metal helmets. . . . Thus, the romantic portraits and murals of the Coronado expedition . . . are little more than fantasy. The Coronado expedition looked for all the world like a massive troop of Mexican Indian warriors, rather than medieval knights. (Flint 2005:206)

In addition, traveling companions were reluctant to bury valuables or even useful items with someone who died along the way. The case of prospector Dave O'Neill reminds us that practicality trumped sentimentality. When Dave died along El Camino near Papago Well in 1916, his companions buried him, but "knew" that he'd want them to have his burros and chewing tobacco, since Dave would no longer be needing them (Nichol 1939).

Historical Inscriptions

On the cliffs and boulders at Tinajas Altas are many historical inscriptions, as well as prehistoric petroglyphs and pictographs. The inscriptions consist primarily of names with a few dates. Unfortunately, few of the names can be connected to known individuals; further research may reveal who these persons were and why they visited Tinajas Altas. The Tinajas Altas Archaeological Survey reports that as many as six may be from the nineteenth century, with one possibly as early as 1834 (Hartmann and Thurtle 2000). None can be attributed to earlier times. One known date was carved on March 11, 1927, by Gale Barrett, who was an inspector for the US Immigration Service (USIS).

Of particular interest is an inscription in black paint that was noted by Kirk Bryan in 1917 and again in 1920. It reads "Josefa Jacoba Martin, Abril 16, 1874." In 1917, Bryan (1925:91, fig. 8) writes, "43 years after it had been written, the inscription was in almost perfect condition except for the last 'a' in Jacoba, which was lost by chipping of the back wall. Three years later, in September, 1920, the 'a' in Josefa had also disappeared and parts of other letters had been obliterated." By 1998, when the inscription was re-photographed most of it had eroded away (Hartmann and Thurtle 2000: fig. J.8).

Some inscriptions suggest that modern persons were buried at Tinajas Altas or had their ashes scattered there. Although the number of gravesites has declined drastically since the early twentieth century, occasionally remains of recently deceased individuals are added. In 1989 the ashes of David F. Roberson were scattered over the area. He was a US Border Patrol pilot who flew for the Yuma Sector, which includes Tinajas Altas in its patrol area. Roberson was killed in July 1989 when his plane crashed in the Yuma Dunes about 35 miles northwest of Tinajas Altas. He loved to fly his Piper Super Cub down Tinajas Altas Canyon, looking for smugglers and undocumented aliens. Because of his dedication and the dedication of many agents like him, many people lived who otherwise would have died of thirst or exposure.

Rescues give agents their highest satisfaction. International politics, economics, and legalities aside, human lives are still the bottom line. The sentiment is that "In rescues we've done some real good." Retired agent Joe McCraw confirms this, "The thrill is trackin' em up before they die. A rough ol' way to go—run outta water in this desert. We wish they'd try another sector . . . for their own sakes" (Broyles 2006). The devils of heat and thirst still prowl El Camino del Diablo.

Modern Military Use

Although Tinajas Altas has no military battle history, it has a strong military connection due to its location within a military training range. Tinajas Altas, as well as nearly all land south of what is now Interstate 8 between Yuma and Ajo, was included in the Luke Air Force Gunnery Range in 1941 and 1942. Thousands of World War II and Korean War pilots learned to fly on the range. Their training included basic flying, air-to-ground strafing of machine guns and rockets, and aerial dogfights with machine gun attacks on towed targets. The presence of spent 2.75-inch rockets in the vicinity of Tinajas Altas does not necessarily mean it was a target; one post–World War II strategy was for interceptor jets to fire salvos of unguided rockets at high-altitude enemy bombers (target drones, in this case), so rockets were widely scattered. Thousands of 2.75- and 5-inch rockets were launched over this period. In addition, from 1953 to 1958, approximately 60 squadrons per year, with 18 to 24 planes in a squadron, fired .50-caliber ammunition at towed aerial targets. These activities were carried out in "a large deserted area extending southeast and east of the base for some 70 miles" (Office of Information Services 1957[?]:3).

In 1947, the Army Air Corps became the US Air Force. The Luke Range was renamed in 1986 as the Barry M. Goldwater Air Force Range in honor of long-serving and popular Arizona senator Barry M. Goldwater, who also served at the Yuma base when it first opened. In 1982, the Air Force ceded management authority of the western half of the range to the US Navy and the US Marine Corps, leaving Tinajas Altas under the immediate jurisdiction of the Yuma Marine Corps Air Station (Tunnicliff et al. 1986:2–6).

Pilot training on the range was largely air-to-air and air-to-ground until the early to mid-1980s, when the Marine Corps expanded its pilot training to include interaction between pilots and ground units. These units involved radar tracking, simulated anti-aircraft guns and missiles, observation, and telemetry. Modern warfare requires continually updated equipment and training, so by 2008 a number of remote electronic emitter posts were positioned on the range to simulate enemy radar installations, and portions of the range were dedicated to laser sighting and attacking of targets. Several training villages, including Yodaville northwest of Tinajas Altas, simulate structures that student pilots may encounter in a real war. Visitors to the range may encounter helicopters as well as strategically inserted ground observers who man cryptic listening and observations posts. Special areas have been designated for large encampments of tent cities for Marines, and most of the Marines who served in the conflict in Iraq received some instruction within the Goldwater Range.

Making a Go of It: Miners, Cowboys, and Farmers

Since the time of the Gadsden Purchase in 1853–1854, no one has figured out how to make a living at or near Tinajas Altas. Even before Congress withdrew the region from mining, grazing, and habitation and designated it for military use, it was too barren of saleable minerals, lacked grazeable plants, saw too few visitors to warrant a store, and lacked enough water to set up housekeeping. Nonetheless, some miners and ranchers have given it a try.

Another Gold Rush

As we have described, during the Gold Rush of 1849 and subsequent rushes in the nineteenth century, thousands of miners did pass by Tinajas Altas. Some of them going west were novices, but many were professional prospectors and miners who understood minerals and ores. They undoubtedly walked and crawled through every canyon in the Tinajas Altas area. They camped, rested, prospected, hunted, got lost, and got found. With the exception of a few tall tales and one adit, no evidence of precious minerals is associated with the immediate Tinajas Altas area. With hindsight, we see that the arduous efforts of the pioneers were poignantly useless;

Grave of Eight, December 14, 1913. Labeled "Grave of man, woman, and two children who died of thirst north of Tule Mountains," eight miles east of Tinajas Altas. (Photo by Charles Sheldon. Courtesy Neil B. Carmony)

professional geologists have surveyed the region and reported nothing promising in the southern Tinajas Altas Mountains. The granite is beautiful but barren. To be sure, flecks of gold or silver may be found in many unexpected places, but it is highly unlikely that any commercial or even hobby production is possible in the Tinajas Altas Mountains.

In 1858 placer gold was discovered at Gila City, now called Dome, just east of Yuma; in 1860 near Ehrenberg, along the Colorado River about 65 miles north of Yuma; and in 1862 at Lagunas, along the Colorado River just north of Yuma. Again, as in 1849–1850, venturers from Mexico marched northward to try their pans, picks, and luck at bringing gold from the earth. Tinajas Altas was again a stopover. Perhaps hundreds or even several thousand prospectors made the trip, but a firm number is impossible to estimate.

One family group reportedly never made it to Tinajas Altas. Their water jug broke, and they died about eight miles east of the tanks at a site now called the Grave of Eight. In "Perils and Wonders of a True Desert," written in 1896 by boundary surveyor David D. Gaillard (1896:603), the tragedy becomes clearer:

> One of the best known and most pathetic cases of death from thirst was that of an entire Mexican family of six or eight persons, who were pushing on toward Tinajas Altas, their total supply of water contained in a wicker-covered glass demijohn. When about eight miles from the tanks their horses gave out, and in unloading the wagon, by some unfortunate accident, their demijohn was broken. Utterly ignorant of the distance to the tanks, or of their location, the husband set out alone on foot to find them, telling his family to await his return where he had left them. Weak and faint, he returned from his unsuccessful search and joined his dying family under a neighboring palo-verde tree where their bodies were found by the next traveler, and buried in a single grave beside the road. Pious hands had piled stones on the grave in the form of a cross, and had encircled the whole by a ring about thirty feet in diameter, formed of stones piled side by side. . . . The wagon tracks made when the poor Mexican drove his exhausted team to the side of the road, were plainly visible thirty years afterward, and at the very spot still remained pieces of glass and wickerwork from the broken demijohn, and the skulls of the two horses."

Today this site consists of a large circle of rocks with the number "8," and a row of rocks that has been aligned to point toward Tinajas Altas which is, coincidentally, eight miles away. Photographs of the site in 1896 and 1913 show a rock-outline circle with a low mound in its center and a linear rock-alignment extending partway across the center with short cross-arms (Gaillard 1896; Sheldon 1993), but sometime after 1913 and before the 1970s someone added an "8" and extended the centerline nearly to both edges. This changed its meaning from a gravesite to a directional and distance marker indicating eight miles to Tinajas Altas.

According to Eldred Wilson (1931), who produced the first geologic map of the area, a few quartz veins locally

Grave of Eight, 2010. (Photo by Bill Broyles, 2010)

Another marker in the desert east of Tinajas Altas, December 1913. (Photo by Charles Sheldon. Courtesy Neil B. Carmony)

stained with copper attracted the attention of a few prospectors, but nothing of value was found. In an inventory of the mineral resources of the military range, geologist Glenn Miller (Miller and Jett 1979:36) confirmed that "past prospecting activities in the Tenajas Alta [*sic*] Mountains uncovered nothing of economic significance."

But hope springs eternal among miners. Irwin (ca. 1970:27) reported a *San Diego Herald* account of Yuma war chief Cebello Pola [Caballo en Pelo] leading a prospecting trip into the Tinajas Altas Mountains in 1853, but the results are unknown. Another prospector, Pablo Valencia, looked for gold southeast of Tinajas Altas in the Pinacate lava field in 1905, but was unsuccessful and nearly died of thirst for his effort (Broyles 1982, 1988; McGee 1906b). In 1910, Carl Lumholtz was startled to find at Tinajas Altas a prospector whose companion would soon follow him in a wagon. They were from Wellton and on their way to Sierra Pinacate to look for gold in the lava. The man showed Lumholtz "a most unusual sample of free gold that literally studded a dark brown piece of rock which, in fact, seemed to be old lava." By evening the partner showed up and they talked about prospecting in "Sierra del Viejo where Cipriano Ortega once worked a silver mine which was lost later and could never be found again. He had also been looking for the lost mission" (Lumholtz 1912:321).

Only three groups of prospect diggings occur in the southern Tinajas Altas Mountains: the Crescent, the Dolores, and the Old Smokey groups. None of these appears in Keith's (1978) study of Yuma County mines, but they have been discussed by others. The Crescent group, just south of Tinajas Altas, includes the Crescent 1 and 2, the Golden King, and the Barragone Star, all in the Fortuna District (Miller and Jett 1979). The minerals of these prospects likely are copper, but the extent and location of the prospects are unrecorded, and the nature of the minerals sought is undefined.

The Dolores, in the Wellton or La Posa District, was southeast of Tinajas Altas in the Lechuguilla Mountains, which some geographers take to be a part of the Tinajas Altas Mountains. Miller and Jett (1979) group it with prospects or claims having no file information other than a name.

The other group near Tinajas Altas was Old Smokey No. 1 in the Fortuna District, south of Raven Butte, but Bureau of Land Management records show no current or valid mining claims in the Tinajas Altas area. Master title plan maps and files list the area as NOE ("Not Open to Entry") due to its military withdrawal by US Public Law No. 99-606.

On the southern side of the border, Mexican topographic maps show a single mine nearby, the Joya Vieja mine, at the west end of Sierra Viejo, Sonora. It may be a silver prospect. And Mina del Desierto along Mexico Highway 2 has been worked to varying degrees for decades, but apparently to little profit. Ironically one low-tech but profitable mine in the limestone peaks of El Capitan, twenty miles west of Tinajas Altas, produces lime for masonry and cement projects in San Luís Río Colorado and beyond.

The only truly profitable mine in the Tinajas Altas region was the Fortuna Mine, along the western branch of El Camino del Diablo, 24 miles (38 km) northwest of Tinajas Altas. It produced more than $2.5 million in gold and silver (Baker 1980; Miller and Jett 1979) and was the richest and largest mine in the what is now the Goldwater Range. The Betty Lee Mine in the Copper Mountains, across the Lechuguilla Valley to the northeast of Tinajas Altas, had extensive tunnels and shafts, a small mill, and up to 30 employees at one time, but it never was profitable (Miller and Jett 1979; Walter Wooley, personal communication 1984).

Winters (1963:47–63) recounts a contorted tale concerning a miner's camp and prospect in Borrego Canyon, a mile south of Tinajas Altas. This camp and prospect are likely part of the Crescent group. The characters are Ol' Charlie Foster (we wonder if this is the same Charlie Foster who was Jeff Milton's friend described by Hornaday [1908]), Rosa Berry, her husband Ray Berry, and a daughter Maria. The year is 1930; the place, Wellton. Charlie lived "in a shack on the edge of Wellton," but also had a camp south of Tinajas Altas where he "eked out an existence of sorts bootlegging liquor to the Sand Papago Indians [Hia C'ed O'odham] who roamed the border" (Winters 1963:48–49). He also prospected, and at least according to the story, was successful in finding gold in the southern Tinajas Altas Mountains. The story contains proposed matrimony, a double-cross, a jilted partnership, a robbery, jail, and a lode twice found . . . and twice lost.

Tangible remains of Ol' Charlie's camp include a "row of rusted tin cans [that] outlined the spot where a tent or canvas house had stood. The trash pile was composed of lard pails, bully beef cans, and several miners' carbide containers. A 1930 license plate dated the place" (Winters 1963:60). At a nearby site in the boulders was Charlie's more secret camp. "Here we found the remains of his smelter, piles of crushed rock (not gold-bearing), broken crucibles, and other signs of mining activity. . . . It appeared that this place had been used as a supply point, assay office and mine headquarters" (Winters 1963:61). Winters figures that the mine connected with this camp is an adit, that is, a horizontal tunnel into the mountain, on the other side of the canyon. (This mine and camp are discussed in more detail in Hartmann and Thurtle 2000.) Interestingly, the Bureau of Land Management has no record of a claim or patent here. The 15' and 7.5' USGS

Kiln at Charlie's Camp. (Photo by Gayle H. Hartmann, 1998)

quadrangles show no other adits, prospects, or mines in the vicinity. The closest is the Smith Mine in the southwestern Copper Mountains. The Smith Mine is described in Olga Wright Smith's (1956) wonderful book *Gold on the Desert.* The Smith's venture failed financially, but she did discover a deep love for the desert.

Where mines fail, tales of lost gold prosper. Such tales are an integral part of the Old West. At least two are associated with the Tinajas Altas region. One yarn features a group of Mexicans finding a vein of gold two feet wide on a small granite hill near Tinajas Altas (Mitchell 1952). The year was 1850. The tale is fraught with problems that detach it farther from reality than usual in such fabrications—it places Tinajas Altas east of Cabeza Prieta Peak instead of west, and a small spring is reported south of Tinajas Altas—it seems that all good treasure stories demand a secret source of something. Needless to say, the group was attacked by "Papagos," got lost, and never could relocate the site.

The most repeated fable of lost gold in this region is that of El Jabonero (the soapmaker) and revolves around clues to a lode found and then lost (Egerton 1974; Penfield 1973; Townsend 1969; Weight 1955; Willson 1948, 1970). The vicinity of the gold is so vague that it may or may not include the Tinajas Altas area, and most seekers place the setting closer to Tule Tank or Quitobaquito, but this tale, too, has little substance and much imagination.

Cowboys

Like mining, grazing livestock at Tinajas Altas was a passing fancy. The forage is sparse, seasonal, and sometimes nonexistent. Barely enough could be found to sustain livestock passing through the area; it was neither feasible nor profitable to pasture them there. Although there is no evidence of long-lived cattle camps, line camps, or ranches, transient and trespass stock as well as personal riding and pack stock have long been part of the Tinajas Altas story. A surprising number of cattle were driven past Tinajas Altas and some were watered there, especially in the late 1800s and early 1900s by Cipriano Ortega, José Redondo, and Alberto Celaya. Later, during the migrations and commutes of the early 1800s, oxen that pulled heavy Mexican carts also drank and fed here. Prospectors of 1849 and the 1860s traveled by horseback and brought livestock with them. Visitors such as Carl Lumholtz, Charles Sheldon, and Kermit Roosevelt had pack and saddle stock, and they packed supplemental feed for their animals.

Tinajas Altas did stand ready as a reserve waterhole for cattle drives from Sonoyta, Altar and Caborca to Mexicali and Yuma along El Camino del Diablo. Anyone traveling the Camino today, or Mexico Highway 2 south of the border, can only marvel at the audacity of the parties who drove herds of livestock along it. In wet winters and springs, when forage was rich, Mexicans drove sheep and cattle from the vicinity of Caborca and Sonoyta to markets in San Luis Río Colorado and Baja California since no road or rail paralleled the route. Longhorn steers cost a few dollars in Mexico and could be sold in California for up to $100 a head.

Accounts of these drives include reminiscences of Cipriano Ortega, who was mentioned in chapter 3 as something of a brigand. Ortega lived in a hacienda named Santo Domingo, on the Sonoyta River west of the town of Sonoyta (Hoy 1990). His presence is remembered in the naming of Cipriano Pass, 10 miles (16 km) northwest of Tinajas Altas and the Cipriano Hills in Organ Pipe Cactus National Monument. Perhaps he is the one who pioneered a string of cattle camps between Sonoyta and San Luis. These camps were at Santo Domingo, Represa Cipriano or Las Playas, Tule Well or Tule Tank, Lechuguilla Tank (with a backup at Tinajas Altas), Laguna Prieta, and then the Colorado River. These made it possible to water at five places in the 120 mile (190 km) journey.

At least one prominent Yuma businessman and politician, José Maria Redondo, made his fortune moving cattle along this route. He had traversed the Yuma Trail as a teenager in the 1849 Gold Rush, and later returned to drive cattle (Beattie 1928). We assume that these livestock drives occasionally passed near Tinajas Altas or that the vaqueros at least refilled their water barrels at the tinajas.

A few American cowboys continued to brave the Camino as late as the 1870s. Rancher John Hohstadt, with his family, drove a herd of 275 cattle from San Luis Obispo, California, to Sonora in 1876. According to the *Douglas Dispatch*, quoted in an anonymous obituary in the *Arizona Historical Review* (Anon. 1929:6–7):

> As they came this way from San Luis Obispo, they traveled down to San Diego and thence east to Yuma where they crossed the Colorado with their herd and wagons on the ferry, about 18 miles from the town of Yuma, that crossing taking them into the state of Sonora. But they turned north into Yuma and rested there about three weeks when they again resumed the movement to the east following up the Gila River to what was then Mission Camp where they passed another three weeks. There they got a Mexican guide and headed south into Sonora with Tinaja Alta [*sic*] as their objective. Arriving there they rested over night watering their cattle from water taken from a tank in solid rock said to be 150 [*sic*] feet deep and of good, pure rain water. This was about 40 miles from the Gila. These waterholes were meccas for the travelers although some of them never reached the water supply. This particular waterhole was said to have been the scene of death of more than 300 persons who perished on their way to the water and the sand was said to be dotted with bodies of the victims.

The Hohstadt family survived the trek. But when they finally reached their ranch in northern Sonora, only 45 head of cattle remained, a loss of 230. A more logical route from Mission Camp would have been by way of Cabeza Prieta Tanks.

Another well-known figure in the history of this area, Alberto Celaya of Sonoyta, was also involved in cattle drives from Caborca to Mexicali before 1910. According to Julian Hayden (2007) who was friends with Celaya, he helped move the cattle of two Mexican businessmen. A Sr. Sandoval, founder of the first bank in Nogales, and a Sr. Gavilondo from Hermosillo had bought 6,000 steers in Chihuahua and then moved them to Sonoyta, where the cattle wintered along the Sonoyta River bottom. Then Celaya and other cowboys moved the herd westward to Laguna Prieta and then on to the Colorado River. This is confirmed in Lumholtz (1912), who adds that the cattle were watered at a tinaja in the Lechuguilla Mountains, eight miles southeast of Tinajas Altas.

Cattle were also driven on the Mexican side of the border passing just south of Tinajas Altas. Sonoran rancher Ramón Sanora Mendez occasionally ran herds of up to 700 head from Sonoyta to Mexicali in the early twentieth century (Oscar Bell, personal communication with Jacquie Gage Kahn 1997). Sanora got the idea of traveling through the desert by observing that Indians could survive in and travel through this arid region. He met the Hia C'ed O'odham hermit, Juan Caravajales, who made his home at Papago Tank in the Sierra Pinacate; he showed Sanora the location of waterholes (see Bell et al. 1980, Lumholtz 1912 for discussions of Caravajales). After a few successful trial runs across the desert with a few cattle to see if these drives were possible, Sanora began making them on a regular basis.

The first part of the journey, from Caborca to Sonoyta, took 20 to 30 days. Sanora would lease fields in Sonoyta for about a month to feed the cattle and fatten them up for the next part of the trip, from Sonoyta to San Luis, which took 40 to 60 days. Pack mules would pull wagons with wooden kegs

Bureau of Animal Industry camp at Tinajas Altas, looking west, about 1947–1949. Rufus and Nadyne Allen's living quarters (at left), a shed (center), and two corrals (right center and bottom center). (Rufus Allen photo courtesy of the Arizona Historical Society/Yuma, Rufus Allen Collection)

of water. These kegs would sometimes be filled at waterholes including Tinajas Altas and brought to the cattle along the trail. The arrival of the Southern Pacific railroad to Arizona between 1877 and 1880 brought an end to the cattle drives, for it could deliver cattle safely to market, without loss of weight or numbers.

Bartolo Ortega, younger brother of Cipriano Ortega, supplied stock and feed for W. J. McGee's expedition in 1900. McGee (1901:141) lamented "the fact that the entire zone traversed by the [Old Yuma] trail was gradually impoverished by the long-continued—and short-sighted—overtaxing of its meager resources." He reported that the earliest European visitors found a "varied flora delicately balanced between hard habitat and dependent fauna through eons of adjustment" and "the sum of living resources was sufficient for reasonable demands. . . . With the third and fourth generation and the gradually increasing number of caucasian travelers, the resources began to suffer; the forage grew scant . . . , with the decimation of plants and the trampling of stock the soil grew less retentive of the scant moisture." Settlers "not only drove herds to devastate the enfeebled flora along the way, but stocked the adjacent ranges with cattle and sheep."

McGee (1901:143) proclaimed that, "of all the examples of destructive contact between pioneers and [Native American] precursors none are more impressive than those so clearly attested by the Old Yuma Trail." He closed with a plea that "adjusting means to conditions in the reclamation of arid lands is the leading lesson of the Old Yuma Trail."

In years of luxuriant winter rains it is possible that some cattle strayed as far south as Tinajas Altas from settlements along the Gila River. Cattle are known to have wandered from the Gila River to the vicinity of Heart Tank in the Cabeza Prieta National Wildlife Refuge. As recently as the winter of 1974 or 1975, a bull from the Guadalupe Victoria Ranch southeast of Tinajas Altas strayed as far west as Laguna Prieta, where it was seen at least 65 miles (105 km) from home (Pancho Begarán, personal communication 1976). Toward the end of spring, and the last of the lush winter forage, it wandered back to the ranch, fat and healthy.

A Cattle Inspection Station

Between 1947 and 1953 an epidemic of hoof-and-mouth disease plagued Mexico. To fight its spread, the US Department of Agriculture established special patrols along the border to intercept and exterminate cattle trespassing from Mexico (Fox 1997). Between Yuma and Lukeville, the Bureau of Animal Industry (BAI) had intermittent line camps at Tinajas Altas, Tule Well, Papago Well, and Quitobaquito.

The camp at Tinajas Altas was on the terrace east of the tanks, where its concrete slabs remain visible today. The camp consisted of a cabin, two corrals, at least one outhouse, and a small shed or two. One of the BAI agents

Bureau of Animal Industry cabin at Tinajas Altas. (Rufus Allen photo, courtesy Rufus Allen Collection, Arizona Historical Society/Yuma)

The cabin from a slightly different perspective. (Rufus Allen photo, courtesy Arizona Historical Society/Yuma, Rufus Allen Collection)

Sign on Rufus and Nadyne Allen's latrine: "Toilet only. Don't shoot me." Nadyne remarked that the toilet had a wonderful view, with no neighbors between there and Ajo. She also expressed dismay at the number of travelers who randomly shot at any target that caught their eye. (Rufus Allen photo, courtesy Arizona Historical Society/Yuma, Rufus Allen Collection)

assigned to Tinajas Altas was Rufus "Rufe" Allen. He and his wife Nadyne lived at Tinajas Altas during 1947–1949, and later at Tule and Papago wells. They regularly drove to town for water and supplies, including extra feed for their horses. They usually took along a horse in the horse trailer, so that they could use the horse to pull out the car when it got stuck in the sand. Allen and other BAI agents patrolled the border on horseback and shot trespass cattle, later burying them in pits dug by a bulldozer (Rufus Allen, personal communication 1986).

The camp fell into disrepair and likely its wood, metal, and other materials were scavenged for use elsewhere. The remains were noted by various visitors, including Madeleine Rodack (1968:26), who describes the ruins of the camp as consisting of "old chicken wire, cracking cement floors, and a few ancient boards and corrugated iron sheets. Rusty tin cans are everywhere and the place is quite a mess."

As recently as 1996, trespass cattle from Mexico have been seen and reported in the vicinity of Tinajas Altas. Presumably they water at the lowest pool. One carcass was observed about 60 feet east of that pool (Bill Broyles, personal observation 1995). Domestic goats from Mexico are sometimes seen in the southern Tinajas Altas Mountains, coming north from the La Joyita truck stop along Mexico Highway 2; they may occasionally reach Tinajas Altas.

Farmers

Bureau of Land Management master title plats show no references to homesteads in the vicinity of Tinajas Altas nor have we found any historical records of farming. The nearest agricultural land is about 23 miles (37 km) north of Tinajas Altas, just south of the town of Wellton. This arable land of the Gila River Valley is river bottom, not bajada or arroyo soil such as that found at Tinajas Altas. The nearest well is at the Mexican truck stop of La Joyita. Its persistence, depth, and water quality are unknown to us. The truck stop of El Sahuaro 5 miles (8 km) south of Tinajas Altas has a hand-dug well or cistern, but again the depth, persistence, and quality are unknown. It is known that water for human consumption and personal use is occasionally hauled to these sites in tank trucks, barrels, and bottles from San Luis Río Colorado.

A communal farm (*ejido*) at Aquiles Cerdán (Serdán), Sonora, 16 miles (26 km) west of Tinajas Altas, was struggling in 1986 and is now abandoned. A few families lived there, eking out a living by working at a roadside *llantería* (tire-repair shop). They raised a few goats on paltry native vegetation and table scraps. The families planted a few citrus trees and small gardens, which were watered with gray water and even "yellow" water (urine); one man laughed that it was beer processed through his bladder that made his trees grow large and the fruit turn sweet (Broyles 2006).

The ejido communal farm system was part of a land reform movement in Mexico—an attempt to settle more people on their own land. Ejidos were even established in seemingly inappropriate places in northern Sonora, such as on the outskirts of the Pinacates southeast of Tinajas Altas, between Sonoyta and Puerto Peñasco (Rocky Point), and around Caborca, Sonora. Sadly in many cases, water proved inadequate and most of these enterprises stand abandoned. Ironically, although the ejido system was an attempt to redress the historical gap between wealthy land-owning families and poor workers—a gap lingering since the time of the Spanish conquerors—some ejido lands deeded to poor families have ended up in the hands of wealthy families (including drug cartels), undermining the whole point of the program.

Water is intermittently available in smaller tinajas in the vicinity of Tinajas Altas (Broyles 1996b, 1997), but these catchments could not have supplied water sufficient for irrigation or domestic animals. Even if full, all the natural tinajas in the Tinajas Altas Mountains would contain not much more than 20,000 gallons; those in the Gila Range would contain at most 17,000 gallons (Broyles 1996b), hardly enough to sustain a garden for one growing season, let alone a farm. However, a sideline industry—making liquor—did prosper nearby during Prohibition (1920–1933), and moonshiners erected stills at tinajas in the Aguila Mountains, 48 miles to the northeast, and Puerto Blanco Mountains, 71 miles to the east-southeast.

Autos on the Camino del Diablo

In 1915, Raphael Pumpelly, the mining engineer who had hair-raising adventures on the Camino del Diablo 54 years earlier as a young man, made a sentimental return at age 77 to Tinajas Altas with his grown children. Ever the adventurous traveler, he had decided that a motor trip from Tucson to Yuma along the Camino would be an interesting sojourn. "It seemed a simple thing to do with three Ford cars and three drivers," said Pumpelly (1918:766) in his memoirs. When they reached Tinajas Altas, Pumpelly noted that the water in the tanks "was alive with polliwogs and mosquito larvae. But strained and boiled, the Tinajas Altas water tasted better than that from Tule Wells" (Weight 1968:34). Pumpelly (1918:774–775) wistfully showed the tanks to his family, recalling how "[i]n 1861 they stood half-way in a waterless stretch of a hundred miles or more, and to find them empty meant almost certain death to man and beast. . . . I remembered now, too well, tales about these potholes to dare peer towards the bottom. . . . It is a matter of history that more than two thousand persons have died of thirst and exhaustion on this part of the Old Yuma Trail." The Pumpelly party spent two days at Tinajas Altas. His children wrote, painted, and photographed, and the drivers repaired the vehicles and patched tires and inner tubes. The group left via Tinajas Altas Pass in order to visit Fortuna Mine.

At the end of their journey, Pumpelly met a man who had recently tried to drive to Tinajas Altas but had to turn back at Tinajas Altas Pass. Within a few weeks Pumpelly received word that three Yuma men tried to match his crossing of the trail, but two died and one was rescued near death (Pumpelly 1918). Pumpelly's appears to be the first record of an automobile trip along the length of the Camino del Diablo.

Raphael Pumpelly III, the grandson of Raphael Pumpelly recalled his grandfather speaking of the trip and reminisced about his two aunts who were on the trip along with his father, Raphael II, and mother, Amelie. His aunt Elise Cabot wrote poetry and his aunt Daisy Smyth was, he recalled, a wonderful painter. He had at his home in California, a watercolor of the Santa Catalina Mountains in Tucson, unsigned but thought to have been painted by his aunt Daisy, Margarita Pumpelly Smyth (Raphael Pumpelly, III, personnel communication with Gayle H. and William K. Hartmann, July 1999).

Shortly after Pumpelly's trip, Charles Sheldon (1993), a hunter who spent time in the vicinity of Tinajas Altas and traveled by horse, reports that in March 1916 a Ford motorcar drove down from Wellton to Tinajas Altas to deliver him a message, indicating again that the road was passable not only for wagons but also for cars.

Geologist Kirk Bryan, who reached Tinajas Altas sometime between 1918 and 1920, noted that few cars had visited Tinajas Altas by way of the Fortuna Mine road between Pumpelly's 1915 visit and Bryan's arrival. Bryan (1925) provided road directions for travelers to the region in an excellent book on the Tinajas Altas region: *The Papago Country, Arizona: A Geographic, Geologic, and Hydrologic Reconnaissance with a Guide to Desert Watering Places.* He lists road signs (some of which he installed), mileages, junctions, and locations where water might be found. He mentions a waterhole at Coyote Water in the center of the Lechuguilla Valley east of Tinajas Altas, and two small tinajas south of Tinajas Altas, one of which may be Surveyors Tank just north of the international boundary. The exact location of these small tanks was not known to Bryan. A "Surveyors Tank" is marked on US Geological Survey topographic maps, and the Tinajas Altas Archaeological Survey (Hartmann and Thurtle 2000) recorded a historical period site, primarily of O'odham affiliation, in association with it, but that location doesn't completely fit several other descriptions of the tinaja.

Bryan (1925:422) advises travelers: "Tinajas Altas lies in a cove in the border of the mountains, and the road comes in around the southern rocky ridge and down a bank into the bed of the arroyo that heads in the canyon above the cove. . . . The best camping ground lies west of the Geological Survey sign, but there is practically no horse feed and no firewood in the vicinity. A few palo fierro or ironwood trees are still standing and may furnish horse feed, but the traveler should bring his own firewood." Also, Bryan (1925:337) advises travelers to boil the water before drinking it. "The water is palatable but there are usually dead bees in it. Occasionally mountain sheep slip and fall into the tanks and contaminate the water. Travelers will find road from Geological Survey sign up arroyo to tanks very difficult for automobiles, and are advised to camp on gravel bench west of sign." The sign post installed by Bryan still stands, but the signs themselves have been stolen or discarded after being nearly obliterated with bullet holes. He (Bryan 1925:422) also mentions that "the number of people who are supposed to have perished at this watering place is doubtless exaggerated, but at least 65 graves can now be counted. There are a number of inscriptions by travelers."

In Bryan's time two routes led north from Tinajas Altas, one to Wellton on the Gila River and one west of the Gila Mountains to Fortuna Mine, then to the small community of Blaisdell, and finally to Yuma. A third trail went west from Tinajas Altas Pass "toward a group of hills in Mexico called El Capitán crossing the international boundary at monument 194. From El Capitán the trail goes across the 'medanos,' or sand dunes, to Laguna Prieta, a pond of slightly salty water 28 miles from Tinajas Altas. Thence it is 30 miles to Colonia Lerdo, a Mexican town on the east bank of Colorado River" (Bryan 1925:422). This may indicate that traffic from Mexico used this trail to Colonia Lerdo, and it may be a remnant of the cattle trail from Sonoyta to Mexicali, as well as a Cocopah trail. Bryan (1925:423) labeled it "part of the so-called all-Mexican route to Lower California, from Sonoita."

Travel in the last decades of the nineteenth century and the first few decades of the twentieth century was slow by modern standards, whether by foot, horse, wagon, or even early automobile. Travelers on roads in this area could expect to average no more than six miles (9.6 km) per hour and to encounter difficulties (Wilson 1931). The road was, however, negotiable by high-clearance passenger cars. As writer Harold Weight noted (1949:10, 14), "The road as far as Tinajas Altas was in good shape. Since the Mexican hoof and mouth epidemic scare, the department of agriculture has maintained a border patrol camp at the [Tinajas Altas] cove and cars travel the route to the highway and Wellton frequently." The Yuma Chamber of Commerce even ran excursions to Tinajas Altas in the cooler months. The road

Kirk Bryan. (Courtesy US Geological Survey; see Whittlesey 1951:88)

southeast from Tinajas Altas toward Tule Well was, however, "still rough and twisty."

James M. Barney, Arizona historian and cartographer, produced a 1908 Yuma County road map. Later he wrote about a visit to Tinajas Altas with notable friends and wrote a review of its history (Barney 1943). The photographs that accompany Barney's article were taken by Benjamin J. Kinsey and one Barry M. Goldwater, for whom the Goldwater Range is named.

Thanks to the Jeep, by the late 1940s adventurous travelers were vacationing all over in the Southwest deserts, having figured out how to visit and camp in relative comfort and safety. Harold Weight (1952:5) reported, "Men no longer stake their lives upon the presence of water in the granite basins of Tinajas Altas. . . . To us, the basins are a curiosity rather than the answer to survival or extinction." He (Weight 1952:11) concludes, "If you have gone without water all the time [during your visit to Tinajas Altas], you will be thirsty when you reach your car again. But I doubt you will be thirsty enough to drink Tinajas Altas' dead, often fouled water. You will be honestly tired. But you will have gained a deeper understanding of the vast slow processes which have made this world habitable."

On the Mexican side of the border, however, change did not come so quickly. In the 1930s much of the road between Sonoyta and San Luis Río Colorado was a treacherous single-track trail through rough lava and shifting sand dunes. Local drivers who knew the ways of the desert (*choferes del desierto*) hired out their cars and drove passengers and cargo across this perilous route, but many novices and tourists perished.

US Geological Survey field vehicle used in the reconnaissance of the Papago country, Yuma County, Arizona, 1920. The crew consisted of a geologist, Kirk Bryan, and a non-technical assistant, mainly Charles Puffer, who were "provided with a Ford automobile, a light camping outfit, a plane table, and other equipment" (Bryan 1925:xvi). (Photo courtesy US Geological Survey Photographic Library, image bry00467)

When two cars met in the sandy portion, the men would lift one of the vehicles and move it out of the road while the other car passed. Then they switched the process and put the first car back on the road. If there were not enough men to move the vehicle, they waited for another vehicle and more manpower (Barrios Matrecito 1977; Julian Hayden, personal communication 1990). San Luis journalist Valdemar Barrios Matrecito (1977) wrote that the road was "really more of a trail" whose severity caused many fatalities and led to many abandoned vehicles along the route.

The portion of Mexico 2 between Sonoyta and San Luis was graded during World War II but was not paved until 1953–1955. At one time it was named Abelardo L. Rodríguez Military Highway, after a Sonoran governor who became president of Mexico (Bill Hoy, personal communication 1996). A few articles have been written about traveling this highway, for example, Murbarger (1960). Occasionally visitors to Tinajas Altas will hear traffic on Mexico Highway 2, the northernmost highway across Mexico and the only one connecting Baja California with the mainland. It approximately parallels the international border a little over four miles south of Tinajas Altas. Since about 1980, plans have been afoot in Mexico to upgrade the road and make it a four-lane divided highway.

Just Having Fun: Campers, Hunters, and Journalists

In our age of popular gusto and entertainment, the fearsomely haunted campsite of Tinajas Altas—scene of deaths and desperation—has become a playground. People from throughout the Southwest have discovered its natural beauty, open space, solitude, and recreational opportunities. Today it is common to find campers, hunters, "explorers," drivers, hikers, photographers, and wildlife watchers enjoying the Tinajas Altas area.

In fact, at present Tinajas Altas is probably the most popular destination in the Goldwater Range. Campers usually stay part of a day or overnight, visit the lowest tank, look at the mortar rocks, perhaps climb to the next accessible tank, and then drive on. Fire rings appear and disappear around Tinajas Altas Canyon. Tree branches, a scarce resource, are often cut for firewood and found scattered on the ground by visitors who may or may not realize that such cutting is illegal and inconsiderate.

Historically, the first visitors who came primarily for recreation were hunters. McGee (1901:138) reported that on his November 1900 trip to Tinajas Altas "we [found] the outfit of a hunting party (now absent on the chase)." Charles Sheldon (1993:23) visited Tinajas Altas in 1913 specifically to hunt desert bighorn sheep. On his 1916 visit he noted that he found "three fine, large bleached skulls of big rams right here near the Tinajas Altas. They were undoubtedly shot by those miserable fellows who come here in summer and watch the waterhole" (Sheldon 1993:110–111).

Kermit Roosevelt visited Tinajas Altas in 1911 while on a bighorn sheep hunting expedition to Mexico. He had heard tales of magnificent bighorn here from Carl Lumholtz, and undoubtedly from his father as well. Theodore Roosevelt, then president of the United States, had been invited to the Pinacate region of Mexico by William T. Hornaday in 1908 following the *Camp-fires on Desert and Lava* expedition, but President Roosevelt went to Africa instead. Kermit Roosevelt (1920) chronicled his own trip in "The Sheep of the Desert," a chapter in his book titled *The Happy Hunting Ground.* He and his party, which included brothers Win and Ike Proebstel, Cipriano Dominguez, and Eustacio Casares, left Wellton the evening of August 10, 1911, by wagon and horseback. They camped en route at a tinaja near Proebstels' mining claim, and later, on August 11, proceeded to Tinajas Altas, or "the Tinah'alta, as they are called colloquially" (Roosevelt 1920:76). He (Roosevelt 1920:73–74) described it as follows:

> Next day we loaded our outfit on our two pack-mules and struck out across the desert for the Tinajas Altas (High Tanks), which lay on the slopes of a distant range of mountains, about four miles from the Mexican border. . . . The usual camping-place is a small plateau, a couple of hundred yards from the lowest tank. This plateau lies in a gulch and is sheltered on either hand by its steep and barren sides. A few hundred feet from the entrance, on the desert and scattered about among the cactus, lie some hundred and fifty graves of men who have died of thirst; for this is a grim land, and death dogs the footsteps of those who cross it. Most of the dead men were Mexicans who had struggled across the deserts only to find the tanks dry. Each lay where he fell, until, sooner or later, some other traveler found him and scooped out for him a shallow grave, and on it laid a pile of rocks in the shape of a rude cross. Forty-six unfortunates perished here at one time of thirst [when a Mexican bandit named Blanco murdered them by thirst and then robbed them].

Most of these early hunters and other early visitors were gentle on the land. Charles Sheldon (1993), for example, prided himself on small camps and small fires. He hunted alone and at night slept without a tent, wrapping himself in a buffalo robe on bare ground.

By the 1920s and 1930s, visitors had switched largely from horses and wagons to automobiles. The early Model T and Model A Fords, with their high clearance, were surprisingly hardy, even though tires blew out frequently. Travelers were able to bring more creature comforts, although their "outfits" were largely homemade and very simple by modern standards. Travel was a chancy affair, road conditions were difficult, and visitors were few. One couple from New York, the Hogners, traveled the Camino del Diablo in their 1929 Ford roadster in the winter of 1936, stopping at both Tule Tank and Tinajas Altas. In her book *Westward, High, Low, and Dry* (illustrated by Nils Hogner), Dorothy Hogner (1938:177) writes,

> We could see the Tinajas Altas Mountains shining white, very, very far away. . . . Was it possible that it was only ten miles across to Tinajas Altas? So said our guide. But distances *are* deceiving in the desert. . . . However, our immediate concern was with the stretch of gravel bench beneath us. It was composed of stones and boulders, and so rough and hard was it that there was scarcely a mark of the road left. Obviously no one had been that way by car, for months.

But it was fun. People ventured there to explore and camp just for the heck of it. Adventuresome people such as Bob Lenon, Tad Nichols, Catherine and Ralph Irwin, Madeleine and Juel Rodack, Joe and Lillie McCraw, and Judy and Glen Franks climbed into their automobiles and visited the area. They enjoyed the healthy climate, met friends, and remained life-long lovers of this desert. Tinajas Altas was at the core of their desert experience; as one said, "It was like our church" (Glen Franks, personal communication 1997). Mary Huntington (1946), for example, wrote about her journey along El Camino del Diablo in *Westways*. Although she provided no new information, her article and others fostered a love of the desert and encouraged travel to such places as Tinajas Altas for post–World War II travelers who no longer faced limited wartime rations of fuel and tires.

Some of the travelers were fueled by articles by Randall Henderson, editor of *Desert Magazine*, first published in 1937 in El Centro, California and, after 1948, in Palm Desert, California. He delighted in taking his readers to out-of-the-way places. In an article titled "Watering Places on the Devil's Highway" (Henderson 1940), he narrates the area's human history and goes exploring to Raven Butte, Tinajas Altas, and the Gran Desierto dunes. He comments on pictographs at Tinajas Altas as well as several other works of rock art both at Tinajas Altas and Raven Butte. He noted that at that date, the cable was in place from tank one to tank two at Tinajas Altas (Henderson 1940). Bryan (1925:337) also noted that a steel cable for travelers stretched between "the second and third tanks . . . up which it is easy to climb the smooth rock face." Today steel rods, but no cables, are in place between the first and third tanks.

According to several families from nearby communities, Tinajas Altas became a popular place for family picnics as early as the 1930s (Glen Franks, personal communication 1997; Ralph Irwin, personal communication 1997; Joe McCraw, personal communication 1997). Evidence of use from this period (or even earlier) can be seen in undated photographs in the Pioneer Museum in Wellton. In one photograph, three young women pose in front of the lowest tank, their dress and hairstyles reminiscent of the 1920s. Another photograph from the same set shows a natural rock shelter a short distance north of Tinajas Altas; campfire soot is visible on the front of the shelter even then. This shelter, known locally as Cave Rock, is in much worse condition today. At present, soot covers most of the interior walls as well as the rock art within the shelter. In the photograph, a paloverde tree near the shelter opening was intact; it has since been chopped down. Other vegetation has also disappeared, a result of the popularity of this spot for camping.

Photographer Tad Nichols (2007) reminisced about a typical visit for both science and pleasure in the 1930s. The group he was with drove touring cars, used gas lanterns and campfires, used a tarp for shelter, and slept wrapped in blankets. In contrast to these hardy travelers, modern visitors bring house trailers and several vehicles, and trample large spaces. One camper denuded an area more than 100 feet in diameter and reduced it to fine-grained dust with his house trailer, vehicles, and large campfires on a bench immediately northeast of Tinajas Altas Canyon (Bill Broyles, personal observation December 1996). Until regulations were implemented late in the 1990s, off-road vehicles were driven to many previously undisturbed spots on the terrace, slopes, valley, and arroyo banks at Tinajas Altas (Franklin 1996; G. Hartmann 1989).

Bathing or swimming in Tinajas Altas is a major breach of desert etiquette, on the level of spitting in the punchbowl. Someone coming for a drink would be angered. The supply of water is limited and is used by wildlife, and it is illegal to camp within one-quarter mile of any desert waterhole or to inhibit game from reaching the water.

Some who visit Tinajas Altas come to write or to photograph. The paired tanks on either side of an enormous boulder, numbers three and four, are especially photogenic. Photography professor Mark Klett saw the scene with a human component and took the image; it is the cover for

Natural rock shelter just north of Tinajas Altas. A paloverde tree that appears in older photos at the front of the opening has been reduced to firewood and the inside walls have been blackened by campfires, obliterating ancient pictographs. (Photo by Bill Broyles, 2010)

Gary Paul Nabhan's *Desert Legends: Re-storying the Sonoran Borderlands*, as well as an interior page (Nabhan 1994: cover, 7). Willard Clay captured the scene in color, as you can see on the cover of this book (Broyles 1993:8–9). Pulitzer award winner Jack Dykinga, a friend of Klett, Nabhan, and Clay, also camped at Tinajas Altas and took a number of color images, including one at those tinajas (Dykinga 2004:112). George Steinmetz, another of the photography fraternity, also selected these tinajas for his photographs of desert bighorn sheep (Chadwick 2006). And award-winning videographer Dan Duncan filmed several episodes of the television series *The Desert Speaks* at Tinajas Altas.

Several novels mention Tinajas Altas, but so far as we know, only one has been set there. Written by a border patrolman, Kent Williams's *Border Gold* (2004) involves agents and smugglers in pitched battles, chases, and endless problems. The names of its characters are thinly disguised respellings of real agents and the routine is realistic, but the events are imaginary. One poet, Robinson Jeffers (1887–1962), wrote "Dead Man's Child" set at Tinajas Altas, although he probably never visited the tanks (Jeffers 1928). Popular accounts such as Gaillard (1896) and Michler (1857–59) fired his imagination.

Others come to paint. Astronomer William K. Hartmann, who wrote the classic book *Desert Heart: Chronicles of the Sonoran Desert* (1989), ranks Tinajas Altas as one of his favorite places to set up his easel. His plein air paintings offer him a "a profound and personal way to interact with the land." Where some artists favor watercolors or oil, Hartmann prefers acrylic, "for in the desert acrylics dry within a few minutes, allowing vigorous repainting of any weak spots, and then the finished painting is soon dry enough to pack in the back of a car. What I strive to record is my experience of the place. The painting has to say, 'The painter of this picture was *there*, and the painter of this picture experienced something special.' The painting has to invite the viewer to experience some of the same things."

And what about painting outdoors in the wind and weather? "I hope that my images project a sense of the place itself, so I often mix dust and even bits of plants into the paint to get texture effects. I don't care too much if a gust of wind blows dust onto the paint. The best plein air paintings have a spontaneous, rough-and-ready quality, and I'm happy if the viewer realizes that this painting was made by a human being working physically, and sometimes uncomfortably, in that place. The painting is part of that place, and the place is part if it" (William K. Hartmann, personal communication December 2008). Two of Hartmann's pictures are shown in this book's color folio.

"Sunday picnic at High Tanks. Good old days before Bombing Range." Three young women, probably from the Wellton area, posing by the lowest tank at Tinajas Altas. The photo is undated, but their dress suggests the 1920s. (Courtesy of Pioneer Museum, Wellton)

Kirk Bryan's 1920s signs at Tinajas Altas, riddled by bullets. The signs disappeared about 1990, though one was later found in a trash pit. Only the post remains. (Photo by Bill Broyles, 1984)

The effects of vehicles traveling off the road near Tinajas Altas. (Photo by Bill Broyles, 1986)

AN INCREASING RATE OF CHANGE

Because of its scenic and recreational values, it is likely that many more people will visit Tinajas Altas in the next few decades. We can predict with sadness that further deterioration of historical and scenic resources will occur at this unique site, but we hope otherwise. Only appreciative visitors who understand the magic spot's significance and beauty can save it.

In the early to mid-1980s, writers begin to note changes in the Camino del Diablo in the vicinity of Tinajas Altas. For example, photographer Tad Nichols (2007:90–91) observed, "Now for some unknown reason in recent years, maybe 1983, somebody sent a bulldozer through there just to the south of the old road. I never saw any need for that at all. There was nothing wrong with the old road. It was sandy and smooth."

Bob Thomas (1980, 1981, 1987, 1989, 1992) practically made a career of visiting Tinajas Altas and filing stories with the *Arizona Republic* (he reports having made at least 25 trips along El Camino del Diablo). In a 1987 article he explains the change in the Camino, chiding the federal agencies for the grading in the Tinajas Altas area (Thomas 1987:S6): "The Border Patrol, Fish and Wildlife Service, and the Air Force have been active in bulldozing roads in the area. . . . In truth, much of the old Camino del Diablo has been ruined by these agencies that have bulldozed and graded 'straight-shot' roads through the entire area for their own convenience and with little attention paid to history or to preservation of local artifacts."

In the late 1970s and early 1980s, the northern portion of the east fork of the Camino del Diablo was widened and graded by the Air Force. This was done in connection with construction of the ISST (ICBM Silo Superhardened Technology) and HAVE HOST military testing and training complexes on the eastern side of the Gila Mountains. In the mid-1980s, roads were improved for Marine Corps training maneuvers, and the Border Patrol created east-west "drag roads," which allowed them to check for foot traffic heading north from Mexico. For example, the Hobbs Drag crosses the Lechuguilla Valley north of Tinajas Altas and the Lower Drag passes near Tinajas Altas to the east. In 2008 the road to Tinajas Altas was again improved to accommodate crew buses and large semi-trucks with trailers used in building the border wall.

Tire tracks indicate that this mature saguaro was knocked over by a large truck, not natural causes. (Photo by Bill Broyles, 1986)

Tinajas Altas, numbers 3 and 4, a popular spot for photography. (Photo by Austin Lenhart)

Consequently, access to Tinajas Altas became a great deal easier. Where once a traveler could expect a difficult trip of several hours from Interstate 8, now the 25 miles (40 km) usually can be covered in less than an hour. This is not always a blessing for those coming here for the solitude, to "feel" the history, or to enjoy the beauty.

Perhaps because access to Tinajas Altas is now so straightforward, some recent writers have been reluctant to publicize the area. For example, although Edward Abbey visited Tinajas Altas (Doug Peacock, personal communication 1994), this is never clear from his writings (Abbey 1984a, 1984b). Abbey may be alluding to Tinajas Altas when he hikes to "Stone Tanks" in his "A Walk in the Desert Hills" (Abbey 1984b). Outdoor writer Doug Peacock (1986, 2005) chronicles hikes through the Tinajas Altas region. Both of these authors considered Tinajas Altas an exceptional, private place and were reluctant to write about it by name. William Hartmann, in his 1989 photo-illustrated chronicle of the area, *Desert Heart: Chronicles of the Sonoran Desert*, reminded readers of the historical and geographical significance of Tinajas Altas, the Sierra Pinacate, and the whole surrounding region, but deliberately avoided detailed maps (W. Hartmann, personal communication 2004).

In recent years the Camino del Diablo has become a travel loop for desert tourists, who enter the Barry M. Goldwater Range at Wellton or Fortuna and drive across the Cabeza Prieta National Wildlife Refuge to exit at Ajo (or vice versa). Visitors include a variety of campers, jeep drivers, history buffs, desert rats, hunters, rock hounds, photographers, and sightseers (e.g., Eckenburg 1984). According to one jeeping couple, "Tinajas Altas remains today the most documented and important camp along El Camino del Diablo" (Hooper and Hooper 1977, 1995:82). In his book *Adventuring in Arizona*, John Annerino (1996a:42–50) provides helpful information on Tinajas Altas and vicinity for the hiker and camper.

Some hike to Tinajas Altas instead of drive. British author Richard Grant hiked to Tinajas Altas in 1994 and wrote "On the Devil's Highway" for *Esquire*, but he includes no detail on Tinajas Altas. His personal observations include, "The 40–pound pack eats into my bruised shoulders, bounces on my kidneys, and I think of pack animals, wagons, horses, mules. Let beasts carry this burden. The problem with this is that beasts tend to die" (Grant 1994:155). Journalist Charles Bowden mentions Tinajas Altas in *Blue Desert* (1986) and *The Sonoran Desert* (1992), as well as "Learning Nothing, Forgetting Nothing: On the Trail of Carl Lumholtz" (Bowden 2007), a look at Lumholtz's trip through Papaguería. And, Tinajas Altas figures prominently in *Sunshot: Peril and Wonder in the Gran Desierto*, a book by Bill Broyles (2006).

Collectors and amateur scientists also visit Tinajas Altas. On a trip to Tinajas Altas and El Camino del Diablo with the Yuma Chapter of the Arizona Archaeological Society, one author raves about the extensive archaeology at Tinajas Altas (Meier 1990). Her own adventure at Tinajas Altas consisted of sliding down a granite ramp and wearing holes in her jeans. It is clear from the paucity of artifactual evidence in the vicinity of Tinajas Altas that, over the decades, many visitors have picked up archaeological souvenirs of their visit.

With communities around the Goldwater Range growing rapidly, Tinajas Altas will undoubtedly see increased waves of visitors. In spite of visitor damage to the natural and cultural resources of the area, the Tinajas Altas experience remains an exciting adventure in a beautiful setting.

Taking Care of Things

In 1853, US ambassador James Gadsden negotiated the purchase of 18,961,920 acres south of the Gila River from Mexico. Southern states sought a transcontinental railroad route to California, and land along or south of the Gila River offered reasonable terrain (Devine 2004). The price paid to Mexico was $10 million, and not everyone believed this was a wise expenditure of tax dollars. From the point of view of Thomas Hart Benton, a member of Congress from Missouri, the purchased land was "utterly desolate, desert and God-forsaken" (Gates 1968:4). Nonetheless, with the acquisition, the lands around Tinajas Altas became US territory and subject to federal land laws. The region became part of the public domain, and there has never been any private ownership in the vicinity of Tinajas Altas.

As we detailed earlier, the international border and the immediately surrounding region were surveyed twice, first in 1854–1856 by a crew led by William H. Emory, then in the early 1890s by a crew led by John W. Barlow (International Boundary Commission 1898). These surveys described Tinajas Altas geographically but produced few consequences in terms of active management. Like all other public domain land in the new territory, Tinajas Altas was placed under the administrative control of the General Land Office, the precursor of the Bureau of Land Management (BLM). It could have been homesteaded under the provisions of various homestead acts or the Desert Land Act of 1877 (Gates 1968:401), but it never was. In fact, the master title plats for these properties maintained by the BLM show that in addition to a lack of homestead entries in the vicinity of Tinajas Altas, there have never been rights-of-way, utility corridors, mining claims, oil or gas leases, railroad easements, official roads, pipeline corridors, or power or water plant sites.

In the early years of the twentieth century, two legislative mandates had a legal effect in the region, although they caused little change on the ground until the construction of the border wall began in 2006. First, in May 1907, all lands within 60 feet of the international border were withdrawn from the public domain by proclamation of President Theodore Roosevelt. The only allowable activity within this strip of land was a public road. Apparently this action was motivated by smuggling and political unrest in Mexico (Tunnicliff et al. 1986:3-1). Second, in June 1910 in anticipation of Arizona statehood on February 14, 1912, Congress promised that in all townships throughout the state, Sections 2, 16, 32, and 36 were to be held in trust for the schools. So, in Township 13 south and Range 17 west, where Tinajas Altas is located, these sections were allocated as school trust lands, although they were later bought by the military.

Not until 1934 did federal attention again affect this region. In that year President Franklin Roosevelt, by means of Executive Order 6910, "withdrew from settlement by homesteaders all unappropriated public domain lands in the far-western states. . . . [H]is order affected most of the acreage of what is now the [Barry M. Goldwater] Range" (Tunnicliff et al. 1986:3-1).

In September 1941, the first million acres of land in this region were withdrawn "from all forms of appropriation under the public land laws and reserved for the use of the War Department as an aerial gunnery range" (Tunnicliff et al. 1986:3-3). Withdrawals continued and in November 1942, an additional 949,000 acres were withdrawn, including Tinajas Altas. All of these lands were also withdrawn from mining and mineral leasing (Tunnicliff et al. 1986:3-4).

When the United States entered World War II, an air base was built on an air field east of Yuma that had been called Fly Field in honor of a Colonel Benjamin Franklin Fly. A short history of this field (Office of Information Services 1957[?]:2) notes that Fly Field became an air base "with the astounding speed which characterized the war effort and by early 1943 Yuma Army Air Base, under its first commander, Colonel James W. Andrew, was graduating classes of pilots." Originally, the mission of the base was to train pilots in AT-6 single-engine advanced trainers, but it soon switched to multi-engine training in AT-17 Bobcats and then to gunnery training in B-17 Flying Fortresses and B-25 Mitchells.

The base was deactivated in 1945, after the end of the war, and then reactivated in 1951 for the Korean War. During the period from 1951 to 1959, the base, which in 1956 became Vincent Air Force Base in memory of Brigadier General Clinton D. Vincent, was used to train fighter-interceptor pilots who practiced with unguided, air-to-air rockets launched from F-86 Sabres. Abundant evidence of those years of air-to-air training exists in the form of ordnance scattered on the ground. Much of the metal debris comes from 2.75-inch practice rockets including fragments of the rockets and rocket motors as well as .50 caliber and 20 mm cartridge casings, ammunition belt links for these cartridges, and parts of aerial illumination flares. Although most of this material is fragmentary and inert, picking up items that are identifiable as military ordnance is both unwise and against range regulations.

Vincent Air Force Base was closed on December 31, 1959, and reopened on January 1, 1960, as Marine Corps Auxiliary Station Yuma, now known as Marine Corps Air Station Yuma. The Marine Corps continues to train over the western portion of the Goldwater Range, although it no longer practices with live-fire ordnance in the Tinajas Altas area. Various legislation authorized continued military use of the Goldwater Range, including *The Military Lands Withdrawal Act of 1986* (US Public Law No. 99-606) and *The Military Lands Withdrawal Act of 1999* (US Public Law 106-65).

Not surprisingly, considering its remoteness, natural beauty, and human sagas, several proposals have urged added protection for Tinajas Altas's historic, scenic, and biological values. Between 1935 and 1937, President Franklin Roosevelt and US Congress considered a proposal for a very large—4,000,000 acres—national monument or game range (wildlife refuge) in southwestern Arizona. This monument to protect both bighorn sheep and organ pipe cactus was promoted by numerous Arizona conservation organizations, including the Boy Scouts of America. The monument would have included what is now Organ Pipe Cactus National Monument and Cabeza Prieta National Wildlife Refuge, plus a considerable amount of land to the north and west of the present-day refuge. Although the boundaries were imprecise, they would have extended toward Yuma and some distance north of I-8. The monument would have included both the Tinajas Altas Mountains and the Gila Mountains. The idea was opposed by Arizona cattle ranchers, the Arizona State Land Department, and the Arizona Colorado River Commission among others (Arizona Colorado River Commission 1938). So, instead of a large monument, the much smaller Organ Pipe Cactus National Monument was created in April 1937.

The Boy Scouts, as well as the Arizona Game Protective Association, continued to press for a wildlife preserve that would protect bighorn sheep (Miller 1938). As a result, in 1939 President Roosevelt set aside an 860,000–acre Cabeza Prieta Game Range, the size it is today. It is now known as Cabeza Prieta National Wildlife Refuge.

In 1965, Morris K. Udall, then congressman from Arizona, and his brother, Stewart Udall, then Secretary of the Department of the Interior, proposed a Sonoran Des-

The US Customs sign at Tinajas Altas "gate," four miles south of the tinajas. Travelers were ordered to report to the nearest Customs Officer for inspection. (Photo by Juel and Madeleine Rodack, 1967. Courtesy Gayle H. Hartmann)

In 2008 a border barrier was built across High Tanks "gate" by the Department of Homeland Security. (Photo by Luke Evans)

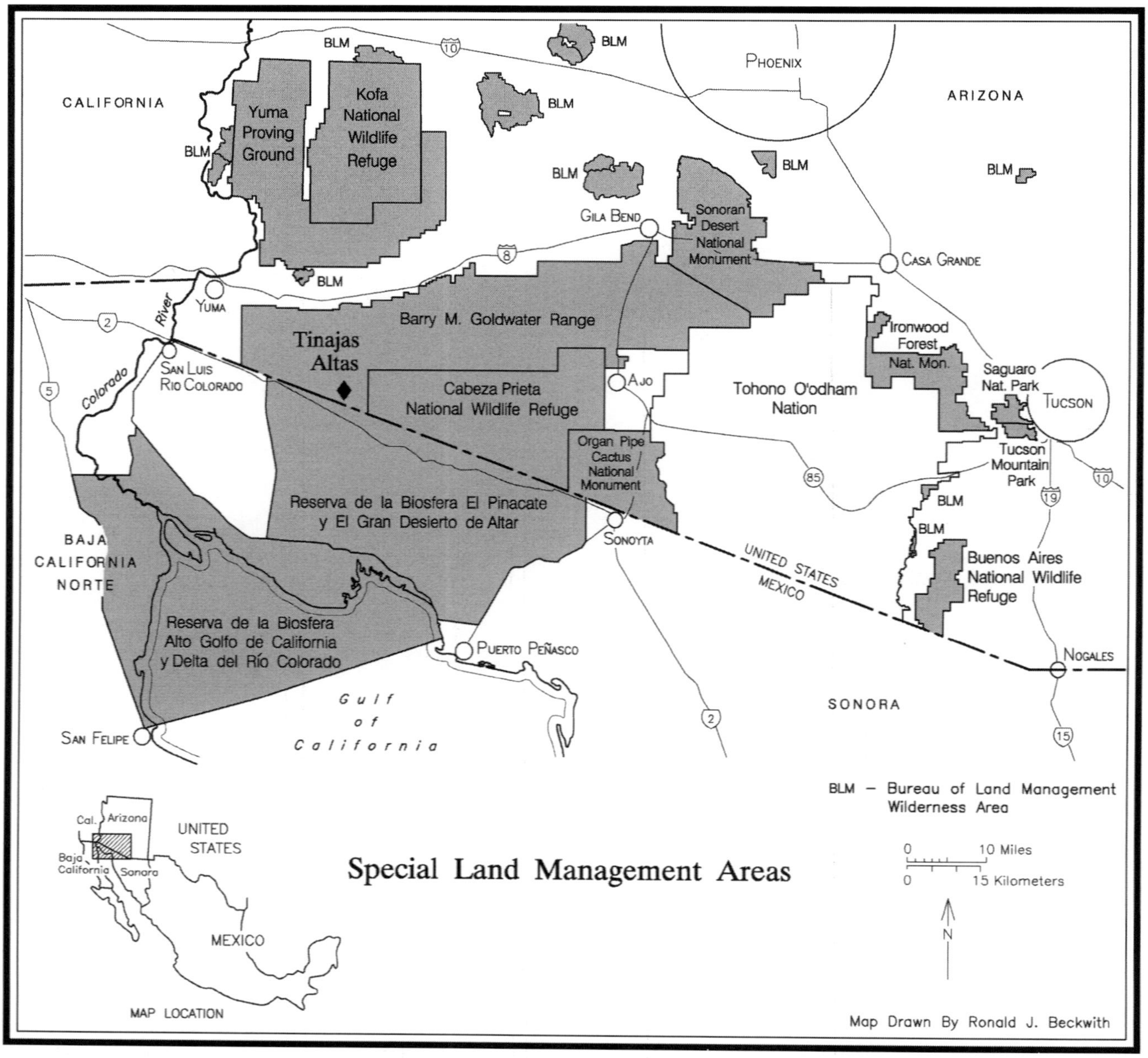

Nominally protected desert lands around Tinajas Altas. (Map by Ronald J. Beckwith)

ert National Park that would have extended from Organ Pipe Cactus National Monument to Tinajas Altas. It would have included the wildlife refuge and been approximately 1,242,000 acres in size, fairly large but still smaller than the area proposed for protection in the 1930s. It would have ranked seventh in size in the national park system at that time. It was to include all lands within the national monument and the Cabeza Prieta Game Refuge, plus the addition of 79,000 acres west of the game range (Felger, Broyles et al. 2007; Udall 2007).

In November 1966, Congressman Udall addressed the Rio Grande Chapter of the Sierra Club on the subject of the Sonoran Desert National Park. He placed extraordinary value on the natural values of this land and firmly believed that it would be suitable for a national park (Udall 1967:108).

> let me talk about the Sonoran Desert National Park for just a moment. . . . [O]ne of the great opportunities this next year . . . is that this is a park that doesn't cost any money. It's all federal land. This park then would extend for sixty miles along the Arizona-Mexico border, and it would include a most unusual part of the country. . . . It's stark; it's very exciting and it's unique and unusual. But one of the things about this proposal is that on the Mexican side of the bor-

> der—I am sure there are some of you who have seen this—there's an area called the Pinacate Lava Fields. This is a million or so acres, with the Pinacate Peak looming off in the distance—and this adds again to this feeling that you are off on some alien planet—with this stark black landscape and really strange types of plants that you find, the remnants of this Sonoran Desert as they penetrate to the north—really strange kinds of cacti and unusual plants.
>
> And there have been some preliminary discussions along the lines that if United States will make this a national park, our part of it, that we could then persuade the Mexican government to set aside the Pinacate area as a Mexican national park, so you would have back-to-back, across sixty miles of this border, two great national parks. You would have an International Park, similar to what we have on the Canadian border up in Montana. This is one of the really nice features; one of the intriguing ideas.
>
> So the reason we should preserve it is that it is a starkly beautiful area. It's different. It's barren, and some of the objectors say, "Well, this is just a barren waste." This is one reason to preserve it. We don't have anything else like this in the United States and a chunk of it ought to be preserved for that very reason.

The Department of the Interior supported the proposal by sending a four-man National Park Service inspection team to the area in January 1965. An article by Weldon Heald, a consultant to the Department of the Interior, summarizes the team's position. He points out that

> to qualify for National Park status an area should exemplify a definite type of natural scenery and be the finest of its kind. So each of our thirty-two national parks is different from the others and are all supreme examples of America's originally rich and diversified wilderness. However, there still remain a number of highly individual and significant types not now represented. One of them is the Sonoran Desert. But Organ Pipe Cactus National Monument and the Cabeza Prieta Game Range [including Tinajas Altas] offer the opportunity of filling this gap by preserving a superb sample of our natural heritage for the benefit and enjoyment of the crowded generations to come. (Heald 1965:3)

For reasons that are not clearly understood, President Lyndon Johnson chose to expand Arches and Capitol Reef national monuments in Utah, but he failed to sign the Sonoran Desert Park proposal (Udall 2007).

A third proposal came in 1974, when the Department of the Interior suggested that the Cabeza Prieta National Wildlife Refuge be expanded by 79,000 acres (31,971 hectares). The same area west of the refuge that was part of the 1965 Sonoran Desert National Park proposal. This expansion would have included the Tinajas Altas Mountains from Raven Butte on the north to the international boundary on the south. The acreage was added to the refuge in 1975 by Public Land Order No. 5493 in a proposed Bureau of Land Management-US Fish and Wildlife Service land exchange, but the "addition was revoked several months . . . [after its passage] due to some technical flaws in the order" (Tunnicliff et al. 1986:3.14). The deal was done on the signature of Interior Secretary Rogers Morton, but Congress nullified it (Ken Voget, personal communication 1987; see Public Land Order 5502). The inclusion of Tinajas Altas into the refuge was again proposed in the mid-1980s as part of a larger BLM-USFWS-Arizona State Land Department land exchange that involved the Kofa refuge, the creation of Buenos Aires National Wildlife Refuge, and the creation of Catalina State Park, but a regional administrator in the Fish and Wildlife Service declined to include the addition (D. Dean Bibles, personal communication 2009). Boundary signs were erected and then taken down, though some of the posts remain. In each of the proposals, military training would have continued, but the land and cultural resources and visitors would have been managed by the Fish and Wildlife Service.

On December 3, 1982, the Tinajas Altas Mountains became an Arizona State Natural Area with the concurrence of the Air Force and Marine Corps. The designated area consisted of the same 79,000 acres that had been briefly added to the Cabeza Prieta National Wildlife Refuge in 1975. Although this natural area designation had no direct legal effect, it recognized the mountains for several special attributes: "[their] relatively undisturbed character, the wildlife and vegetation indigenous to the location, the several tinajas (or natural water tanks) located in the mountains, and the fact that the tinajas were an important stopping place along the Camino del Diablo" (Tunnicliff et al. 1986:sec. 14, p. 10).

In the late 1990s, a binational agreement was signed between Mexico and the United States that affected the border region of northwest Mexico and southwest Arizona. It was inspired by the Biosphere Reserve designation on the Mexican side. The agreement encouraged cooperative actions

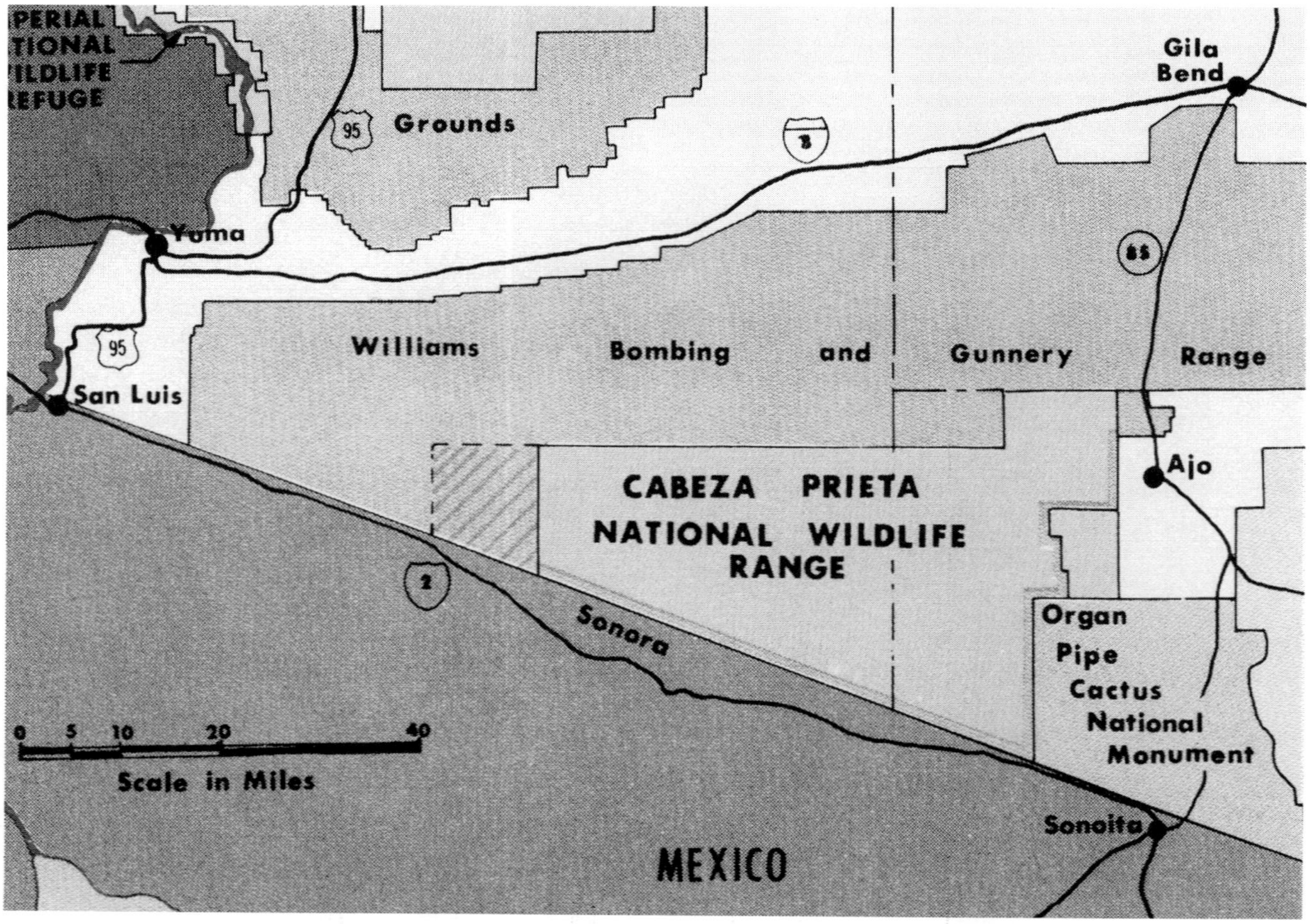

The proposed Tinajas Altas addition to Cabeza Prieta National Wildlife Refuge (the diagonally hatched area immediately west of the refuge). (Map from Cabeza Prieta National Wildlife Range 1970)

with a goal of natural resource protection. BLM's Tinajas Altas Area of Critical Environmental Concern was one of the listed areas (US Department of the Interior 1997).

Since Heald's report the United States has added 30 new national parks, bringing the total to 57 as of 2011, and Mexico has created two enormous biosphere reserves just across the border south of Tinajas Altas. In 1993 Mexican President Carlos Salinas de Gortari proclaimed the Reserva de la Biosfera El Pinacate y Gran Desierto de Altar in Sonora and the Reserva de la Biosfera Alto Golfo de California y Delta del Río Colorado in Sonora and Baja California (Chester 2006; Felger, Broyles et al. 2007).

In 1998, a proposal surfaced to create a Sonoran Desert National Park and Preserve. This grassroots citizens' proposal—endorsed by Stewart Udall, former secretary of interior, Dean Bibles, former state director of the Bureau of Land Management for Arizona, and Exequiel Ezcurra, former president of Mexico's National Institute of Ecology, among others—asked to combine Organ Pipe Cactus National Monument, Cabeza Prieta National Wildlife Refuge, and the Barry M. Goldwater Range. The goal of the proposal was to have Organ Pipe managed as a national park, and the Cabeza Prieta and the Goldwater Range portions managed as a national preserve, with military uses continuing. The "national preserve" designation increases natural and cultural resource protection, but allows exemptions for such activities as military training and big game hunting. The Department of Defense would have been co-manager with the National Park Service, much like existing arrangements with White Sands National Monument and Channel Islands National Park.

The park and preserve concept was seen as a way to combine military training with natural and cultural resource protection, keeping intact an irreplaceable and significant portion of the Sonoran Desert ecosystem. When associated with the Pinacate and Upper Gulf biospheres of Sonora, it would create the world's premier desert park. It would be America's first borderland sister park with Mexico. A similar arrangement with Canada has been highly successful since 1932, and that park—Waterton-Glacier International Peace Park—is also a renowned World Heritage site.

Touch the past. Two sets of bedrock mortars have been moved from Tinajas Altas. This group is on the south lawn of the Arizona Historical Society in Tucson. (Photo by Bill Broyles)

Cross-border agreements and sister parks are increasingly seen as ways to encourage conservation while retaining national sovereignty and protecting national security. World-wide, dozens of international, national, and local groups are working to designate special lands for protection, conservation, enjoyment, and understanding. These groups include governments, corporations, non-profit corporations, and citizen coalitions. Conservation International, International Conservation Union, The Nature Conservancy, Sierra Madre, The Wild Foundation, Wildlife Conservation Society, and World Wildlife Fund are especially active and effective. Several presentation books have compellingly made the case for bi-national parks. These include *Celebration of the Land: The Last Sanctuaries* (Lawrence 2000), *The Gulf of California: A World Apart* (Gil et al. 2001), and *Transboundary Conservation: A New Vision for Protected Areas* (Mittermeier et al. 2005). As Charles Chester reminds us, "Protecting life on Earth will ultimately require an international approach" (Chester 2006:3).

Frederick Gehlbach, a biologist from Texas, studied mammals and lizards along the US-Mexico border. He learned much about the land and its species, but he also learned about the bigger needs for open places, natural places, places where humans can go to appreciate and understand nature and themselves. He wished that his students and friends could see the world as he had.

> Above all, I support a Gran Desierto International Park, connecting Organ Pipe Cactus and Cabeza Prieta preserves with the Gila and Tinajas Altas ranges in Arizona and the extraordinary wilderness southward, the Desierto de Altar, Sonora. This oft-mentioned but untried wildland could be North America's greatest desert museum, at least twelve thousand square miles in extent, matched only by the Sierra Madre in its vastness. (Gehlbach 1981:277)

If you come to love a place, as we openly love Tinajas Altas and this region of desert and gulf, you hope that it will be protected so that future generations will have the opportunity to love it too.

An ironwood tree that has shaded travelers for centuries at Tinajas Altas. (Photo by Bill Broyles, 2010)

Desert bighorn ram. (Photo by Bill Broyles)

Native American grinding holes at Tinajas Altas (© Jack Dykinga)

Tinajas Altas (© William K. Hartmann).

Bighorn sheep at Tinajas Altas (© George Steinmetz).

Pool at Tinajas Altas (© Jack Dykinga).

Sunset Glow Across the Valley (© William K. Hartmann).

5 Natural History

Bill Broyles

Although by some standards Tinajas Altas is in an extreme desert, the region contains a wealth of plants and animals that somehow survive and thrive despite high summer temperatures and little rain. Sometimes we wonder how anything can live here at all. But, long after their camping trips are over, visitors to Tinajas Altas and El Camino del Diablo are apt to smile and remember bighorn sheep, foxes, Kearney sumac, ocotillos, saguaros, many-headed barrel cactus, ravens, turkey vultures, goldfinches, kangaroo rats, Pinacate beetles that stand on their heads, comic road runners, and beargrass as tall as trees. Life can be good.

Old hands returning to Tinajas Altas usually stop a quarter mile from the cliffs and scan the slopes with binoculars, looking for bighorn sheep, hoping to see a golden eagle perched on a promontory, and dreaming of seeing a mountain lion. At dusk or dawn, campers may be startled to see flocks of pelicans or great blue herons flying high overhead, as they navigate the axis of the mountains between the Colorado River and the Gulf of California, and more than one groggy sleeper has been puzzled to hear the honking of ducks or geese migrating at night.

The hot, dry climate plays an important role in determining which mammal species can live here and how they live (see appendix 2; Henry 2007). Desert bighorn sheep (*Ovis canadensis mexicana*), the largest mammals here, inhabit the region and are plentiful enough in the Tinajas Altas area to be hunted. Their distinctive horns and statuesque poses make them favorites to see (Krausman et al. 1999; Lee 1989; Monson and Sumner 1980). Bighorn primarily stay in the mountains, but visitors may discover cloven tracks in sandy washes where a bighorn stood browsing tender ironwood leaves a mile or more from the security of craggy slopes.

Coyotes are ubiquitous. Their trails can be seen along arroyos, across the flats, and leading toward waterholes, where they hunt, scavenge, and sometimes drink. Coyotes are omnivorous, eating everything from rabbits to cactus fruit, grasshoppers to carrion. Coyotes also figure prominently in Tohono and Hia C'ed O'odham myths and tales, frequently as a trickster.

Gray foxes may be seen at dusk among the rocky cliffs, while kit foxes prefer sandy flats where they hunt rodents, especially kangaroo rats. Bobcats are scattered throughout the valleys as well as the hills, and subsist on rabbits, rodents, reptiles, and birds. Coyotes, foxes, and bobcats can derive sufficient moisture from their prey to enable them to live independently of any waterhole.

A few mountain lions (*Puma concolor*, formerly *Felis concolor*) inhabit the region but are rarely seen and may be transients, tourists like ourselves, just passing through. Lions, especially younger males, may roam a hundred miles seeking a home range not already occupied by another lion (Logan and Sweanor 2001; McIvor et al. 1994). Given the sparse prey here, there likely are no more than a partial handful of lions in the entire mountain range at any given time. There has been some speculation that these are a subspecies, *Puma concolor browni* (Yuma puma), but the evidence so far is inconclusive (McIvor et al. 1994).

A herd or two of javelina (*Pecari tajacu sonoriensis*) currently roam the area, but javelina are recent residents, likely arriving sometime between 1934 and 1945. In the past century javelina have expanded their western and northern ranges in Arizona and New Mexico, although the reasons why are unclear. Javelina did not appear in the archaeological evidence dating back 11,000 years at Ventana Cave, about 110 miles east of Tinajas Altas (Haury 1975), although they are abundant in that area today; a single javelina-like tooth from one of the cave's oldest horizons represents a different genus, *Platygonus* (Knipe 1956). Neither Edgar Mearns (1907), William Hornaday (1908), Carl Lumholtz (1912), Charles Sheldon (1993, for the period 1913–1916), Kermit Roosevelt (1920), nor Goode P. Davis (1982, for the period 1824–1865) reported javelina west of Sonoyta, Sonora, or Ajo, Arizona. By 1934 javelina were reported near Tule Well and Tule Tank, 18 miles east of Tinajas Altas. Before the end of World War II, a herd was spotted near Tinajas Altas (Halloran 1945), and between 1956 and 1959 specimens appeared in the Kofa National Wildlife Refuge 40 miles north of Tinajas Altas (Hoffmeister 1986). Javelina eat a wide variety of things including roots, fruits, stems, seeds, insects, reptiles, and even carrion and cactus stems (Sowls 1997).

Sonoran pronghorn (*Antilocapra americana sonoriensis*), an endangered subspecies inhabiting sandy plains and rocky bajadas, have been seen within 10 miles of Tinajas Altas in recent years. Pronghorn have home ranges that may cover between 16 and 1,109 mi^2 (43 and 2,873 km^2; Hervert et al. 2005) and they are capable travelers. For example, in 2009 a young buck pronghorn released from a captive-breeding

Bighorn ewes, lambs, and young rams at Tinajas Altas. (Photo by Bill Broyles)

pen ten miles west of Ajo was tracked by its radio-collar. The buck traveled north of the Crater Range, then west past the Mohawk Mountains, farther west to within a few miles of Tinajas Altas, south across the border to the west side of the Pinacate Mountains, and then east of the Pinacate to the vicinity of the Sierra San Francisco, a distance of several hundred miles and all within few weeks (James Atkinson, US Fish and Wildlife Service, personal communication 2010). Historically their range extended into southeastern California and northeastern Baja California (Thompson-Olais 1994). Pronghorn are the desert's fastest runners and because of their extremely good eyesight and wariness, they are very elusive, causing some folks to call them the ghosts of the desert. Neither mule deer nor white-tailed deer live in the Tinajas Altas area.

Small mammals include the desert cottontail, black-tailed jackrabbit, and numerous species of rodents (Henry 2007; Hoffmeister 1986). The kangaroo rat (*Dipodomys* spp.) is one mammal truly adapted to the desert, for it can metabolize seeds to make sufficient moisture for its needs. Most of the mice found in the area are seed eaters, but one, the southern grasshopper mouse (*Onychomys torridus torridus*), devours beetles, grasshoppers, and scorpions, as well as other mice. When researchers use live-traps to survey rodents, they usually can identify grasshopper mice by their odor before even opening the trap. A couple of interesting books, Frederick R. Gehlbach's *Mountain Islands and Desert Seas* and Michael A. Mares's *A Desert Calling*, touch on rodent surveys in this Sonoran Desert borderland.

Ecologists point out that most mammals here have adapted to high daytime summer temperatures by spending much of the day underground (Gehlbach 1981; Turner and Brown 1982). Temperatures a foot underground may be as much as 60°F cooler than the blistering surface at midday. Burrows dug by rodents, badgers, tortoises, lizards, and coyotes are used by these animals themselves, as well as other species such as snakes and burrowing owls (Merlin 1999). Woodrats (*Neotoma* spp.), also called packrats, construct homes from twigs, cactus joints, and debris to protect themselves from sun, rain, and predators, but even these rude houses are shared, uninvited, by lizards, tortoises, and snakes. Bighorn sheep rest in tafoni caves and shaded places during the hottest part of summer days. Even the bird population is influenced by the sparseness and openness of the vegetation (see appendix 3; Griffin 2007). Around the pools, one is apt to see mourning or white-winged doves, ravens, or goldfinches. Occasionally a prairie falcon drops from the cliffs and devours an unwary bird. Gambel's quail call to each other as they scratch for seeds in the underbrush. Roadrunners, thrashers, cactus wrens, and Gila woodpeckers are all here (see Phillips et al. 1963; Corman and Wise-Gervais 2005).

During warm weather snakes and lizards may be out and about throughout the day, except for the hottest hours of mid-summer, and in cooler weather they are most active after warming themselves in sunshine (see appendix 4; Rosen 2007). Geckos and night lizards are active after sunset. Large chuckwallas squeeze into the rock crevices in Tinajas Altas Canyon when they feel threatened, whiptail lizards nose in the dirt looking for bugs, horned lizards sit beside ant hills waiting for food to walk by, and Gila monsters venture out of their hidden burrows only a few weeks a year to eat quail eggs and unwary lizards (Beck 2005). Rattlesnakes generally hunt at night, when they have the advantage of heat-sensitive "vision," but they may sun themselves near dens, even in winter. For more information on reptiles see Brennen and Holycross (2006), Lowe et al. (1986), and Shaw and Campbell (1974).

Animals that were likely important food for Native Americans include desert iguanas, chuckwallas, bighorn sheep, various species of woodrats and mice, jackrabbits, cottontails, ground squirrels, and desert tortoises (see appendix 1; Felger 2007; Rea 1997, 1998). Some animals generally were not eaten because of cultural taboos (coyotes), taste (skunks), association with physical or spiritual illnesses (gophers, vultures), or insufficient meat (pocket mice).

Chuckwalla. (Photo by Bill Broyles)

Prominent invertebrates include the giant desert hairy scorpion, apt to be seen hunting at night. Both tarantulas and their nemeses, tarantula wasps can be seen, usually near dusk. Seasonally, blister beetles, millipedes, velvet mites, pallid-winged grasshoppers, velvet "ant" wasps, and harvester ants may "bloom" in great numbers (Jandrey 2001; Werner and Olson 1994). In wetter years there may be sufficient annual plants (evening primroses, sand verbenas, mallows, and bladderpods) to allow bumper crops of butterflies (especially painted ladies) and sphinx moths (Stewart et al. 2001).

In broad daylight, when animals may be resting out of sight or avoiding midday heat, visitors can always track down unusual plants (see appendix 5; Felger and Van Devender in press). The most common plants are bursage and creosotebush. They are key plants for holding soil together, providing shade for seedlings of other species as well as animals, and furnishing food. Bursage, both triangle leaf and white bursage, thrives in sandy soil somewhat better than creosotebush, and both species are common on the broad valley floors with creosotebush communities continuing up the upper bajadas and into the mountains. Creosotebush, that hardy and widespread symbol of deserts, ranges throughout most of the deserts of both North and South America. After dew or rain, creosotebush gives the desert its signature fragrance—"the desert smells like rain" (Nabhan 1982). If you inspect creosotebush closely, you'll find that it has its own entourage of green-colored crab spiders, clickers (a type of grasshopper), and other grasshoppers. If you erase creosotebushes and bursages from a photo of the desert landscape, you can see how empty it would look without them. International explorer Carl Lumholtz (1912:223), who passed by Tinajas Altas, wrote, "To me the [creosotebush] is a symbol for health and an example of cheerful existence under adverse circumstances."

Other typical species may seem more interesting because there are relatively fewer of them. These include ocotillo and several species of cacti, including saguaro, beavertail cactus, teddybear cholla, diamond cholla, and the impressive many-headed barrel cactus. In the better watered confines of the canyon itself and on the slopes, leguminous trees, especially ironwood, and the aromatic elephant tree, a relative of frankincense, are notable, as well as large clumps of desert lavender and the bright-red-orange flowering chuparosa. Often mistaken for a yucca, desert tree-beargrass (*Nolina bigelovii*) is visible on the steep slopes of the canyon. High in sheltered, north-facing alcoves Kearney sumac (*Rhus kearneyi*) grows, one of only a few places where it occurs in Arizona. A few western honey mesquite are present in Tinajas Altas Canyon, but they are uncommon throughout the rest of the mountain range. Each of these plants adds something to our enjoyment and understanding of this desert.

*Desert tree-beargrass (*Nolina bigelovii*) at Tinajas Altas canyon. (Photo by Bill Broyles)*

Following bountiful winter rains, the desert may bloom with wildflowers spanning the entire valley and rolling up the mountain slopes. Colorful primroses, sand verbenas, desert sunflowers, and stately ajo lilies dazzle our eyes and we spend hours recording the splendor with our cameras. The plants provide food for rabbits, caterpillars, and rodents, while the flowers set seeds for birds, rodents, and even ants. By biomass, the combined weight of all ants and termites may outweigh all of the mammals, birds, and reptiles combined (Jones and Nutting 1989; MacKay 1991; MacKay et al. 1989). Termites are essential to the ecosystem by recycling nitrogen and other nutrients back into the soil—if not for termites, we'd soon be smothered by dead wood, leaf litter, and dung, and living plants would die for lack of nitrogen and trace minerals (Phillips and Comus 2000). Solar radiation also plays a major role in plant litter decomposition (Austin and Vivanco 2006).

*Many-headed barrel cactus (*Echinocactus polycephalus*) and creosotebushes surrounded by biological soil crust with the Tinajas Altas Mountains in the background. (Photo by Gayle H. Hartmann, 1998)*

It is easy to overlook one of the very most important living organisms at Tinajas Altas, biological soil crust, but it is indispensable. This crust community of living organisms catches rainfall and dew and holds that moisture in the soil, reducing evaporation. These organisms bind the soil together, stalling erosion. They also contribute plant-essential nutrients to the soil (Belnap 2007). The crust goes by several names: cryptogamic, cryptobiotic, microphytic, or microbiotic crusts, and even BOB (for bio-organic buddies).

> Biological soil crusts are a complex mosaic of living organisms—cyanobacteria (formerly called blue-green algae), green algae, lichens, mosses, liverworts, bacteria, and fungi—that weave throughout the top few millimeters of the soil surface. Most of the organisms in soil crusts are invisible, especially in hot deserts such as the Sonoran. Despite their small size, they provide many essential ecosystem services, including providing the glue that holds the desert soils together. (Belnap 2007:343; see also Belnap and Lange 2001)

The moss, algae, and lichen partners of the soil crust can green-up and resume photosynthesis within hours after receiving rain (Evenari et al. 1982). When pounded by too much foot, hoof, or vehicle traffic, these soft crusts disintegrate and die, leaving the soil vulnerable to erosion and depleting both moisture and nutrients. The crust is like the desert's skin, and when it is flayed we are left with "moon dust"—bare, lifeless, drifting ground. Roger Rosentreter, Mathew Bowker, and Jayne Belnap (2008) have provided a handy guidebook to the most common lichens and bryophytes.

Don't bust the crust. (Courtesy of Jayne Belnap, US Geological Survey)

But tinajas are the real attraction here. These marvelous pools are a curiosity by virtue of being an occasional source of water for humans but, to the careful observer, a rich diversity of life exists at and under the surface of their waters. These pools may look simple, but they are both complex and fascinating.

Rock pools are temporary and unpredictable habitats for aquatic invertebrates, but generally they form a rich and diverse neighborhood. For example, invertebrate communities in a similar series of perennial and temporary rock pools in the White Tank Mountains west of Phoenix, Arizona, contain more than 95 taxa (species or subspecies). Aquatic insects (65 taxa) were the major group, followed by crustaceans (13 taxa) and rotifers (10 taxa). Among the aquatic insects, the orders Diptera (true flies, mosquitoes, gnats; 30 taxa), Coleoptera (beetles; 19 taxa), and Hemiptera (true bugs; 11 taxa) were particularly well represented (Kubly 1992). Two semi-aquatic vertebrates, the red-spotted toad (*Bufo punctatus*) and the Sonoran Desert toad (*Bufo alvarius*, formerly called Colorado River toad), also inhabit rock pools, but only the former commonly reproduce in them. Similar numbers of species may occur at Tinajas Altas, though a thorough study is needed.

When rains come and flow into the tinajas, the rising water signals the invertebrate inhabitants that it is time to reproduce, hatch, emerge, and disperse. Their responses vary according to the time of year, the magnitude of the flows, the chemistry and temperature of the water, and the life cycle of the species. Considering these complex relationships, we can almost view the tinaja itself as a living organism.

We marvel how invertebrates adapt to survive in this harsh environment. Take temperature and evaporation, for example. May and June are drought months during which relative humidity declines to 10 percent or less, maximum air temperatures often exceed 105°F (40°C), and monthly pan evaporation rates reach 17 inches (43 cm). As water evaporates, salts are concentrated and the dwindling water mass is increasingly susceptible to solar warming. Summers are fast paced and impose harsh deadlines, so aquatic animals must race against the drying of their pools.

Winters are mild by many standards. Freezing air temperatures rarely occur at Tinajas Altas, as shown by thermometer readings in the closest communities (Wellton, Yuma) and by the presence of plant species that are easily damaged by frost, such as *Bursera microphylla* and *Jatropha cuneata*. We have no reports of surface ice forming on the pools during winter months, but cooler air and water temperatures do affect invertebrates by slowing productivity of zooplankton and algae (including diatoms and photoplankton), which are key links in the food chain. Average winter monthly pan evaporation falls to 4.3 inches (11 cm), but the annual average (98 inches; 250 cm) still exceeds the cumulative yearly precipitation by a factor of 20. Living in such a changing and potentially hostile environment is quite challenging for any species.

In the best of years, winter storms emanating from the Pacific Ocean bring gentle, soaking rains, often lasting several days. The first storm provides water to a dry hole or adds water to a waterhole that otherwise would dry up. Within hours some species start hatching from last year's eggs or emerging from intermediate life stages. Within a few days, the replenished pools are populated by a diversity of rotifers, copepods, and cladocerans, microscopic open-water invertebrates (zooplankton) that characteristically make a living by grazing on their plant-world counterparts. Their bottom-dwelling cousins, the ostracods or seed shrimp, begin life as open-water swimmers, but most soon descend to bottom sediments for their food. Some of these creatures can complete their life cycles in a few weeks; others may take over a year to mature, but their life-stages are tuned to the presence or absence of water in the tinajas.

Algae remnants or spores quickly respond to water and sunlight. Typically, algae and blue-green algae (cyanobacteria) run through seasonal or annual blooms and die-offs. Common filamentous green algae may include the divisions *Chlorophyta*, *Cyanophyta*, *Chrysophyta*, *Euglenophyta*, and *Pyrrophyta*. The blue-green algae may include *Oscillatoria*, *Lyngbya*, *Microcystis*, *Nostoc*, *Phormidium*, *Anabena*, *Gleotrichia*, and *Schizothrix* (Rosenstock et al. 2004). Human visitors sometimes are surprised to see entire pools covered with green blankets of algae.

Additions to the tinaja invertebrate community arrive in due time, some carried in by flowing water, some walking overland from protected damp refugia, some arriving as eggs—like fairy shrimp—and spores of algae carried in soil on birds' feet. Others fly in from neighboring or distant pools (see Jackson and Fisher 1986). These flying insects include biting and non-biting midges, mosquitoes, predaceous diving beetles, water boatmen, backswimmers, damselflies, and dragonflies. Some of the airborne colonists become residents in the tinajas, while others simply lay their eggs and move on to other haunts or, having reproduced, die. Many of the species that arrive by air are predators that do not have drought-resistant stages in their life cycles. Their dilemma, in response to the never-ending cycle of drying and wetting of tinajas, is not whether to leave, but when to leave. This uncertainty benefits species that are able to remain through periods of dryness, because they are able to respond rapidly to the next filling of the pool, thereby getting a head start on their many enemies and competitors.

The montage of tinaja invertebrates not only uses different food sources, but they do so within different parts of the rock pools (Kubly 1990, 1992). Large (*Gerris* sp.) and small (*Microvelia* sp.) water striders and gyrinid beetles cruise the surface of the water, whereas the backswimmer *Notonecta* sp. is typically at or just below the surface, lying on its back with its abdomen just above the surface so it can breathe. As indicated by its common name, the backswimmer sees the world while swimming upside-down on its back; it does not swim backward. Its large eyes and back-swimming view of the pool bottom help it scout for a variety of prey and dodge other predators. A smaller backswimmer, *Buenoa* sp., lives closer to the bottom, below its larger relative—and for good

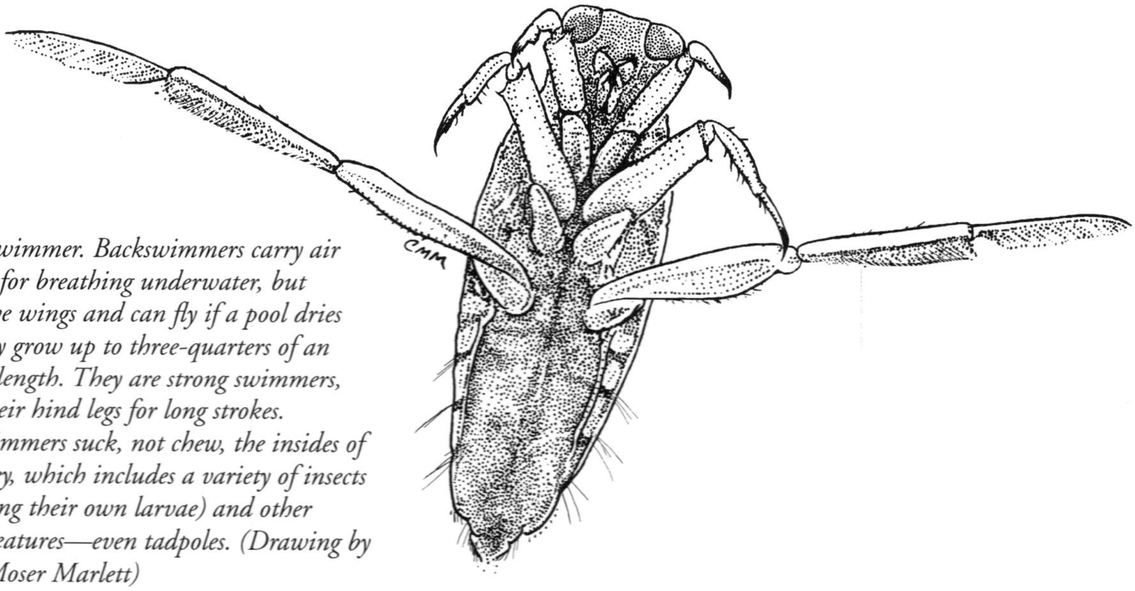

A backswimmer. Backswimmers carry air bubbles for breathing underwater, but they have wings and can fly if a pool dries up. They grow up to three-quarters of an inch in length. They are strong swimmers, using their hind legs for long strokes. Backswimmers suck, not chew, the insides of their prey, which includes a variety of insects (including their own larvae) and other small creatures—even tadpoles. (Drawing by Cathy Moser Marlett)

reason. Both are predators who pierce their prey with stiletto-like beaks, and without hesitation the larger backswimmer will use its size and speed to capture and impale its smaller relative. Some predators shred their prey, but others pierce their prey and then "drink" them.

Predaceous diving beetles (*Laccophilus* sp., *Deronectes* sp., and *Rhantus* sp.), water scavenger beetles (*Tropisternus* sp., whose larvae are consummate predators), water boatmen (*Graptocorixa* sp.), and creeping water bugs (*Ambrysus* sp.) spend most of their time in bottom sediments, but they must rise to the surface periodically to replenish their supply of oxygen. Some of them actually gather bubbles of air to extend their stay below the surface, much like scuba divers refilling their tanks. In the course of their sojourns across the bottom or to and from the surface, they risk meeting other predators, such as the nymph of the dragonfly *Libellula* sp., which does not depend on surface air and makes its living by ambushing unfortunate victims.

Poolside observers can enjoy its residents by noting which feeding strategy a species uses. Common strategies include filtering water for microscopic food, collecting food in a retractable "net," grazing, scraping food off of rocks, shredding organic matter into digestible sized pieces, or preying on other species.

Open-water predators and parasites may dwell on the bottom, but some float or swim in open water, and the rest swim or float on the surface. The predators and parasites include the rotifer *Asplanchna* sp., hemipterans (backswimmers and water striders), and the whirligig beetle *Gyrinus* sp. The whirligig beetle and backswimmers rely largely on

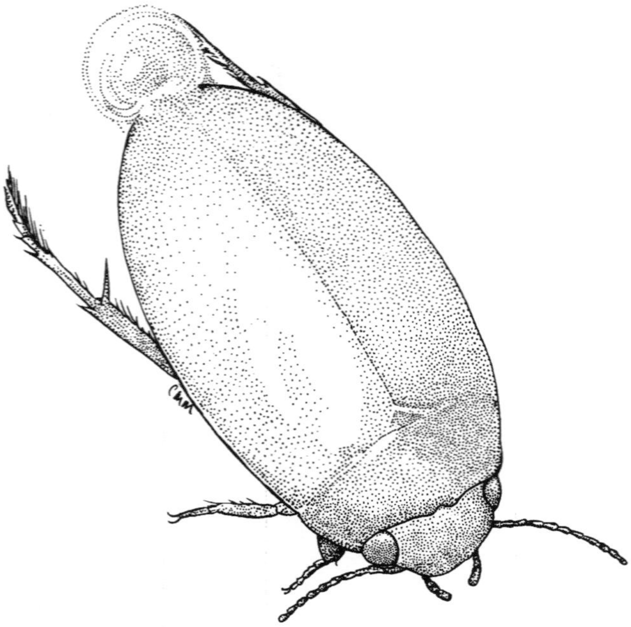

A predaceous diving beetle. These insects can dive for extended periods of time by breathing from air bubbles trapped underneath their wing covers. They feed on insects including mosquito larvae, microscopic animals, detritus, and carrion. These beetles grow up to an inch in length, but look larger in the water. They have wings and can fly if they need to leave the pool. Typically their six legs swim in unison, and they come to the surface tail-first. Their larvae are slender, soft-bodied predators known as water tigers. Similar-looking water scavenger beetles come to the surface head-first and swim by paddling their six legs alternately. (Drawing by Cathy Moser Marlett)

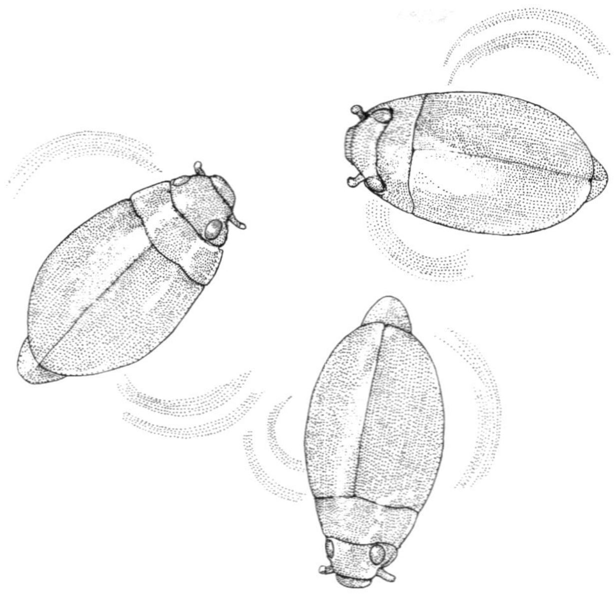

Whirligig beetles. These small dark-metallic-colored beetles feed on the surface by swimming in small circles, giving them their name "whirligig" and their family name Gyrinidae (from the same root word as gyre and gyroscope). Although it might seem as if they swim aimlessly, their movements in open water are tuned to feeding on microscopic animals and avoiding predators. They are usually seen in groups that have complex social structures, where hungrier beetles take position on the outer circle, in effect wagering greater access to food against greater risk of being eaten by predators. They grow to about one-half-inch long and have divided eyes (roughly shaped like figure 8s) that allow them to see above and below the surface simultaneously, keeping track of predators and their own food, which consists of other insects and organic bits. (Drawing by Cathy Moser Marlett)

waterways or aerial pathways to colonize other temporary rock pools, whereas the water striders, *Gerris* sp. and *Microvelia* sp., are semiaquatic but can travel overland for short distances. *Asplanchna* sp. is adapted to withstand extended periods of dryness.

Predaceous beetles, hemipterans, dragonflies, damselflies, and dipterans largely lack drought-resistant stages in their life cycles. Instead, they rely on permanent-water pools during drought periods, and their flying adults colonize temporary waters. One damselfly, *Archilestes grandis*, lays its eggs within the cambium of willow trees, which occur along the distant Gila and Colorado rivers. For survival it depends upon well-developed flight capabilities, selection of perennial or longer-lived temporary pools, relatively rapid development, and reproduction or hatching periods that coincide with long-lasting pools. Predaceous beetles have semi-terrestrial pupal stages that survive in cells dug in sediments forming the banks of their habitats; however, they are susceptible to mortality from high summer temperatures found in desert soils and from rapid immersion in water entering pools as runoff. Ironically, successful reproduction for beetles may be restricted largely to temporary pools.

The open water herbivores (feeding on plants and algae) and detritivores (feeding on dead organic matter) comprise most rotifers and crustaceans, and all mosquito larvae. All taxa within the group possess some mechanism for remaining in or near temporary pools during the period of dryness. Crustaceans, except possibly the copepod *Eucyclops agttis*, have desiccation-resistant eggs or embryos. This same means of surviving dryness also is found among the mosquitoes, for most species have drought-resistant eggs, although some species also have aestivating (dormant) adults. Rapid development to sexual maturity, an adaptive feature of some temporary pond inhabitants, is particularly noticeable for the mosquito *Psorophora signipennis*. Identifiable larvae of this species may be present within two days of summer rainfall, and pupae may appear within one week following inundation.

Bottom-dwelling herbivores and detritivores include representatives from several taxonomic groups. The clam shrimp, all ostracods, the ceratopogonid midge, and some chironomids have desiccation-resistant stages in their life cycles. Insects that remain at the pool—the mayfly *Callibaetis*, corixids, beetles, and other dipterans—generally must colonize temporary pools through flight as adults. Only during rare occasions when rock pools are connected by flowing water can they move between pools. The pulmonate snail, a species of *Physella*, can move only during these brief periods. It attains a limited capacity to withstand dryness by burrowing into bottom sediments and secreting a mucoid plug or epiphragm across the shell aperture. The ostracods are able to persist in the dry sediments of intermittent pools.

Aquatic insects lacking drought-resistant stages and the snail generally prefer larger pools. The ceratopogonid midge, a species of *Dasyhelea* that withstands drying in the larval stage, prefers pools of lower volume because its larvae drown if submerged completely. The highest number of taxa may not occur in a perennial pool, but rather in intermediate-sized temporary pools. Some invertebrates—the tardigrade *Hypsibius* sp., the clam shrimp *Eulimnadia texana*, the fairy shrimp *Streptocephalus mactini*, the caddisfly *Limnephilus*, and the mosquitoes *Psorophora signipennis*, *Culex tarsalis*, and *Anopheles pseudopunctipennis franciscanus*—may be found only in ephemeral pools that dry up after a month or so. For example, fairy shrimp eggs require a dry period before they can hatch when the pool refills with water.

When the winter rainy season ends and temperatures rise, evaporation begins for the tinaja invertebrate communities.

Reproduction of most invertebrates is evident by virtue of the appearance of young individuals, all of which must pass through a series of instars or life stages before reaching the adulthood. During this period of high biological activity the invertebrates become structured into a "trophic hierarchy" of those being eaten and those doing the eating. Regardless of their places on the food chain, survivors must still face the prospects of increasing water temperature or evaporation of the entire pool.

The onset of spring and summer accelerates the pace of life in the tinaja. Activity is intense and competition fierce. Transformations from aquatic to terrestrial stages are common. Mosquito pupae rise to the surface for a final time, wriggle out of their pupal cases, and emerge to the atmosphere where they inflate with fluid and air to fly away in search of a blood meal or plant sap. Mayfly nymphs with noticeably darkened wing cases suddenly rise to the surface and in an instant erupt into air. After a short flight to a more protected place, the last transformation, a molt from subimago to imago or adult, is necessary before the newly emerged mayfly is free to fly away. After all that work, it seems sad to say that the adult's chief function in life is to return and lay eggs, beginning the next generation, and that within hours or days its own life is over.

Other insects leave the drying tinaja at different times, and each individual that does so faces the challenge of finding a more favorable place in which to wait out the remaining time until the summer monsoons begin. As the cumulus clouds build, and barometric pressure plummets in anticipation of summer thunderstorms, the number of insects emerging from drying pools becomes even more pronounced, but the vast variety and number of invertebrates flying through the air above us may go unnoticed except by insect-eating bats and birds such as lesser nighthawks and ash-throated flycatchers. Especially at dusk you may observe birds and bats feeding several hundred feet above the ground, giving an idea of how full the air is with insects.

The number of pools containing water declines under the relentless summer sun, and summer rains may or may not replenish them. Pools may or may not persist through the fall or winter. Pool size is not a reliable predictor of duration, because shading, leakage, the amount of sediments, size and permeability of the immediate drainage area, and presence of bordering riparian vegetation can greatly influence pool persistence (see Brown and Johnson 1983). For example, a pool may be exposed to sunlight and wind, may have relatively deep sand or gravel sediments, or may be bordered by alluvial sediments that support acacia and mesquite. While these plants provide some shade to the pool, their roots remove water from the basin. Smaller, but longer-lived pools, such as numbers 7 and 8 at Tinajas Altas, typically are well-protected from sun and wind by sheer rock faces and overhangs.

Summer thunderstorms may appear to douse the countryside, but in reality they are highly variable and spotty. A common lament is that "it rained everywhere but here" (McDonald 1959). This variability means that only some pools may receive significant water. If and when major monsoon storms finally come, they often are deluges that wash large amounts of sediment and organic debris from the surrounding slopes into the pools. The water standing in pools is soon stained brown by organic residue.

Monsoon storms may flash flood the canyon bottoms and scour the pools, but most species have evolved strategies to prevent being swept downstream. Algae attaches itself to rocks. Invertebrates may use hooked feet or suckers to cling to rocks or sticks; some hide under rocks, in crevices, or burrow in sand and gravel; some are streamlined and huddle within the thin boundary layer where water flows more slowly over rocks or the pool bottom. At least one black fly species (of the family *Simuliidae*) anchors itself to a rock with a silk safety line. Still, many individuals are flushed downstream beyond the life-giving pools or crushed by the tumbling rocks and sand.

Surprisingly, the desiccation-resistant stages of ephemeral pond invertebrates are able to withstand this onslaught of roaring water and, as in winter, within a few days following a thunderstorm, the water contains early hatches of rotifers, copepods, and cladocerans. Soon to follow are the larger crustaceans, the fairy shrimp and clam shrimp. The former is recognizable by the way it swims gracefully about on its back, using its fringed feet to filter food particles from the water. Fairy shrimp have eleven pairs of legs and subsist on a diet of microscopic animals and organic scraps. Adults grow up to one-inch in length, and you may see females carrying eggs sacs behind her gills. The clam shrimp looks like a small, swimming clam, but its shell, or carapace, is comprised of chitin rather than calcium salts. Both grow quickly and pass through molts in short order to become adults within a few weeks, develop eggs and sperm, and reproduce before the rapid vanishing of their aquatic world. They race to escape the myriad flying predators that enter from above in their quest for food. The predators, in turn, are forced to find their prey in diminishing pools as the monsoon season comes to an end.

Tinajas in desert canyons can be viewed as aquatic islands in a terrestrial ocean (see MacArthur and Wilson 1967). These "islands" are unreliable and scattered, providing habitat in some years and none in others, and then drying and forcing invertebrates to skip a cycle or to find new pools in

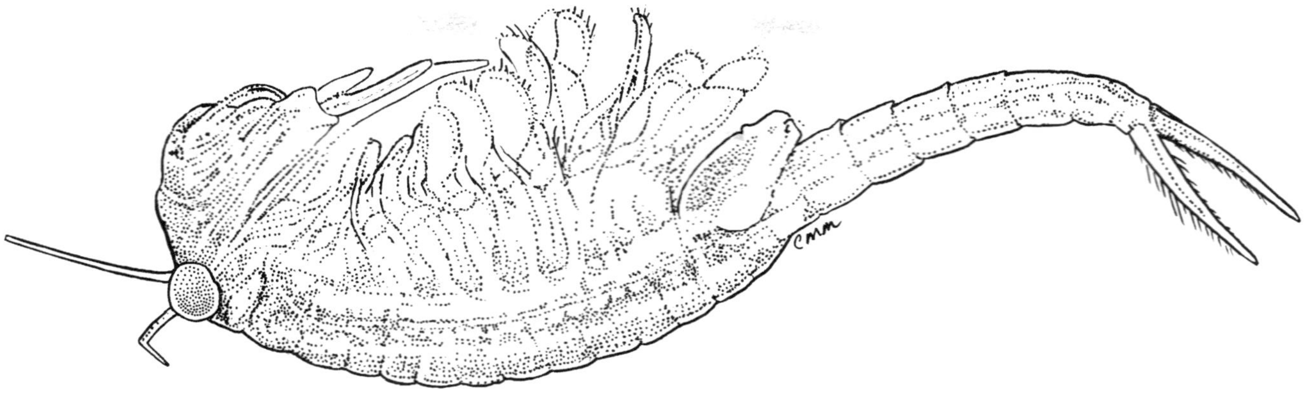

A fairy shrimp. Various species of fairy or brine shrimp occur throughout the West, appearing as if by magic in temporary or vernal pools following rains. (Drawing by Cathy Moser Marlett)

other canyons. We suspect that some pools occasionally "die" and remain vacant until colonizers arrive from neighboring tinajas. This means that, over time, the entire network of desert pools is crucial for survival of these hardy species, just as a network of pools was essential for human travelers. Both aquatic and terrestrial invertebrates are strong bioindicators of environmental health (Hodkinson and Jackson 2005), and long-term studies of freshwater invertebrates are crucial if we are to understand local biological relations, regional ecosystems, and global changes (see Jackson and Füreder 2006). The pools at Tinajas Altas are a remarkable repository of knowledge waiting to be better understood.

In Western movies thirst-wracked travelers knew not to drink from desert waterholes if they couldn't see bugs and green scum, for their absence meant that the water was poisonously alkali or laced with arsenic. Today we appreciate that an active, lively pool is a sign of a healthy ecosystem. The rich and fascinating life within and around desert waters reminds us how rare and special they are, and how remarkable and fragile life itself really is.

Saguaro (Carnegiea gigantea). *(Photo by Joan E. Scott)*

Conclusion
Reflections on an Exceptional Place

Tinajas Altas is a place of contradictions and amazement. It is a place whose beauty exceeds all expectation. It is a place many have visited but where few are ever seen. It is a place vibrantly alive but where death is close by—in memory, in names etched on rock, in graves. Tinajas Altas both lures and repels. In spring the flaming chuparosa and the pale desert lavender are aglow with color and abuzz with bees. After a summer storm, life-giving water runs in the arroyo, sometimes to a depth of several feet, but then quickly hides beneath the sands. Summer nights are star-filled and refreshingly cool, while dawn brings calm light and whispers of immortality, but by mid-afternoon the canyon may become too hot for us to walk from the arroyo up to the second tank. Our faces turn red from the slightest exertion and we hide in the shade of venerable ironwood trees lining the wash, as have countless visitors before us. The rocks hold heat like a smelting oven, and when not a whisper of breeze cools the land, the fever paralyzes everyone, even the desert bighorn and vultures. In the winter, the air is invigorating, while slanting light glances off the stark white granite of the canyon, and shadows play in thousands of rock niches created over the millennia by water and wind.

Tinajas Altas is archaeologically rich—it speaks of millennia of human use—but many of its artifacts have been taken to distant shelves, burned in campfires, or trampled under foot and tire. Yet its history lives in countless stories told by the travelers, scientists, soldiers, miners, and cowboys who have watered and rested in the canyon's confines. Virtually anyone who ever traveled through this region went to Tinajas Altas. Personal diaries, letters, and stories remain to be collected or shared. Where did the people go? What did they seek? Who were they? Many riddles linger unanswered.

And a book by Native Americans is yet to be written. As Goldwater Range archaeologist Adrianne Rankin puts it, "The most important piece of future work should be a proposed volume on Tinajas Altas written by Native American informants. . . . Finally, the end result of future work should be a National Register nomination for Tinajas Altas as a Traditional Cultural Property" (Hartmann and Thurtle 2000:xxxv).

Tinajas Altas is a place of great value. Scientifically, it is a key to understanding the lifeways of early people, the climates of millennia past, and the birth of the Sonoran Desert itself. Ecologically, it is a hub of life in this region. Thick vegetation along the washes, in some seasons colorful and lush, meets the stark creosotebush flats. The steep granitic mountain slopes descend dramatically to the Lechuguilla Valley on the east and the Gran Desierto on the west. It is a place that provides opportunities both for recreation and reflection: it is a place to hike, photograph, study, sketch, observe, and relax. Its grand and pure scenic value is national park quality. Educationally, it may teach generations the value and mysteries of life and rock and sky. It stands as a vivid reminder of how precious water is.

Even though Tinajas Altas has been compromised by a few inconsiderate visitors, it retains enormous significance. It is a signature waterhole on a heritage trail that covers millennia of Southwest history. It features and focuses wildlife and people alike. It lends spiritual and personal power to more than a dozen Native American groups, as well as being a source of inspiration and solitude for hosts of "desert rats." It sparks our courage and our imagination. It speaks of an Old West not far past, where human ambition and endurance were pitted against the full intensity of the Sonoran Desert. Tinajas Altas is a national treasure, deserving our fullest allegiance.

Perhaps Randall Henderson (1940:34) spoke for all of us when he said of Tinajas Altas, "It is the kind of place that makes poets want to write poetry, and the others wish they could."

The desert kit fox still listens, ever alert. (Photo by Robert S. Henry)

When You Go

Tinajas Altas is located within the Barry M. Goldwater Range–West, an active military training facility under the direct control of the US Marine Corps at Yuma. Visitation is allowed under restricted conditions, and permits *must* be obtained from the USMC Range Management Office. Call 928-343-4393 for details.

Tinajas Altas is a sacred place for many reverent people, and certainly holds sacred as well as historical interest. Each stick, stone, and scrap of pottery at Tinajas Altas is protected by federal law, desert etiquette, and spirits past. At least four agencies—the US Marine Corps, Bureau of Land Management, US Customs and Border Protection (Border Patrol), and Arizona Game and Fish Department—monitor and defend this site. A cadre of citizen volunteers also watches and studies here. If you see suspicious activity or find damage, immediately notify the Range Management Office or the nearest officer.

Because of its remoteness, wildness, and distance from conveniences, visitors must contact the Range Management Office for the latest conditions and permissions. You may confront flashfloods, thirst, bandits, breakdowns, venomous animals, precipitous slopes, and any number of seen or unforeseen dangers. This is not a stroll in your own backyard. All-wheel-drive vehicles with high ground clearance are preferred, and should be self-contained with gas, water, and food. There are NO services once you leave Interstate 8. The roads are dusty, sandy, narrow, brushy, rough, and at times muddy. You may be stranded for days.

Visitors commonly arrive by vehicle or, occasionally, on foot. Some enjoy a day-trip from Yuma or Wellton. More adventurous people may elect to travel the length of El Camino del Diablo from Ajo, and they usually allot two to four days for this lifetime adventure.

Backpackers and hikers can plot any number of routes cross-country from Yuma, Wellton-Tacna, or even Ajo. Because they are limited in the quantity of water they can carry, virtually all hike in the cool seasons and then only after careful planning. Some rendezvous periodically with another person in a vehicle to replenish water and food. Walking across the desert, even for a few miles, has its own memorable rewards.

If planning to travel through the Cabeza Prieta National Wildlife Refuge, you must contact their office at 520-387-6483. They too require a permit and can alert you to the latest conditions as well as suggesting ways to enjoy your visit safely.

History has proven time after time that you should NEVER rely on any desert waterhole for your survival.

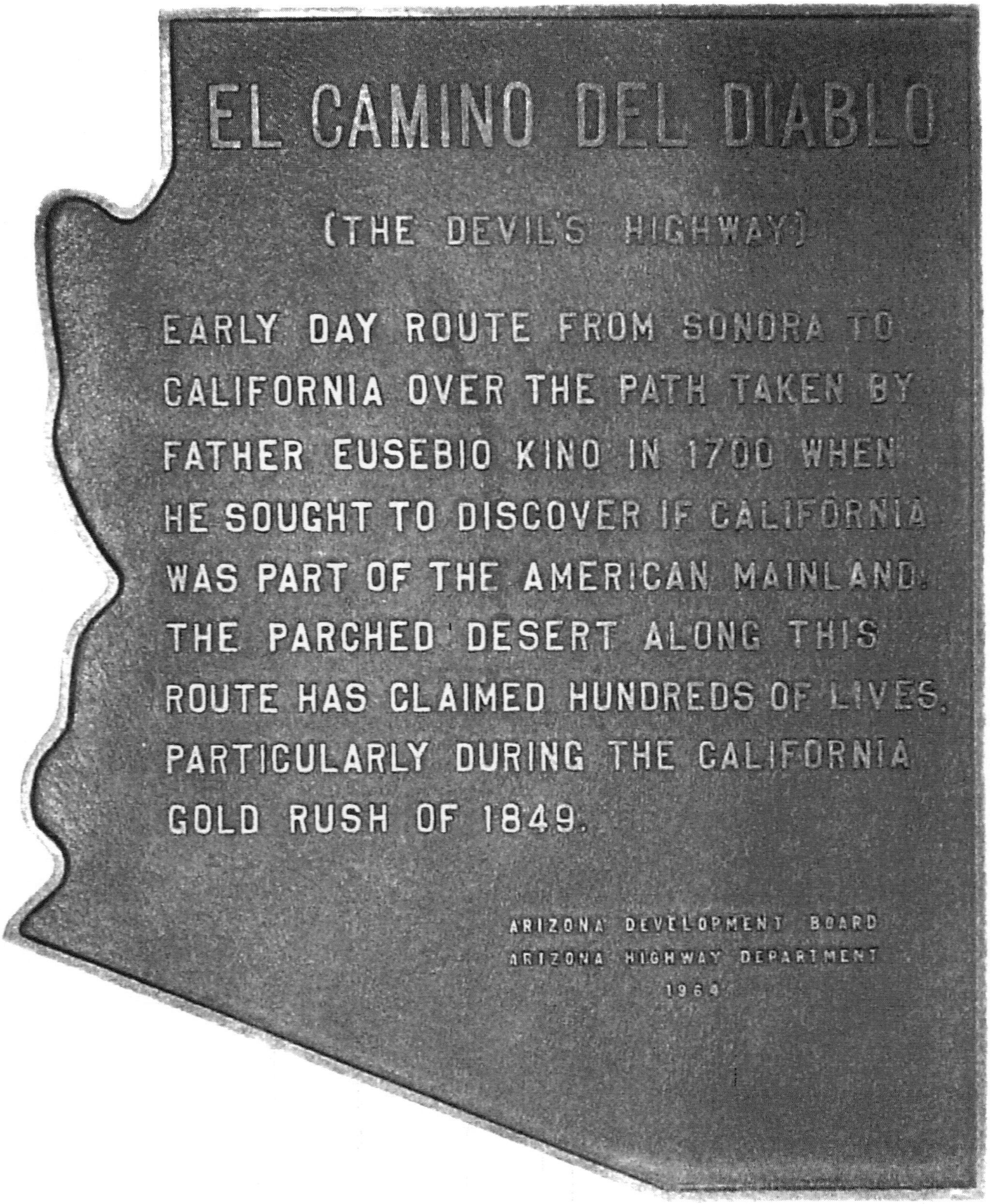

El Camino del Diablo commemorative plaque, Fortuna, Arizona. (Photo by Bill Broyles)

Acknowledgments

Much of this book grew from Gayle Harrison Hartmann and Mary Charlotte Thurtle's report on a historical and archaeological project about Tinajas Altas and the surrounding area. It was titled *The Only Water for 100 Miles*, vols. 1 and 2 (Phoenix: Arcadis: Geraghty and Miller, and Tucson: SWCA Environmental Consultants, July 2000), Prepared for 56th Range Management Office, Luke Air Force Base, Arizona. Project number AZ000686.00. We are very grateful to the United States Air Force and United States Marine Corps, especially Ron Pearce (USMC), Adrianne G. Rankin (USAF), and W. Bruce Masse (USAF).

Gayle Harrison Hartmann and Mary Charlotte Thurtle were part of an archaeological crew whose work they enthusiastically applaud. The crew included Lee Fratt, Chris Hogan, Jane Kolber, Troy Marino, Scott Seibel, Mark Slaughter, Rick Ward, and Greg Whitney. In addition, volunteers Kyle Bajakian, Beth Humstone, Kevin Price, Courtney Russell, Claudine Scoville, and Peter H. Smith supplied information and encouragement. Special thanks go to Tom Euler and Melissa Keane(SWCA), Ken Houser (ARCADIS), Sgt. William Adams (USMC), and Colonel R. Child (US Border Patrol).

In the historical portion of the SWCA report (vol.1), Gary Paul Nabhan, assisted by Jacquie Gage Kahn, interviewed living Native Americans and researched their uses of plants and animals; Thomas E. Sheridan researched and wrote the section covering the Spanish Entrada and Mexico from 1540-1854; and, Bill Broyles delved into the more recent history and maps. In the archaeological portion (vol. 2), although Hartmann and Thurtle were the editors and principal authors, numerous individuals contributed various sections of text. They are Richard V. N. Ahlstrom, Thomas A. Biggs, Karen A. Demsey, Lee Fratt, Linda M. Gregonis, S. Jerome Hesse, Richard G. Holloway, Mark Klett, Jane Kolber, Annick Lascaux, Austin Lenhart, Meg McDonald, M. Steven Shackley, Mark C. Slaughter, Jennifer G. Strand, and Arthur W. Vokes.

The current volume is the result of blending all this material plus additional contributions. Many willing and capable hands aided our effort, including Ronald J. Beckwith (maps), Luke Evans (mileage chart, mapwork, and field information), William K. Hartmann (information and use of his paintings), Wilton "Bill" Hoy (historical information), Austin Lenhart (photos and information), Robert "Bob" Lenon (map and history information), and Joan Scott (photo preparation). US Bureau of Reclamation biologist Dennis J. Kubly helped prepare part of the "Natural History" chapter, and sections of it were thoughtfully reviewed by John Alcock (Arizona State University), Richard S. Felger, and William J. Matter (University of Arizona), who each offered helpful suggestions for its improvement. Geologist James T. Gutmann helped us better understand geological processes. US Geological Survey biologist Jayne Belnap helped us with information on the desert soil crust. Appendices were graciously provided by Arizona Game and Fish Department biologist Robert S. Henry, wildlife biologist and naturalist David J. Griffin, University of Arizona research scientist of renewable natural resources Philip C. Rosen, and botanist Richard S. Felger (Herbarium, University of Arizona Botany Department, and San Diego Natural History Museum) and deep ecologist Thomas R. Van Devender of Sky Island Alliance. At the Arizona State Museum in Tucson, Rebecca Donnelly, former Curator of the Photographic Collections, Mike Jacobs, Curator of Archaeological Collections, and Curator of the Photographic Collections Jannelle Weakly were exceptionally helpful. Julie Hoff, map librarian at the Arizona State Archives in Phoenix, was an enormous help on old maps.

We are proud to present color photos by Willard Clay, George Steinmetz, and Jack Dykinga, and paintings by William K. Hartmann. Others, such as Neil Carmony and Erik Enderson, generously shared photos, and are noted in the captions. We have enjoyed many discussions with history buffs, such as John Annerino, Oscar and Fillman Bell, Charles Bowden, Carol Brooks, Neil Carmony, Charles Conner, Roger DiRosa, Lorraine Marquez Eiler, Patricia Etter, Paul Ezell, Don Fedock, Richard Felger, Bernard L. "Bunny" Fontana, Julian Hayden, Mark Haynes, Patrick Haynes, Floyd Herbert, Mike Holt, Steve Holt, Freeman Hover, Ronald Ives, Richard Laugharn, Bob Lenon, Francisco "Paco" Manzo Taylor, Joe McCraw, Al McGinnis, Ami Pate, Adrianne Rankin, David Rankin, David Roberson, Mark Santiago, Dennis Segura, Norman and Hilah Simmons, Dana and Betty Smith, Christine Szuter, David Taylor, Jeanne Williams, and Lyle Williams. We are grateful to all.

We thank the entire staff at the University of Arizona Press including Kathryn Conrad, Allyson Carter, Al Schroder, Lisa Stallings, Holly Schaffer, and Scott De Herrera.

Linda Gregonis ably handled our copyedits and indexing. Bill Benoit at Simpson and Convent designed our book. We are grateful to them all.

Gary Paul Nabhan and Jacquie Gage Kahn extend special thanks to the Hia C'ed O'odham Alliance, Lorraine Marquez Eiler, Joe Joaquin, Frank Jim, and Lorey Cachora.

Thomas E. Sheridan, Gary Paul Nabhan, and Bill Broyles are affiliated with the University of Arizona's Southwest Center and gratefully acknowledge Dr. Joseph C. Wilder's and the university's support and encouragement. Joe is a living monument to Southwest research and books.

And as you find "new" history, please share it with your local historical society. As tomorrow becomes today, and today becomes the past, we need to remember and record our history, for only then can we understand who we are, where we've been, and where we're going.

Appendix 1

Some Native American Names and Uses of Plants, Mammals, and Reptiles of the Tinajas Altas Region

Gary Paul Nabhan and Jacquie Gage Kahn

Native American names for living things are very interesting to many people, but also a bit confusing, due to unusual diacritical marks and variant spellings. For example, where Tohono O'odham may use a 'w' (*wispismal*, the word for chuparosa), the Hia C'ed O'odham use a 'v' (*vispismal*). For more information see Gary Paul Nabhan, Wendy Hodgson, and Frances Fellows's "A Meager Living on Lava and Sand? Hia C'ed O'odham Food Resources and Habitat Diversity in Oral and Documentary Histories" (1989); Amadeo Rea's *At the Desert's Green Edge* (1997) and *Folk Mammalogy of the Northern Pimas* (1998); Richard S. Felger's "Living Resources at the Center of the Sonoran Desert: Regional Uses of Plants and Animals by Native Americans (2007) and Felger et al. "Botanical Diversity of Southwestern Arizona and Northwestern Sonora" (2007); and Donald F. Hoffmeister's *Mammals of Arizona* (1986).

Compare to Henry (this volume, appendix 2), Rosen (this volume, appendix 4), and Felger and Van Devender (this volume, appendix 5). Alternate O'odham spellings from Amadeo Rea's 1997 volume are shown as (AR) and those from Richard Felger 2007 as (RF). Native names and uses are indicated by the following: (H) = Hia C'ed O'odham, (T) = Tohono O'odham, (Q) = Quechan, and (C) = Cocopah.

Tohono O'odham woman harvesting saguaro fruit. (Photo by Edward S. Curtis. Courtesy the Library of Congress, image no. LC-USZ62-111942)

Family name Scientific name (Common English name)	Hia C 'ed O'odham (H) and Tohono O'odham (T) Name	Cocopah Name (C), or Quechan Name (Q)	Uses/Significance
	PLANTS		
Acanthaceae (Acanthus family)			
Justicia californica (chuparosa; desert hummingbird-bush)	vispisimal, vispismal jeej (AR)		
Agavaceae See **Asparagaceae**			
Amaranthaceae (Amaranth family) includes			
Chenopodiaceae (Goosefoot family)			
Atriplex polycarpa (desert saltbush)	'oñik 'i:vaki	P?ay (C)	Ash; seeds eaten
Chenopodium murale (netleaf goosefoot)	'oñik 'i:vaki		Stalks, greens, seeds eaten
Asparagaceae (Asparagus family), includes **Agavaceae** (Agave family); **Nolinaceae** (Beargrass family)			
Agave deserti subsp. *simplex* (desert agave)	a'ud, 'a'ud (RF)	m?aly	Leaf bases and hearts roasted as food; leaf fiber used for sandals and bags; stalks and thorns used as utensils; medicine for infected wounds (T;H)
Nolina bigelovii (desert tree-beargrass)	moho		Fiber for baskets; seed and flowers eaten
Asteraceae (Composite family)			
Ambrosia ambrosioides (canyon ragweed)	nunui je:j, ñuñui jeej (AR)		Medicine for arthritis (H)
Ambrosia [*Hymenoclea*] *salsola* var. *pentalepis* (white burrobush, cheese bush)	ivatod, iivdhat (AR), 'i:vadhod (RF)		Indicator plant
Bebbia juncea var. *aspera* (sweetbush, chuckwalla delight)	hauke'es, hauk 'u 'us (RF)		
Encelia farinosa var. *farinosa* (brittlebush)	Tohaves		Gum chewed, sap used as adhesive indicator plant (H,T)
Peucephyllum schottii (desert fir, pygmy-cedar)	pihul		
Porophyllum gracile (*hierba del venado*)	do'i-uv u'us (AR)		Possible tea

Family name Scientific name (Common English name)	Hia C 'ed O'odham (H) and Tohono O'odham (T) Name	Cocopah Name (C), or Quechan Name (Q)	Uses/Significance
Brassicaceae (Mustard family)			
Descurainia pinnata (tansy mustard)	su'uvad	kasen	Seeds eaten, eye and stomach medicine
Lepidium lasiocarpum (sand peppergrass)	ka:kovani, avhad (AR)		Medicine
Burseraceae (Torchwood family)			
Bursera microphylla (elephant tree)	'usabkam		Medicine
Cactaceae (Cactus family)			
Carnegiea gigantea (saguaro)	bahidaj (fruit), ha:san (plant)	?a?a	Fruit eaten, made into wine (T, H); seeds eaten; stems used as wood
Cylindropuntia acanthocarpa var. *coloradensis* (buckhorn cholla)	ciolim, hannam (AR)	xpa	Food
Cylindropuntia bigelovii (teddybear cholla)	hadsadkam	cac qw?i s i	
Cylindropuntia ramosissima (diamond cholla)	vipinoi; vihoi	atot (Q)	Food
Echinocerus engelmannii var. *chrysocentrus* (strawberry hedgehog cactus)	i:svig, iisviki (AR), 'isvigi (RF)		Food
Ferocactus cylindraceus [acanthodes] (desert barrel cactus; mountain barrel cactus)	jiavuli, chiavul (AR), jiavul (RF)	mo ycac	Fruit eaten; stems boiled down for candy
Mammillaria grahamii (Arizona fishhook cactus)	ban:ban ha-'i:swig; ban mauppa; ban cekida		Fruit eaten
Opuntia basilaris var. *basilaris* (beavertail cactus)	gisoki	xpa	Food and pottery paint
Peniocereus greggii var. *transmontanus* (night-blooming cereus)	ho'ok va'o		Food
Chenopodiaceae see **Amaranthaceae**			
Cucurbitaceae (Gourd family)			
Cucurbita palmata (coyote gourd; coyote melon)	adavi	xa mas	Medicine
Ephedraceae (Joint-Fir family)			
Ephedra aspera (Boundary ephedra; Mormon tea)	ko'okpalk, ku'upok	?i siw	Medicinal tea
Euphorbiaceae (Spurge family)			
Jatropha cuneata (limberbush)	va:s		Red dye, basketry, medicine for heart and sores

Family name Scientific name (Common English name)	Hia C 'ed O'odham (H) and Tohono O'odham (T) Name	Cocopah Name (C), or Quechan Name (Q)	Uses/Significance
Fabaceae (Leguminosae) (Legume family)			
Acacia greggii (catclaw acacia)	u:pad, 'u:pad		Food, wood
Olneya tesota (ironwood)	hoi'idkam	kwnya; ʔi wir; axpal(Q)	Seeds eaten; flowers used as medicinal tea; wood used for utensils and posts
Parkinsonia [*Cercidium*] *florida* (blue paloverde)	ko'okomadk; kalisp	susá	Food, wood
P. microphylla (foothill paloverde)	ke:k cehedagi, kuk cehedagi	susá	Flowers used in a drink; stems in construction, seeds eaten
Prosopis glandulosa (western honey mesquite)	kui, kui chepelk (AR)	ʔa [pods] (C); ʔanya y; eva (Q)	Pods eaten; gum used as medicine (sores, sore throat), hair dye, eaten; wood for fuel, utensils, construction; roots used for rope
Fouquieriaceae (Ocotillo family)			
Fouquieria splendens (ocotillo)	melhog, melok (AR)	i nyáy	Building material; drink; hardened nectar eaten as candy
Krameriaceae (Ratany family)			
Krameria grayi (white ratany)	'edho, e:d, eedho (AR), edho, he:d (RF)		Roots pounded for red dye for basket fiber, tanning agent for deer hides; leaves used medicinally for sores and diabetes
Malvaceae (Mallow family)			
Sphaeralcea ambigua (desert globemallow)	ñiatum, hadam tatkam (AR)		No use
Nolinaceae. See **Asparagaceae**			
Nyctaginaceae (Four-O'clock family)			
Boerhavia spp. (spiderling)	makkom ha-jeved, makkum jeej (AR)		No use
Orobanchaceae (Broom-rape family)			
Orobanche cooperi [*ludoviciana*] (desert broomrape)	mo'oktak, mo'o stas, mo'otadk (AR), mo'otask (RF)		Roasted and eaten
Plantaginaceae (Plantain family)			
Plantago ovata [*insularis*] (woolly plantain; Indian wheat)	mumsa, muumsh (AR)		Food; medicine

Family name Scientific name (Common English name)	Hia C 'ed O'odham (H) and Tohono O'odham (T) Name	Cocopah Name (C), or Quechan Name (Q)	Uses/Significance
Chiroptera (Bats)			
Myotis sp. (bat)	nanakimel, nanakmel (RF)		
Macrotus sp. (bat)	nanakimel		
Felidae (cats)			
Puma [Felis] concolor (mountain lion)	mavid	nymca, yimca, nymca xanycku y (male)	Not eaten (C, Q)
Lynx [Felis] rufus (bobcat)	gevho	nymi	Not eaten (C, Q)
Heteromyidae (Kangaroo rats and pocket mice)			
Dipodomys spp. (kangaroo rat)	dahivo, dahiwua (AR)		
Perognathus sp. (pocket mouse)	Nahagia nahagui (AR)		
Leporidae (Rabbits and hares)			
Lepus californicus (black-tailed jackrabbit)	cuk cu:vi, chuk chuuvi (AR)	k?uy	Meat eaten; blankets
Sylvilagus audubonii (desert cottontail)	tobi	xly?a	Meat eaten; blankets
Muridae (Mice and rats)			
Neotoma albigula (white-throated woodrat, packrat)	kosoñ		Meat eaten
N. lepida (desert woodrat, packrat)	kosoñ	su ny	Meat eaten; bones pounded and eaten (Q, C)
Peromyscus spp. (cactus mouse, deer mouse)	nahagia		Meat eaten (H)
Mustelidae (Skunks and allies)			
Taxidea taxus (American badger)	kaavi, ka:vi	mxwa	Eaten if no other food available (H)
Sciuridae (Squirrels)			
Ammospermophilus harrisii (Harris' antelope ground squirrel)	cupi, chuchupi (AR)		Possible food
Spermophilus tereticaudus (round-tailed ground squirrel)	selik	cis	Food (H); possible food
Tayassuidae (Peccaries)			
Pecari tajacu (javelina, collared peccary)	miseiñ ko:ji, mischiñ kooji (AR)	kucyayak	Ceremonially hunted (H); meat eaten (H); bones cremated (H). Arrived at Tinajas Altas after 1934

Family name Scientific name (Common English name)	Hia C ‘ed O‘odham (H) and Tohono O‘odham (T) Name	Cocopah Name (C), or Quechan Name (Q)	Uses/Significance
	ANURA (Toads and Frogs)		
Pelobatidae (Spadefoot toads)			
Scaphiopus couchii (Couch's spadefoot)	olas babad		
Bufonidae (Toads)			
Bufo alavarius (Sonoran desert toad)	babad	xanyá	Poisons dogs (H)
	TESTUDINES (Turtles)		
Testudinidae (Tortoises)			
Gopherus agassizii (desert tortoise)	ce'o komakaced		Food (H); never food (Q)
	SAURIA (Lizards)		
Eublepharidae (Eyelidded geckos)			
Coleonyx variegatus (western banded gecko)		kwanyúk pkwyaw	Taboo (H)
Xantusiidae (Night lizards)			
Xantusia vigilis (desert night lizard)			Taboo (H)
Iguanidae (Iguanas and related lizards)			
Callisaurus draconoides (zebra-tailed lizard)	o'otkol		Sure can run fast (H)
Dipsosaurus dorsalis (desert iguana)	obmad		Food (H)
Phrynosoma mcallii (flat-tailed horned lizard)	cemamagi	cxnyas; cxyas	Taboo (H)
P. solare (regal horned lizard)	cemamagi		Taboo; need a kulan from a makai if it makes you sick if you've touched it (H); when touched, they hunch shoulders (H)
Sauromalus ater [formerly *obseus*] (common chuckwalla)	cenased	xanys?í y	Food (H, Q)
Uta stansburiana (side-blotched blizard)	hujed		Eats flies and bugs (H)
Teiidae (Whiptails and allies)			
Aspidocelis (*Cnemidophorus tigris*) (tiger whiptail)	wajelo		
Helodermatidae (Beaded lizards)			
Heloderma suspectum (Gila monster)	ciadag	xlkwaw	Makes you sick, biting causes bloated belly

Family name Scientific name (Common English name)	Hia C ‘ed O‘odham (H) and Tohono O‘odham (T) Name	Cocopah Name (C), or Quechan Name (Q)	Uses/Significance
	SERPENTES (Snakes)		
Colubridae (Northern colubrid snakes)			
Lampropeltis getula (common kingsnake)		xnypuk	All snakes taboo to Hia C‘ed O‘odham, believed all snakes at one time were humans
Phyllorhynchus spp. (leaf-nosed snake)		?awíyacác	Snakes not eaten by Quechan or Cocopah
Viperidae (Vipers)			
Crotalus spp. (rattlesnake)	koi		

Appendix 2
Mammals at Tinajas Altas

Robert S. Henry

Desert bighorn ram. (Photo by Bill Broyles)

The following list is based on Henry 2007 and Hoffmeister 1986.

Order Chiroptera—Bats

Phyllostomidae—Leaf-nosed Bats

Macrotus californicus. California leaf-nosed bat

Vespertilionidae—Plain-nosed Bats

Antrozous pallidus subsp. *pallidus.* Pallid bat.

Eptesicus fuscus subsp. *pallidus.* Big brown bat.

Myotis californicus subsp. *stephensi.* California myotis.

Parastrellus hesperus subsp. *hesperus.* Western pipestrelle, canyon bat.

Molossidae—Free-tailed Bats

Eumops perotis subsp. *californicus.* Western mastiff bat.

Nyctinomops femorosaccus. Pocketed free-tailed bat.

Nyctinomops macrotis. Big free-tailed bat. Expected but no records.

Tadarida brasiliensis subsp. *mexicana.* Brazilian free-tailed bat, Mexican free-tailed bat.

Order Carnivora—Carnivores

Canidae—Dogs and Allies

Canis latrans subsp. *mearnsi.* Coyote.

Urocyon cinereoargenteus subsp. *schottii.* Gray fox. Especially in rocky or brushy areas.

Vulpes velox subsp. *macrotis (= Vulpes macrotis).* Kit fox. In open habitats.

Felidae—Cats

Lynx rufus subsp. *baileyi.* Bobcat.

Puma concolor subspp. *azteca.* Mountain lion, cougar, puma.

Mustelidae—Skunks and Allies

Spilogale putorius subsp. *leucoparia.* Spotted skunk.

Taxidea taxus subsp. *berlandieri.* American badger.

Procyonidae—Raccoons and Allies

Bassariscus astutus subsp. *yumanensis.* Ringtail.

Order Artiodactyla—Even-toed Hoofed Mammals

Tayassuidae—Peccaries

Pecari tajacu subsp. *sonoriensis.* Collared peccary. In canyons and washes. Newcomer to area since mid-twentieth century.

Antilocapridae—Pronghorns

Antilocapra americana subsp. *sonoriensis.* Pronghorn. no recent records, but expected. Occurs in nearby Mohawk Valley and Tule Desert.

Bovidae—Sheep, Cattle and Allies

Ovis canadensis subsp. *mexicana.* Desert bighorn sheep. In mountains and on bajadas, occasionally in washes near mountains.

Order Rodentia—Rodents

Sciuridae—Squirrels

Ammospermophilus harrisii. Harris' antelope squirrel. In rocky or gravelly habitats.

Spermophilus tereticaudus subspp. *neglectus, tereticaudus.* Round-tailed ground squirrel. In valleys; subsp. *tereticaudus* only in Colorado and lower Gila river valleys.

Geomyidae—Pocket Gophers

Thomomys bottae subsp. *pusillus.* Botta's pocket gopher. In fine-grained soils; many variants.

Heteromyidae—Kangaroo Rats and Pocket Mice

Dipodomys deserti subsp. *deserti.* Desert kangaroo rat. In fine-grained valley soils.

Dipodomys merriami subsp. *Merriami.* Merriam's kangaroo rat. In fine-grained soils.

Chaetodipus baileyi subsp. *baileyi.* Bailey's pocket mouse.

Chaetodipus intermedius subsp. *phasma.* Rock pocket mouse. Rocky or gravelly habitats.

Chaetodipus penicillatus subspp. *pricei, penicillatus.* Desert pocket mouse. Valley floor.

Harris' antelope squirrel (Ammospermophilus harrisii). *(Photo by Bill Broyles)*

Perognathus amplus subsp. *amplus.* Arizona pocket mouse. Valley floor.

Muridae—Mice and Rats

Neotoma albigula subsp. *mearnsi.* White-throated woodrat, packrat.

Neotoma lepida subsp. *auripila.* Desert woodrat, packrat. Prefers rocky habitats.

Onychomys torridus subsp. *torridus.* Southern grasshopper mouse.

Peromyscus crinitus subsp. *disparilis.* Canyon mouse. In rocky habitats.

Peromyscus eremicus subsp. *eremicus.* Cactus mouse.

Order Lagomorpha—Rabbits and Hares

Leporidae—Rabbits and Hares

Lepus californicus subsp. *eremicus.* Black-tailed jackrabbit. In open habitats.

Sylvilagus audubonii subsp. *arizonae.* Desert cottontail. In brushy habitats.

Appendix 3
Birds at Tinajas Altas
David J. Griffin

Western screech-owl (Otus kennicottii). *(Photo by David J. Griffin)*

This appendix is based on Griffin 2007. The birds listed here are known, likely, or expected to occur at Tinajas Altas itself or within five miles of Tinajas Altas. This includes the open water of Tinajas Altas and Borrego Tank, the Tinajas Altas Mountains, Tinajas Altas Pass, the gravelly outwash plains and bajadas, numerous small, dry washes, the creosotebush-bursage flats, areas of desert pavement, and the xeroriparian area of Coyote Wash. The list is based on direct observations (by DJG and others, including Henry Detwiler, who graciously reviewed this list) and the author's experience and understanding of birds in the region.

Status and Abundance

Status and abundance codes were developed and modified from a number of sources including Monson and Phillips (1981), Unitt (1984), Groschupf et al. (1988), Andrews et al. (1992), Witzeman et al. (1997), Rappole (2000), Corman and Wise-Gervais (2005) and Griffin (2007), but most closely follow those of Russell and Monson (1998). As Russell and Monson (1998) noted, when using codes for the relative abundance of a species it is important to remember that bird numbers can vary within a species' range. Additionally, some species can be highly localized in distribution and/or restricted to habitats with limited extent. It is also important to remember that many species' numbers can fluctuate with changing seasons and from year-to-year, especially in desert communities. The following status and abundance codes are subjective and intended to be used to represent the results in a single day's effort by a practiced observer familiar with the area, in appropriate habitat. An asterisk (*) beside a species' name indicates that it is known to breed in the area. An exclamation point (!) beside a species' name indicates that it is an introduced (exotic) species. Nomenclature and organization follow that of the American Ornithologists' Union *Check-list of North American Birds* and supplements (American Ornithologists' Union 1998, 2000, 2010).

Status

Permanent resident. Birds typically present in area all year, usually breeding.
Summer resident. Birds typically present in area from late March through October, usually breeding.
Summer visitor. Birds that typically breed south of the area and migrate north in late summer or fall then return south in winter; or birds that remain in the area during summer, but do not breed.
Winter resident. Birds that remain in the area during winter, typically from October through March or later.
Winter visitor. Birds that occur temporarily in the area during winter.
Passage migrant. Birds that typically pass through the area in spring (late February to mid-June) and/or fall (late July through October); individuals may remain for one or more days, and rarely more than a week.

Abundance

Abundant. Invariably seen, often in conspicuous flocks (>100 birds/day)
Common. Frequently encountered, either as individuals or in pairs or small groups (25–100 birds/day)
Fairly common. A few individuals or pairs may be seen (10–25 birds/day)
Uncommon. Present in area (region, habitat), but may not be found in one or two days of observations (1–10 birds/day)
Rare. Present in area, but infrequently seen; several days may be required to find it (1–5 birds/day and 1–10 birds/season)
Casual. Out of usual range, but could be expected every year or two (5 or more records within the region)
Irregular. Within its range, but not present every year
Accidental. Far from usual range and not expected again (1–5 records within the region)

HABITATS

For each species, habitats (i.e., places and components that make up where the species forages, nests, rests, etc.) are listed in accordance to their relative importance. Major habitats are listed before the slash and minor habitats after the slash; habitats in italics are where breeding most often occurs. Habitats are designated by the following codes:

Dune. Sand dune (and/or sand sheets)
Creosote. Creosotebush–white bursage scrub (and/or Saltbush Scrub)
Riparian. Desert riparian/xeroriparian
Paloverde mixed. Paloverde-mixed cactus
Cholla. Cholla (Teddy Bear and/or Chainfruit "forests")
Cliffs. Rocks and cliffs
Pavement. Desert pavement
Open water. Open water (i.e., the tinajas at Tinajas Altas and Borrego Tank)

CHECKLIST OF THE BIRDS

Anseriformes

Anatidae—Ducks, Geese, Swans

Anas platyrhynchos. Mallard. Passage migrant (spring and fall)/winter visitor. Casual to irregular. Open water/creosote.
Anas cyanoptera. Cinnamon teal. Passage migrant (spring and fall). Casual to irregular. Open water/creosote.
Anas clypeata. Northern shoveler. Passage migrant (spring and fall). Casual to irregular. Open water/creosote.
Anas crecca. Green-winged teal. Passage migrant (spring and fall). Casual to irregular. Open water/creosote.

Galliformes

Phasianidae—Partridges, Grouse, Turkeys

Callipepla gambelii. Gambel's quail.* Permanent resident. Fairly common to abundant. Creosote/*riparian*/*paloverde mixed*/open water.

Pelecaniformes

Ardeidae—Herons, Bitterns

Ardea herodias. Great blue heron. Passage migrant (spring and fall)/summer and winter visitor. Rare to casual. Open water/riparian/creosote.
Ardea alba. Great egret. Passage migrant (spring and fall). Rare to casual. Open water/riparian/creosote.
Egretta thula. Snowy egret. Passage migrant (spring and fall). Rare to casual. Open water/riparian/creosote.
Bubulcus ibis. Cattle egret. Passage migrant (spring and fall)/winter visitor. Casual to accidental. Open water/riparian/creosote.

Red-tailed hawk (Buteo jamaicensis). *(Photo by Bill Broyles)*

Threskiornithidae—Ibises, Spoonbills

Plegadis chihi. White-faced ibis. Passage migrant (spring and fall). Irregular. Open water/creosote.

Accipitriformes

Carthartidae—New World Vultures

Cathartes aura. Turkey vulture.* Permanent resident/passage migrant (spring and fall). Uncommon. Riparian/*cliffs*/paloverde mixed/creosote.

Padionidae—Ospreys

Pandion haliaetus. Osprey. Passage migrant (spring and fall). Casual to irregular. Creosote/open water.

Accipitridae—Hawks, Kites, Eagles

Elanus leucurus. White-tailed kite. Winter/summer visitor. Accidental. Creosote/paloverde mixed.

Circus cyaneus. Northern harrier. Passage migrant (spring and fall)/winter visitor. Uncommon to casual. Creosote/dune.

Accipiter striatus. Sharp-shinned hawk. Passage migrant (spring and fall)/winter visitor. Uncommon to irregular. Creosote/riparian/paloverde mixed.

Accipiter cooperii. Cooper's hawk. Passage migrant (spring and fall)/winter visitor. Uncommon to casual. Creosote/riparian/paloverde mixed/open water.

Parabuteo unicinctus. Harris's hawk. Summer/winter visitor. Uncommon to irregular. Paloverde mixed/riparian.

Buteo swainsoni. Swainson's hawk. Passage migrant (spring and fall). Rare to irregular. Creosote/paloverde mixed/dune/riparian.

Buteo jamaicensis. Red-tailed hawk.* Permanent resident/winter visitor/passage migrant. Uncommon to fairly common. Paloverde mixed/ creosote/*riparian/cliffs*/open water (see Broyles 2006, for an enchanting tale of a desert wanderer's encounter with a red-tailed hawk at Tinajas Altas).

Buteo regalis. Ferruginous hawk. Passage migrant (spring and fall)/winter visitor. Rare to irregular. Creosote/paloverde mixed/riparian.

Aquila chrysaetos. Golden eagle.* Permanent resident/passage migrant (spring and fall). Uncommon to irregular. Creosote/ paloverde mixed/*cliffs*/open water.

Falconiformes

Falconidae—Falcons

Falco sparverius. American kestrel.* Permanent resident/winter visitor. Fairly common to rare. *Paloverde mixed/creosote/riparian/cliffs.*

Falco columbarius. Merlin. Passage migrant (spring and fall)/winter visitor. Rare to irregular. Creosote/paloverde mixed/riparian/open water.

Falco peregrinus. Peregrine falcon. Passage migrant (spring and fall). Rare to irregular. Open water/paloverde mixed/cliffs.

Falco mexicanus. Prairie falcon.* Permanent resident/winter visitor/passage migrant. Uncommon to rare. Creosote/ paloverde mixed/*cliffs*/open water.

Charadriiformes

Charadriidae—Lapwings, Plovers

Charadrius vociferus. Killdeer. Winter visitor/passage migrant (spring and fall). Uncommon to rare. Open water/creosote.

Scolopacidae—Sandpipers, Phalaropes

Tringa melanoleuca. Greater yellowlegs. Passage migrant (spring and fall). Rare. Creosote/open water.

Tringa flavipes. Lesser yellowlegs. Passage migrant (spring and fall). Rare. Creosote/open water.

Tringa solitaria. Solitary sandpiper. Passage migrant (spring and fall). Rare. Open water/creosote.

Family name Scientific name (Common English name)	Hia C 'ed O'odham (H) and Tohono O'odham (T) Name	Cocopah Name (C), or Quechan Name (Q)	Uses/Significance
Rhamnaceae (Buckthorn family)			
Zizyiphus obtusifolia (gray thorn; white crucillo)	'us jevedpad, a:gos, u:supad, u'us chevadbad (AR)		Fruits eaten; roots boiled for stomach aches
Santalaceae (Sandalwood family) includes **Viscaceae** (Mistletoe family			
Phoradendron californicum (desert mistletoe)	hakovad, to:ki, haakvod (AR), ha:kvad, toji (RF)	x yus (C)	Berries dried and stored
Solanaceae (Potato family)			
Lycium macrodon (wolfberry)	s-cuk kuavuli	Xcuc (C)	
Lycium parishii (Parish wolfberry)	kuavuli		Berries eaten
Nicotiana obtusifolia [*trigonophylla*] (desert tobacco; coyote tobacco)	ban ha-vivga, ban vivga (AR), o'odham ha-vivga (RF)	mela' 'v (Q)	Leaves smoked as tobacco
Viscaceae (Mistletoe family). See **Santalaceae**			
Typhaceae (Cattail family)			
T*ypha* sp. (cattail)	'uduvad, udvak (AR), 'uduvhag (RF)	xnya y; xanycplak (root); atpil (Q)	Used in baskets (H); rhizomes eaten; shoots eaten; pollen eaten
Zygophyllaceae (Caltrop family)			
Larrea divaricata subsp. *tridentata* (*L. tridentate*) (creosotebush)	segai, segoi, shegoi (AR),	tksi, xpsi	Leaves used medicinally in dozens of ways; ceremonially
	MAMMALS		
Antilocapridae (pronghorns)			
Antilocapra americana (pronghorn)	ku:vid, kuuvid (RF)	m?u	Meat eaten (H, T, Q); bones cremated ceremonially (H); not hunted (C). Historical range but in modern times
Bovidae (Sheep, cattle, and allies)			
Ovis canadensis (bighorn sheep)	cesoiñ, cheshoñ (AR)	mu yayák	Meat eaten; bones cremated ceremonially (H); horns taboo (H)
Canidae (Dogs and allies)			
Canis latrans (coyote)	ban	txpa; xtpa	
Vulpes macrotis (*Vulpes velox* subsp. *macrotis*) (kit fox)	kaso, cuavi	prxa w	Possibly eaten (T, H, Q)

Actitis macularia. Spotted sandpiper. Passage migrant (spring and fall). Rare. Open water/creosote.

Numenius americanus. Long-billed curlew. Passage migrant (spring and fall). Uncommon to rare. Creosote/open water.

Calidris mauri. Western sandpiper. Passage migrant (spring and fall). Uncommon to rare. Creosote/open water.

Calidris minutilla. Least sandpiper. Passage migrant (spring and fall). Uncommon to rare. Creosote/open water.

Limnodromus scolopaceus. Long-billed dowitcher. Passage migrant (spring and fall). Rare to casual. Open water.

Gallinago gallinago. Wilson's snipe. Passage migrant (spring and fall). Rare. Open water/creosote.

Columbiformes

Columbidae—Pigeons, Doves

Columba fasciata. Band-tailed pigeon. Passage migrant (spring and fall)/winter visitor. Irregular/accidental. Paloverde mixed/riparian.

Zenaida asiatica. White-winged dove.* Summer resident. Fairly common to abundant. *Paloverde mixed/riparian/* open water/creosote.

Zenaida macroura. Mourning dove.* Permanent resident/winter visitor. Fairly common to abundant. Creosote/*riparian/paloverde mixed/* open water.

Cuculiformes

Cuculidae—Cuckoos, Roadrunners, Anis

Geococcyx californianus. Greater roadrunner.* Permanent resident. Uncommon. Creosote/*riparian/paloverde mixed.*

Coccyzus americanus. Yellow-billed cuckoo. Passage migrant (spring). Rare. Riparian/paloverde mixed.

Strigiformes

Tytonidae—Barn owls

Tyto alba. Barn owl.* Permanent resident. Rare to irregular. Riparian/*paloverde mixed/cliffs.*

Strigidae—Typical owls

Otus kennicottii. Western screech-owl.* Permanent resident. Uncommon. *Riparian/paloverde mixed.*

Bubo virginianus. Great horned owl.* Permanent resident. Uncommon. *Riparian/paloverde mixed/cliffs.*

Micrathene whitneyi. Elf owl.* Summer resident/winter resident. Uncommon to fairly common/casual. *Riparian/paloverde mixed.*

Athene cunicularia. Burrowing owl. Winter resident. Rare to irregular. Creosote/paloverde mixed/dune/pavement.

Asio otus. Long-eared owl. Winter resident. Rare to irregular. Riparian/ paloverde mixed/creosote.

Asio flammeus. Short-eared owl. Winter resident. Rare to irregular. Creosote/dune/riparian.

Caprimulgiformes

Caprimulgidae—Nighthawks, Nightjars

Chordeiles acutipennis. Lesser nighthawk.* Summer resident/passage migrant (spring and fall). Uncommon to fairly common. *Creosote/pavement/* dune/riparian/*paloverde mixed/* open water.

Phalaenoptilus nuttallii. Common poorwill.* Summer resident/winter resident. Uncommon to fairly common. *Paloverde mixed/creosote/* riparian/*pavement* /open water.

Apodiformes

Apodidae—Swifts

Chaetura vauxi. Vaux's swift. Passage migrant (spring and fall). Uncommon. Creosote/riparian/paloverde mixed/open water.

Aeronautes saxatalis. White-throated swift.* Permanent resident/passage migrant (spring and fall). Uncommon to common. *Cliffs/* creosote/paloverde mixed/riparian/open water.

Trochilidae—Hummingbirds

Archilochus alexandri. Black-chinned hummingbird. Passage migrant (spring and fall). Uncommon. Riparian/paloverde mixed/open water.

Calypte anna. Anna's hummingbird. Summer/winter visitor. Casual to rare. Riparian/paloverde mixed/open water.

Calypte costae. Costa's hummingbird.* Permanent resident/winter resident. Uncommon to common. *Riparian/paloverde mixed/* creosote.

Stellula calliope. Calliope hummingbird. Passage migrant (spring). Rare. Riparian/paloverde mixed.

Selasphorus rufus. Rufous hummingbird. Passage migrant (spring and fall). Uncommon. Riparian/paloverde mixed.

Piciformes

Picidae—Woodpeckers

Melanerpes uropygialis. Gila woodpecker.* Permanent resident. Uncommon to common. *Riparian/paloverde mixed.*

Sphyrapicus nuchalis. Red-naped sapsucker. Passage migrant (spring). Rare. Riparian/paloverde mixed.

Picoides scalaris. Ladder-backed woodpecker.* Permanent resident. Uncommon to common. *Riparian/paloverde mixed/* creosote/cholla.

White-winged doves (Zenaida asiatica). *(Photo by Bill Broyles)*

Colaptes auratus. Northern flicker. Winter resident. Uncommon to common. Paloverde mixed/riparian/creosote.

Colaptes chrysoides. Gilded flicker.* Permanent resident. Uncommon to common. *Paloverde mixed/riparian/* creosote.

Passeriformes

Tyrannidae—Tyrant flycatchers

Contopus cooperi. Olive-sided flycatcher. Passage migrant (spring and fall). Rare to uncommon. Riparian/paloverde mixed.

Contopus sordidulus. Western wood-pewee. Passage migrant (spring and fall). Uncommon. Riparian/paloverde mixed/creosote.

Empidonax traillii. Willow flycatcher. Passage migrant (spring and fall). Rare to uncommon. Riparian/paloverde mixed/open water.

Empidonax minimus. Least flycatcher. Passage migrant (fall). Accidental. Riparian/paloverde mixed.

Empidonax hammondii. Hammond's flycatcher. Passage migrant (spring and fall). Uncommon. Riparian/paloverde mixed/creosote.

Empidonax wrightii. Gray flycatcher. Passage migrant (spring and fall)/winter resident. Uncommon/casual. Riparian/paloverde mixed/creosote.

Empidonax oberholseri. Dusky flycatcher. Passage migrant (spring and fall)/winter visitor. Uncommon/rare. Riparian/paloverde mixed.

Empidonax difficilis. Pacific-slope flycatcher. Passage migrant (spring and fall). Uncommon to fairly common. Riparian/paloverde mixed/creosote.

Sayornis nigricans. Black phoebe. Passage migrant/winter visitor (spring and fall). Rare to uncommon/rare. Open water/riparian.

Sayornis saya. Say's phoebe.* Permanent resident/winter resident. Uncommon to fairly common. Creosote/riparian/paloverde mixed/cliffs.

Myiarchus cinerascens. Ash-throated flycatcher.* Permanent resident/winter resident. Uncommon to fairly common/uncommon. *Riparian/paloverde mixed/* creosote.

Myiarchus tyrannulus. Brown-crested flycatcher. Passage migrant. Rare/uncommon. Riparian/paloverde mixed.

Tyrannus verticalis. Western kingbird.* Summer resident/passage migrant (spring and fall). Rare to irregular/uncommon. *Riparian/*open water/*paloverde mixed.*

Laniidae—Shrikes

Lanius ludovicianus. Loggerhead shrike.* Permanent resident/winter resident. Uncommon/fairly common. Creosote/*riparian/paloverde mixed/*cholla/pavement/dune.

Vireonidae—Vireos

Vireo bellii. Bell's vireo. Passage migrant (spring and fall). Rare to uncommon. Riparian/creosote/paloverde mixed/open water.

Vireo vicinior. Gray vireo. Passage migrant (spring and fall)/winter resident. Rare/uncommon. Riparian/ paloverde mixed.

Vireo plumbeus. Plumbeous vireo. Passage migrant (spring and fall). Uncommon. Riparian/paloverde mixed/creosote.

Vireo cassinii. Cassin's vireo. Passage migrant (spring and fall). Uncommon. Riparian/paloverde mixed/creosote.

Vireo gilvus. Warbling vireo. Passage migrant (spring and fall). Uncommon. Riparian/paloverde mixed.

Corvidae—Crows, Jays

Corvus corax. Common raven.* Permanent resident. Uncommon. Creosote/ riparian/*paloverde mixed/cliffs.*

Alaudidae—Larks

Eremophila alpestris. Horned lark.* Permanent resident/ winter resident. Uncommon to abundant. *Dune/creosote/pavement.*

Hirundinidae—Swallows

Tachycineta bicolor. Tree swallow. Passage migrant (spring and fall). Uncommon. Creosote/riparian/paloverde/ open water.

Tachycineta thalassina. Violet-green swallow. Passage migrant (spring and fall). Rare. Riparian/paloverde mixed.

Stelgidopteryx serripennis. Northern rough-winged swallow. Passage migrant (spring and fall). Uncommon to fairly common. Creosote/riparian/open water/paloverde mixed.

Riparia riparia. Bank swallow. Passage migrant (spring and fall). Rare. Open water/creosote/riparian.

Petrochelidon pyrrhonota. Cliff swallow. Passage migrant (spring and fall). Uncommon to common. Creosote/ riparian/open water.

Hirundo rustica. Barn swallow. Passage migrant (spring and fall). Uncommon to fairly common. Creosote/riparian/open water/paloverde mixed.

Remizidae—Verdins

Auriparus flaviceps. Verdin.* Permanent resident. Fairly common. Creosote/ *riparian/paloverde mixed/*cholla.

Troglodytidae—Wrens

Campylorhynchus brunneicapillus. Cactus wren.* Permanent resident. Fairly common to common. *Paloverde mixed/riparian/creosote/cholla.*

Salpinctes obsoletus. Rock wren.* Permanent resident. Common. *Cliffs/*creosote/*paloverde mixed.*

Catherpes mexicanus. Canyon wren.* Permanent resident. Uncommon to fairly common. *Cliffs/*riparian/paloverde mixed.

Thryomanes bewickii. Bewick's wren. Passage migrant (spring and fall)/winter resident. Uncommon/rare to uncommon. Riparian/paloverde mixed/creosote.

Troglodytes aedon. House wren. Passage migrant (spring and fall)/winter resident. Rare/rare. Riparian/paloverde mixed/creosote.

Polioptilidae—Gnatcatchers

Polioptila caerulea. Blue-gray gnatcatcher. Passage migrant (spring and fall)/winter resident. Rare/rare. Creosote/ riparian/paloverde mixed.

Polioptila melanura. Black-tailed gnatcatcher.* Permanent resident. Fairly common. Creosote/*riparian/paloverde mixed/*cholla.

Regulidae—Kinglets

Regulus calendula. Ruby-crowned kinglet. Winter resident/ passage migrant (spring and fall). Uncommon to fairly common/fairly common. Riparian/paloverde mixed/ creosote.

Turdidae—Thrushes

Sialia mexicana. Western bluebird. Winter visitor/passage migrant (spring). Rare/irregular to uncommon. Creosote/riparian/paloverde mixed.

Sialia currucoides. Mountain bluebird. Winter visitor/passage migrant (spring). Rare to uncommon. Creosote/ paloverde mixed/riparian.

Myadestes townsendi. Townsend's solitaire. Winter visitor/ passage migrant (spring and fall). Casual to uncommon. Paloverde mixed/riparian.

Catharus ustulatus. Swainson's thrush. Passage migrant (spring and fall). Rare to uncommon. Riparian/ paloverde mixed.

Catharus guttatus. Hermit thrush. Passage migrant (spring and fall)/winter visitor. Uncommon to fairly common/ rare. Riparian/creosote/paloverde mixed.

Turdus migratorius. American robin. Passage migrant/Winter visitor/winter resident. Uncommon to fairly common/rare to uncommon. Riparian/paloverde mixed.

Mimidae—Mockingbirds, Thrashers

Mimus polyglottos. Northern mockingbird.* Permanent resident/winter resident. Uncommon to fairly common/ fairly common. *Riparian/paloverde mixed/*creosote.

Oreoscoptes montanus. Sage thrasher. Winter visitor/winter resident. Uncommon/rare. Creosote/riparian/paloverde mixed.

Toxostoma bendirei. Bendire's thrasher.* Summer resident/passage migrant. Irregular/rare to uncommon. Riparian/*paloverde mixed/creosote/*cholla.

Toxostoma curvirostre. Curve-billed thrasher.* Permanent resident. Fairly common. *Paloverde mixed/riparian/cholla/*creosote.

Toxostoma crissale. Crissal thrasher.* Permanent resident. Rare to uncommon. *Riparian.*

Toxostoma lecontei. Le Conte's thrasher.* Permanent resident. Uncommon. *Creosote/dune/*riparian/pavement.

Sturnidae—Starlings

Sturnus vulgaris. ! European starling. Summer/winter visitor. Accidental to casual. Paloverde mixed/riparian.

Motacillidae—Wagtails, Pipits

Anthus rubescens. American pipit. Passage migrant (spring and fall). Rare to uncommon. Open water/creosote/riparian.

Anthus spragueii. Sprague's pipit. Passage migrant (spring and fall)/winter visitor. Accidental. Open water/riparian/creosote.

Bombycillidae—Waxwings

Bombycilla cedrorum. Cedar waxwing. Passage migrant (spring and fall)/winter visitor. Rare/irregular. Riparian/paloverde mixed.

Ptilogonatidae—Silky-flycatchers

Phainopepla nitens. Phainopepla.* Summer resident/winter resident. Uncommon/fairly common to common. *Riparian/paloverde mixed.*

Parulidae—Wood-warblers

Oreothlypis celata. Orange-crowned warbler. Passage migrant (spring and fall)/winter visitor. Uncommon to fairly common. Riparian/paloverde mixed/creosote.

Oreothlypis ruficapilla. Nashville warbler. Passage migrant (spring and fall). Uncommon. Riparian/paloverde mixed/creosote.

Oreothlypis virginiae. Virginia's warbler. Passage migrant (spring). Casual to rare. Open water/riparian/paloverde mixed.

Oreothlypis luciae. Lucy's warbler.* Summer resident. Rare to uncommon. *Riparian/paloverde mixed.*

Dendroica petechia. Yellow warbler. Passage migrant (spring and fall). Uncommon. Riparian/paloverde mixed/open water.

Dendroica coronata. Yellow-rumped warbler. Passage migrant (spring and fall). Uncommon to fairly common. Riparian/paloverde mixed/creosote/open water.

Dendroica nigrescens. Black-throated gray warbler. Passage migrant (spring and fall). Uncommon to fairly common. Riparian/paloverde mixed.

Dendroica townsendi. Townsend's warbler. Passage migrant (spring and fall). Uncommon to fairly common. Riparian/paloverde.

Dendroica occidentalis. Hermit warbler. Passage migrant (spring and fall). Uncommon. Riparian/paloverde mixed.

Oporornis tolmiei. MacGillivray's warbler. Passage migrant (spring and fall). Uncommon to fairly common. Riparian/paloverde mixed.

Geothlypis trichas. Common yellowthroat. Passage migrant (spring and fall). Uncommon. Open water/riparian/paloverde mixed.

Wilsonia pusilla. Wilson's warbler. Passage migrant (spring and fall). Fairly common to common. Riparian/open water/paloverde mixed/creosote.

Icteria virens. Yellow-breasted chat. Passage migrant (spring and fall). Rare. Riparian/open water.

Emberizidae—Emberizids

Pipilo chlorurus. Green-tailed towhee. Winter resident/winter visitor/passage migrant (spring and fall). Uncommon. Riparian/paloverde mixed/creosote.

Pipilo maculatus. Spotted towhee. Winter resident/passage migrant (spring and fall). Uncommon to fairly common. Riparian/paloverde mixed/creosote.

Spizella passerina. Chipping sparrow. Winter visitor/passage migrant (spring and fall). Uncommon. Riparian/paloverde mixed/creosote.

Spizella breweri. Brewer's sparrow. Winter resident. Fairly common to abundant. Creosote/dune/riparian.

Spizella atrogularis. Black-chinned sparrow. Winter resident. Rare to irregular. Riparian/paloverde mixed/cliffs/open water.

Pooecetes gramineus. Vesper sparrow. Winter resident. Uncommon to abundant. Creosote/dune/riparian/paloverde mixed.

Chondestes grammacus. Lark sparrow. Passage migrant (spring and fall). Rare to uncommon. Riparian/creosote/paloverde mixed.

Amphispiza bilineata. Black-throated sparrow.* Permanent resident. Fairly common to abundant. *Creosote/*dune/*riparian/paloverde mixed/*cholla/cliffs.

Amphispiza belli. Sage sparrow. Winter resident. Uncommon to abundant. Creosote/dune/riparian/pavement/cholla.

Calamospiza melanocorys. Lark bunting. Winter resident/winter visitor. Uncommon to abundant. Creosote/dune/riparian/pavement.

Passerculus sandwichensis. Savannah sparrow. Winter visitor/passage migrant. Rare to uncommon. Riparian/creosote/open water.

Melospiza lincolnii. Lincoln's sparrow. Passage migrant/winter visitor. Rare to uncommon. Riparian/open water.

Zonothrichia leucophrys. White-crowned sparrow. Winter resident. Fairly common to abundant. Creosote/riparian/paloverde mixed/cholla/dune/open water.

Junco hyemalis. Dark-eyed junco. Winter resident. Rare to fairly common. Creosote/riparian/paloverde mixed/cliffs.

Cardinalidae—Cardinals, Tanagers

Piranga rubra. Summer tanager. Passage migrant (spring). Casual. Riparian/creosote/paloverde mixed/open water.

Piranga ludoviciana. Western tanager. Passage migrant (spring and fall). Uncommon to fairly common. Riparian/paloverde mixed/creosote/open water.

Cardinalis sinuatus. Pyrrhuloxia. Permanent/summer resident. Rare. Riparian/paloverde mixed.

Pheucticus melanocephalus. Black-headed grosbeak. Passage migrant (spring and fall). Uncommon to fairly common. Riparian/paloverde mixed/creosote.

Guiraca caerulea. Blue grosbeak. Passage migrant (spring and fall). Rare. Riparian/open water/paloverde mixed.

Passerina amoena. Lazuli bunting. Passage migrant (spring and fall). Uncommon to fairly common. Riparian/creosote/paloverde mixed.

Icteridae—Blackbirds

Agelaius phoeniceus. Red-winged blackbird. Passage migrant (spring and fall). Rare to uncommon. Open water/riparian.

Sturnella neglecta. Western meadowlark. Winter resident/summer resident. Uncommon/irregular. Creosote/dune/open water.

Xanthocephalus xanthocephalus. Yellow-headed blackbird. Passage migrant (spring and fall). Rare to uncommon. Open water/riparian.

Euphagus cyanocephalus. Brewer's blackbird. Passage migrant (spring and fall). Rare. Open water/riparian/paloverde mixed.

Quiscalus mexicanus. Great-tailed grackle. Passage migrant/summer visitor. Rare. Open water/riparian.

Molothrus ater. Brown-headed cowbird.* Summer resident/winter resident. Uncommon to fairly common. *Riparian/paloverde mixed/creosote/cholla/* open water.

Icterus cucullatus. Hooded oriole. Passage migrant. Uncommon. Riparian/open water/paloverde mixed.

Icterus bullockii. Bullock's oriole. Passage migrant (spring and fall). Uncommon. Riparian/open water/paloverde mixed.

Icterus parisorum. Scott's oriole.* Summer resident/passage migrant (spring and fall). Uncommon to fairly common. Riparian/*paloverde mixed.*

Fringillidae—Finches

Carpodacus cassinii. Cassin's finch. Winter visitor. Irregular. Riparian/ paloverde mixed/creosote/open water.

Carpodacus mexicanus. House finch.* Permanent resident. Fairly common to abundant. *Creosote/riparian/paloverde mixed/cliffs/cholla/* open water.

Spinus pinus. Pine siskin. Passage migrant (spring and fall). Rare. Riparian/paloverde mixed/creosote.

Spinus psaltria. Lesser goldfinch. Winter resident/summer visitor. Uncommon to fairly common/rare. Riparian/paloverde mixed/open water/creosote.

Spinus lawrencei. Lawrence's goldfinch. Passage migrant/winter visitor. Rare. Riparian/paloverde mixed/creosote/open water.

Spinus tristis. American goldfinch. Passage migrant (spring)/winter visitor. Rare. Creosote/riparian/paloverde mixed/open water.

Appendix 4
Amphibians and Reptiles at Tinajas Altas

Philip C. Rosen

*Sonoran Desert Toad (*Bufo [Ollotis] alvarius*), formerly Colorado River Toad. (Photo courtesy Erik F. Enderson)*

These species are expected within a five-mile radius of Tinajas Altas (updated from Rosen 2007):

X certain
P occurrence probable or occasional
Q occurrence possible

Of the 39 species, 29 species are certain, 5 are probable, and 5 are possible. A pound sign (#) indicates that a species is venomous.

Class Amphibia—Amphibians

Order Anura—Frogs and Toads

Family Pelobatidae—Spadefoot Toads

Scaphiopus couchii. Couch's spadefoot. X

Family Bufonidae—Toads

Bufo (Ollotis) alvarius. Sonoran desert toad. Q
Bufo (Anayxrus) cognatus. Great plains toad. Q
Bufo (Anaxyrus) punctatus. Red-spotted toad. X

Class Reptilia—Reptiles

Order Testudines—Turtles

Family Testudinidae—Tortoises

Gopherus agassizii (morafkai). Sonoran Desert tortoise. Desert tortoise. X

Order Squamata—Lizards and Snakes

Suborder Sauria—Lizards

Family Helodermatidae—Beaded Lizards

Heloderma suspectum. Gila monster. X #

Family Eublepharidae—Eyelidded Geckos

Coleonyx variegatus. Western banded gecko. X

Family Xantusiidae—Night Lizards

Xantusia vigilis. Desert night lizard. X

Family Teiidae—Whiptails and Allies

Aspidoscelis (Cnemidophorus) tigris. Tiger whiptail. X

Family Iguanidae—Iguanas and Related Lizards

Subfamily Iguaninae—Iguanas

Dipsosaurus dorsalis. Desert iguana. X

Sauromalus ater. Common chuckwalla. X

Subfamily Crotaphytinae—Collared and Leopard Lizards

Crotaphytus nebrius. Sonoran collared lizard. X

Gambelia wislizenii. Long-nosed leopard lizard. X

Subfamily Phrynosomatinae—Other Iguanids

Callisaurus draconoides. Zebra-tailed lizard. X

Phrynosoma mcallii. Flat-tailed horned lizard. P

Phrynosoma platyrhinos (*goodei*). (Southern) desert horned lizard. X

Sceloporus magister. Desert spiny lizard. X

Uma notata (*rufopunctata*). Yuman fringe-toed lizard. P

Urosaurus graciosus. Long-tailed brush lizard. X

Uta stansburiana. Side-blotched lizard. X

Suborder Serpentes—Snakes

Family Leptotyphlopidae—Slender Blind Snakes

Leptotyphlops humilis. Western blind snake (threadsnake). X

Family Boidae—Boas

Lichanura trivirgata. Rosy boa. P

Family Colubridae—Northern Colubrid Snakes

Arizona elegans. Glossy snake. X

Chilomeniscus cinctus (*stramineus*). Banded (bandless) sand snake. Q

Chionactis occipitalis. Western shovel-nosed snake. X

Lampropeltis getula. Common kingsnake. Q

Masticophis (*Coluber*) *flagellum.* Coachwhip. X

Phyllorhynchus decurtatus. Spotted leaf-nosed snake. X

Pituophis catenifer. Gopher snake. X

Rhinocheilus lecontei. Long-nosed snake. Q

Salvadora hexalepis. Western patch-nosed snake. X

Sonora semiannulata. Western ground snake. Q

Trimorphodon biscutatus. Lyre snake. P

Family Xenodontidae—Neotropical Colubrids

Hypsiglena torquata (*chlorophaea*). Desert night snake. X

Family Elapidae—Cobras and Their Relatives

Micruroides euryxanthus. Sonoran coralsnake. Q #

Family Viperidae—Vipers

Crotalus atrox. Western diamondback rattlesnake. Q #

Crotalus cerastes. Sidewinder. X #

Crotalus mitchellii. Speckled rattlesnake. X #

Crotalus molossus. Black-tailed rattlesnake. P #

Crotalus scutulatus. Mohave rattlesnake. X #

Appendix 5
Vascular Plants of the Tinajas Altas Region

Richard S. Felger and Thomas R. Van Devender

Desert tree-beargrass (Nolina bigelovii). *Some specimens may stand more than 15 feet tall. (Photo by Charles Sheldon. Courtesy Neil B. Carmony)*

The Tinajas Altas region is generally considered extreme desert. During the usual conditions of drought you won't see much in the way of greenery and even the creosotebushes (*Larrea divaricata*) may be dried up, but go there after a generous rainy time and there is a richness of flowers and greenery and buzzing insects and birds and other creatures. For example, in late October 2004, following hurricane-fringe rains, there was abundant greenery and a full complement of summer-season annuals (ephemerals) as well as the first winter-spring annuals. The plant life molded into the spectacular scenery at Tinajas Altas fits easily into the concept of Forrest Shreve's (1951) Lower Colorado Valley subdivision of the Sonoran Desert. Although some nonbotanist authors claim that the Tinajas Altas Mountains support Arizona Upland vegetation, the entire region is part of the Lower Colorado Valley.

We present a listing of the present-day vascular plants from the Tinajas Altas region. The core flora area is the 80,000 acres of the Tinajas Altas region encompassing the Tinajas Altas Mountains from the Tinajas Altas Pass to the Mexico border at Frontera Canyon and the south end of the Lechuguilla Valley drained by La Jolla Wash. We also include some plants from immediately adjacent areas when we expect them to occur in the core area. Specific geographic information is given in the regional gazetteer by Broyles et al. (2007).

The total flora in this annotated checklist includes 226 species and 2 hybrids in 164 genera and 44 families. Although there has been extensive documentation of plants from this region, intrepid botanists should be able to discover some additional species. The most diverse families in the flora are the Asteraceae, Poaceae, Boraginaceae, Cactaceae, Brassicaceae, Solanaceae, Euphorbiaceae, Fabaceae, Polygonaceae, and Nyctaginaceae, and the most diverse genera are *Cryptantha, Ambrosia, Eriogonum, Boerhavia, Cylindropuntia, Euphorbia,* and *Lycium.* There are 12 non-native species in the flora, which is only 5 percent of the total flora. Across most of the Sonoran Desert region this extremely low percentage of non-natives is matched only in the Gran Desierto dune fields of northwestern Sonora (Felger 2000) and some Gulf of California islands (Felger et al. 2007; Felger and Wilder 2011; Wilder and Felger 2010). Sahara mustard (*Brassica tournefortii*) and Arabian grass (*Schismus barbatus* subsp. *arabicus*) are the only invasive species that are likely impacting native plants in the Tinajas Altas region, although buffelgrass (*Cenchrus ciliaris* [*Pennisetum ciliare*]) is a potential threat (e.g., Felger 2000; Felger et al. 2003; Franklin and Molina-Freamer 2010).

The vegetation and flora are dynamic, changing even now, as they have changed dramatically during the past, along with shifting climate and human presences. Biologists have been collecting herbarium specimens from Tinajas Altas since Edgar Mearns' collections in 1894 and packrats (*Neotoma* spp.) have inadvertently collected and curated Tinajas Altas plants for more than 43,000 years. Identifications, dating, and analysis of these fossil specimens recovered from ancient packrat middens are discussed by Betancourt et al. (1990) and Van Devender (1990, 2007).

The floristic listing presented here is based largely on the specimen-based works of Felger, Rutman et al. (2007) and Felger and Van Devender (in press); see Felger 1980 and 2000, and Felger et al. 2003). Authors of plant names are in those references and on-line in many places. In the following list the plants are grouped as ferns and lycophytes, cone-bearing plants (gymnosperms), and flowering plants (angiosperms). Within these categories the plants are listed alphabetically by family, genus, and species. Designations of plant families follow the APG III format (Stevens 2011). Non-native plants are indicated with an asterisk (*). Our opinions of the accepted scientific names are in **bold**. Selected synonyms of scientific names are in *italics* within brackets [ ]. English-language common names represent our best attempt at reality. For plants lacking a widely accepted common name, none is given. Hia C'ed O'odham and Spanish-language names, when available, can be found in Felger, Rutman, et al. (2007). Flower color, given for most of the plants listed here, refers to the dominant color and not necessarily to a specific flower part or structure. Brief mention of ethnobotanical uses is provided for some of the more culturally important plants—more extensive local and regional information can be found in Felger (2007), which includes documentation and references. Extensive listings of the regional flora can be found at Southwest Environmental Information Network (2011), and other herbarium databases and regional floristic references.

Ferns and Lycophytes

Pteridaceae—Brake Fern Family

Astrolepis cochisensis subsp. ***cochisensis*** [*Notholaena cochisensis*] Small star-scaled cloak fern. Small, tufted ferns; leaves slender and once-pinnate. Among rocks at higher elevations.

Cheilanthes parryi [*Notholaena parryi*] Parry's lip fern. Small tufted ferns; leaves twice pinnate and gray-haired. Among rocks in mountains. This species extends farther into dry places than any other Sonoran Desert fern.

Notholaena californica subsp. ***californica*** [*Cheilanthes deserti*] Indian fern, California cloak fern. Small tufted ferns with beadlike leaf divisions. Sheltered among rocks in the mountains.

Desert sumac (Rhus kearneyi). *(Photo by Bill Broyles)*

Selaginellaceae—Spike-Moss Family

Selaginella eremophila. Desert spikemoss. Matlike plants with scale leaves. Rocky, mostly north- and northeast-facing slopes, especially at higher elevations. Scarce near waterholes, probably due to bighorn grazing.

Cone-Bearing Plants—Gymnosperms

Ephedraceae—Joint-Fir Family

Ephedra aspera. Boundary ephedra, Mormon tea. Shrubs with green twigs and brown scale leaves. Mostly on rocky slopes. Ephedra is a significant medicinal plant.

Flowering Plants—Angiosperms

Acanthaceae—Acanthus Family

Justicia californica [*Beloperone californica*] Desert hummingbird-bush. Shrubs with tubular, red-orange flowers attracting hoards of hummingbirds. A favorite bighorn food plant. Mostly along washes and canyons, less often on lower rocky slopes.

Agavaceae, see Asparagaceae

Amaranthaceae—Amaranth Family (includes Chenopodiaceae)

Amaranthus crassipes var. ***crassipes.*** Spreading amaranth. Summer annuals. Inflorescence branches thick and corky. Localized along Coyote Wash; discovered in October 2004, following exceptional rains.

Amaranthus fimbriatus. Fringed amaranth. Summer annuals. Washes, bajadas, and canyon bottoms. The seeds are edible.

Atriplex polycarpa. Desert saltbush. Shrubs with scurfy-grayish foliage. Widespread; washes, sandy valley plains, bajadas, pediments, and lower slopes.

****Chenopodium murale.*** Netleaf goosefoot. Winter-spring annuals. Locally scarce but well established on the Mesa del Muerto. Young herbage and seeds are edible. Reported as native to the Old World.

Tidestromia lanuginosa subsp. ***eliassoniana.*** Honeysweet. Summer annuals, the herbage generally scurfy-whitish; flowers minute, yellow. Often seasonally abundant; washes, bajadas, canyons, and less common on rocky slopes.

Anacardiaceae—Cashew Family

Rhus kearneyi subsp. ***kearneyi.*** Desert sumac. Hardwood shrubs; evergreen except in extreme drought. Flowers white to pink. North-facing slopes, canyons, and cliff bases to near peak elevations. The type collection is from Tinajas Altas. A refreshing, tart drink can be made from the fruits.

Apocynaceae—Dogbane Family (includes Asclepiadaceae)

Asclepias albicans. White-stem milkweed. Shrubs with tall reedlike stems. Flowers cream-white. Exposed rock slopes and occasional on the valley-plain. Fruits edible when roasted in coals.

Canyon ragweed (Ambrosia ambrosioides). *(Photo by Joan E. Scott)*

Asclepias erosa. Giant sand-milkweed. Robust perennials, the leaves large and thick. Flowers greenish white. Often growing and flowering in late spring and early summer when nothing else seems alive. Sandy soils of Lechuguilla Valley.

Asclepias subulata. Desert milkweed, reedstem milkweed. Perennials with many reedlike stems. Flowers waxy, cream- and yellow-white. Occasional in the Lechuguilla Valley.

Funastrum hartwegii [*Sarcostemma cynanchoides* subsp. *hartwegii*] Climbing milkweed. Perennial vines, sometimes growing on saguaros. Flowers maroon and white. Washes and canyon bottoms. The fresh flowers can be eaten as snacks and chewing gum was made from the milky sap.

Asparagaceae—Asparagus Family (includes Agavaceae, Nolinaceae, and Themidaceae)

Agave deserti subsp. ***simplex.*** Desert agave. Medium-sized agaves. Flowers yellow, April to June. Mountains and occasional on upper bajadas. A significant food and utilitarian plant; uses included the "hearts" pit-roasted for food and leaves processed for fiber.

Hesperocallis undulata. Ajo lily, desert-lily. Growing and flowering during cool seasons from a single, rather large bulb. Flowers white and showy. Sandy soils of Lechuguilla Valley and sometimes on rocky pediments. The bulbs are edible but slimy.

Nolina bigelovii. Desert tree-beargrass, Bigelow beargrass. Yuccalike plants with a woody trunk and tall flowering stalks. Flowers small, white and green, in June. Canyons and slopes including cliffs and rock faces, from lowest to high elevations. Rare giants have trunks three to four meters tall. Simmons (1966) reports that bighorn eat the leaves.

Triteleiopsis palmeri [*Brodiaea palmeri*] Blue sand-lily. Cooler seasons from a cluster of cormlets (bulblets). Flowers deep blue, on tall flowering stalks. Sand flats and dunes west of Tinajas Altas Mountains; widespread in nearby Gran Desierto in Sonora.

Asteraceae (Compositae)—Aster or Daisy Family

Ambrosia ambrosioides [*Franseria ambrosioides*] Canyon ragweed. Shrubs with slender stems and large triangular leaves. Seeds in hook-spined burs. Growing with warm weather, the herbage frost-sensitive. Washes and canyon bottoms. An important medicinal plant.

Ambrosia deltoidea [*Franseria deltoidea*] Triangleleaf bursage. Small shrubs, summer dormant. Seeds in small burs. Slopes, canyon bottoms, valley floor rivulets, and wash banks. Widespread but not extending into such extremely xeric niches as does *A. dumosa.*

Ambrosia dumosa [*Franseria dumosa*] White bursage. Small shrubs, the herbage often whitish. Summer dormant; seeds in small burs. Valley plains, bajadas, and rocky slopes to higher elevations, and extending into harsh, dry sites. One of the most common perennials in the region.

Ambrosia dumosa × ***A. ilicifolia.*** One plant of this presumed hybrid found among plants of the parent species on a steep slope north of the tinajas.

Ambrosia ilicifolia [*Franseria ilicifolia*] Hollyleaf burrobush. Broad shrubs with hollylike leaves. Summer dormant. Seeds in burs. Washes, canyon bottoms, and sometimes on rocky slopes. The persistent, dry, dead leaves rustling in the wind may sound like a rattlesnake.

Ambrosia salsola var. ***pentalepis*** [*Hymenoclea salsola* var. *pentalepis*] White burrobush, cheese bush. Globose shrubs. Flowers straw-colored, the seeds in small, winged burs. Flowering March and April. Lechuguilla Valley and mountain canyons.

Baccharis brachyphylla. Shortleaf baccharis. Low shrubs with inconspicuous flowers mostly in late spring to early fall. Canyon bottoms and higher elevations in the mountains; not common.

Baccharis sarothroides. Desert broom. Broomlike shrubs. Flowers whitish, the fruits minute and become airborne by means of copious white pappus; flowering and fruiting mostly October and November. Coyote Wash; not common.

Bahiopsis parishii [*Viguiera parishii. V. deltoidea* var. *parishii*] Parish's goldeneye. Shrubs with bright yellow flowers. Canyons and slopes to the summit.

Baileya pleniradiata. Woolly desert-marigold. Nonseasonal annuals, but mostly seen in spring. Flowers yellow and showy. Sandy-gravelly soils of washes, bajadas, valley plains, and canyon bottoms at lower elevations.

Bebbia juncea var. ***aspera.*** Sweetbush, chuckwalla delight. Shrubs with fragrant yellow flowers attracting butterflies and other insects; flowering nonseasonally. Bajadas, washes, canyons, and rocky slopes.

Brickellia atractyloides var. ***atractyloides.*** Spiny-leaf brickell-bush. Subshrubs; leaves firm with spinescent teeth. Flowers pale yellow and purplish; February and March. Rocky slopes, mostly north facing, and in canyons.

Calycoseris wrightii. White tackstem. Cool-season annuals with milky sap. Rays of flower heads white with pinkish brown streaks on the lower surfaces. Coyotes Wash and apparently not common.

Chaenactis carphoclinia var. ***carphoclinia.*** Pebble pincushion. Spring annuals; flowers white. Sandy flats, washes, bajadas, and rocky slopes.

Chaenactis stevioides. Desert pincushion. Spring annuals, generally slightly more robust than *C. carphoclinia.* Flowers white or pinkish white. Sandy soils of valley bottoms and rocky slopes.

Encelia farinosa var. ***farinosa.*** Brittlebush. Small shrubs with bright yellow rays and yellow or brownish disks; mostly spring and also with summer-fall rains. Widespread; dry washes, valley floor, canyons, and rocky slopes. The yellowish resin was used as chewing gum, an all-purpose adhesive and sealant, and for medicinal purposes.

Brittlebush (Encelia farinosa). *(Photo by Joan E. Scott)*

Encelia farinosa var. ***farinosa*** × ***E. frutescens*** var. ***frutescens.*** Small shrubs; flower heads with bright yellow rays. Leaves intermediate in shape from the presumed parent species. Scattered in the Lechuguilla Valley.

Encelia frutescens var. ***frutescens.*** Button encelia. Small shrubs. Flowers heads yellow, with disk florets only; reproductive during warmer months. Sandy soils of Lechuguilla Valley.

Erigeron lobatus. Desert fleabane. Spring annuals and sometimes also in summer. Disk florets yellow, the rays pale lavender. Along Coyote Wash.

Filago, see ***Logfia***

Geraea canescens. Desert sunflower, desert gold. Coarse spring annuals; sunflowerlike heads with yellow-orange disk florets and bright yellow rays. Not documented from the Tinajas Altas area, but common in nearby areas on sandy, gravelly, and rocky soils.

Gymnosperma glutinosum. Gumhead. Small shrubs or subshrub perennials with resinous herbage. Flowers heads small and bright yellow. Mostly on rocky slopes

Desert sunflower (Geraea canescens). *(Photo by Bill Broyles)*

and mountain arroyos or canyons. Often bright green and even flowering in extended drought.

Hymenoclea salsola, see ***Ambrosia salsola***

****Lactuca serriola.*** Prickly lettuce, compass plant. Annuals, sometimes to 2+ m tall; flower pale yellow. Locally in clayish soil in dense mesquite brush at the south end of Coyote Wash at the Camino del Diablo crossing.

Laennecia coulteri [*Conyza coulteri*] Coulter's horseweed. Warm-weather annuals; flowers inconspicuous. Infrequent along Coyote Wash.

Logfia arizonica [*Filago arizonica*] Arizona fluffweed. Small cool-season annuals, often white-woolly; flowers minute. Washes, canyon bottoms, floodplains, and bajadas and perhaps soil pockets on rocky slopes.

Logfia filaginoides [*Filago californica, Logfia californica*] California fluffweed. Small, cool-season annuals, often white-woolly with erect stems; flowers minute. Washes, canyons, bajadas, and among rocks on slopes.

Machaeranthera pinnatifida, see ***Xanthisma spinulosus***

Monoptilon bellioides. Mohave desertstar. Low, cool-season annuals. Ray florets white to lavender, the disk yellow. Sandy to rocky soils, washes, bajadas, desert pavements, canyons, and lower slopes.

Palafoxia arida var. ***arida.*** Spanish needles. Spring annuals and sometimes with summer-fall rains; herbage silvery-haired. Flower dull white to pale pink. Not recorded from Tinajas Altas, but found in western Cabeza Prieta Refuge and near the Butler Mountains.

Pectis papposa var. ***papposa.*** Desert chinchweed. Widespread summer-fall annuals; pungently aromatic with conspicuous oil glands. Flowers bright yellow. Sandy to rocky soils, washes, plains, and less common on slopes.

Perityle emoryi. Desert rock-daisy. Cool-weather annuals. Flower heads with white rays and yellow disk florets. Sandy to rocky soils, washes, plains, and slopes.

Peucephyllum schottii. Desert fir, pygmy cedar. Woody shrubs resembling a small conifer, with twisted trunks and branches, and shredding bark. Herbage resinous. Flowers fragrant, of yellow disk florets. Rocky and often exposed mountain slopes.

Desert-fir, pygmy-cedar (Peucephyllum schottii). *(Photo by Bill Broyles)*

Pleurocoronis pluriseta [*Hofmeisteria pluriseta*] Arrowleaf. Small shrubs or subshrubs. Leaf blades narrowly arrow-shaped, often scarcely wider than the petiole. Often in crevices on rock faces, canyons, cliffs, and slopes.

Porophyllum gracile. Hierba del venado. Herbaceous perennials, pungently aromatic. Stems and leaves bluish green, the foliage sparse and quickly deciduous. Flower heads with pinkish-white disk florets; growing and flowering at various seasons. Widely scattered; rocky slopes, canyons, arroyos, and upper bajadas. It was a significant medicinal plant.

Prenanthella exigua. Brightwhite. Diminutive spring annuals. Flower heads with several small rays—white with violet tips. Higher elevations in the mountains.

Psathyrotes ramosissima. Desert velvet. Spring annuals; compact, woolly, and strongly scented. Leaves velvety with deeply incised veins. Flower heads of yellow disk florets. Desert plains west of the Tinajas Altas range.

Rafinesquia neomexicana. Desert chicory. Cool-season annuals; flower heads showy, white tinged with pale rose-purple and yellow. Often growing through small shrubs. Sandy to rocky soils; washes, plains, and rock slopes to higher elevations.

Senecio mohavensis. Mojave groundsel. Cool-season annuals, often small and delicate. Flowers yellow, rather inconspicuous. Often in shaded places among rocks in mountain canyons and on north-facing slopes.

****Sonchus asper*** subsp. ***asper***. Spiny sowthistle. Cool-season annuals; larger plants spiny-prickly and sometimes robust. Flower heads pale yellow. Not common; mostly washes and canyon bottoms.

****Sonchus oleraceus***. Sow thistle. Cool-season annuals, the plants not robust. Flower heads pale yellow. Widely scattered but seldom common; washes and canyons, sometimes on slopes.

Stephanomeria pauciflora. Brownplume wirelettuce. Mound-shaped bushy perennials with sparse foliage. Flowers pale pink; flowering nonseasonally. Washes, canyons, and rocky slopes.

Stylocline micropoides. Desert neststraw. Diminutive, woolly, cool-season annuals; flowers minute. Locally common in the Tinajas Altas Mountains on upper bajadas, benches, and washes, and sometimes in soil pockets on rocky slopes and in canyons.

Trichoptilium incisum. Yellow head. Annuals; mostly growing and flowering during cooler seasons, especially in spring, and sometimes flowering into early summer. Leaves woolly; flowers yellow. Mostly in rocky or gravelly soils; washes, bajadas, and rocky slopes.

Trixis californica var. ***californica***. California threefold. Small shrubs with brittle stems and thin, upright leaves. Flowers yellow and raylike (bilabiate). Canyons, bajadas, and rocky slopes including extremely arid hills to higher elevations in the mountains.

Viguiera parishii, see ***Bahiopsis parishii***

Xanthisma spinulosum var. ***gooddingii*** [*Machaeranthera pinnatifida* var. *gooddingii*] Spiny goldenweed. Herbaceous perennials and also flowering in the first season as nonseasonal annuals. Ray and disk florets bright yellow, mostly in spring, but also with summer rains. Mostly canyons and slopes to the summit, and sometimes along washes and on the desert floor.

Boraginaceae—Borage Family (includes Hydrophyllaceae)

Amsinckia intermedia var. ***echinata***. Devil's lettuce, fiddleneck. Cool-season annuals. Plants rough-haired, five calyx lobes about equal in size, and nutlets rough with ragged-edged ornamentations. Flowers yellow-orange. Widespread, from lowest to higher elevations.

Amsinckia tessellata. Fiddleneck. Cool-season annuals resembling *A. intermedia*, but often more robust, the calyx lobes 3 or 5 and unequal in size, the nutlets with smooth-edged bumps (tessellated). Flowers yellow-orange. It occurs nearby at the western boundary of Cabeza Prieta in the Lechuguilla Valley.

Cryptantha angustifolia. Narrowleaf cryptantha, desert cryptantha. One of the most widespread and abundant cool-season annuals in the region. Flowers white. This is one of those itchy plants that gets in your socks and sleeping bag. Valley bottom and plains, canyons, and rocky slopes.

Cryptantha barbigera. Bearded cryptantha. Cool-season annuals; flowers white. The plants are sometimes sterile with abnormal growth due to infestations of mites. Sandy soils of washes and plains, and on rocky slopes to higher elevations.

Cryptantha maritima [includes var. *maritima* and var. *pilosa*] White-haired cryptantha. Cool-season annuals; flowers minute. As with *C. barbigera*, the plants are sometimes deformed by mites. The varieties do not seem worthy of recognition; both occur intermixed. Common and widespread; washes and sand flats, and bajadas to rocky slopes at higher elevations.

Cryptantha micrantha subsp. ***micrantha***. Dwarf cryptantha. Cool-season annuals; plants diminutive with very slender stems, the roots staining red-purple; flowers minute and white. This is the smallest cryptantha in the region. Common in immediately adjacent regions and expected within the core flora area in sandy soils of valley bottoms and desert plains.

Cryptantha pterocarya var. ***cycloptera***. Wingnut cryptantha. Cool-season annuals. Readily recognized by the broad sepals, green when fresh and only moderately hairy, relatively large fruits, and relatively large nutlets with broad, ornamented wings. Widespread throughout the flora area; desert flats, bajadas, washes, canyons, and rocky slopes to highest elevations.

Cryptantha racemosa. Bushy cryptantha. Annuals or short-lived perennial herbs or subshrubs, growing only during cooler seasons; flowers white. This species and *C. holoptera*, known from nearby northwestern Sonora, are the only cryptanthas in the region that survive more than one season. Rocky slopes and sandy-gravelly washes in the Tinajas Altas Mountains from low to high elevations.

Eucrypta micrantha. Delicate cool-season annuals, aromatic and viscid-glandular; leaves pinnately divided; flowers small and pale blue. Often in protected niches beneath shrubs or sheltered among rocks. Widespread; wash, bajadas, canyons, and rocky slopes.

Nama demissum var. ***demissum***. Purplemat. Cool season annuals, low-growing and often matted. Flowers bright lavender-pink. Washes, and gravelly soils of bajadas and plains; often on open, otherwise barren areas such as desert pavements. The plants are more compact, generally smaller, and with darker-colored flowers than the more common *N. hispida*.

Nama hispida [*N. hispida* var. *spathulatum. N. coulteri*] Bristly nama. Cool-season annuals, the branches often ascending, or with age sometimes spreading-prostrate but not matted. Flowers pale lavender to purple. Widespread and common, especially washes, sandy flats, bajadas, and canyons.

Pectocarya heterocarpa. Mixed-nut comb-bur. Cool-season annuals; flowers minute, white with a yellow throat. The four nutlets spread open like tiny, open jaws with marginal teeth. Widespread; sandy soils of washes and the valley floor to the mountains.

Pectocarya platycarpa. Broad-winged comb-bur. Cool-season annuals. Similar to *P. heterocarpa* but the nutlets are noticeably larger with broader and usually yellowish wings. Widespread including washes, desert flats, and bajadas to rocky slopes.

Phacelia ambigua [*P. crenulata* var. *ambigua. P. ambigua* var. *minutiflora*] Desert heliotrope. Cool-season annuals, stinky and glandular pubescent, and can cause unpleasant dermatitis. Inflorescence branches curved (helicoid). Flowers lavender-purple. Seasonally abundant and widespread, in many habitats including washes, canyons, bajadas, plains, and rocky slopes to their summits. We use a broad taxonomic interpretation for this species.

Phacelia neglecta. Alkali phacelia. Cool-season annuals; plants short, stubby, and semisucculent. Flowers white. Unlike most *Phacelia* in the region, the plants are not stinky and the inflorescences are not curved (helicoid). Generally not common and localized on nearly barren desert pavements and pediments near mountain bases.

Phacelia pedicellata. Pedicellate phacelia. Cool-season annuals, the plants often robust, conspicuously glandular-pubescent with an offensive odor, and can cause unpleasant dermatitis, the herbage often pale green. Flowering branches helicoid. Flowers pale lavender-blue. Canyon bottoms and among rocks, especially on north- and east-facing mountain slopes, from base to higher elevations.

Tiquilia palmeri. Palmer's crinklemat. Low-growing perennials from very slender roots issuing from even deeper, long, thick, black roots. Flowers lavender-pink, following spring and summer-fall rains. Sand flats and dunes near the Butler Mountains and expected near the Tinajas Altas Mountains and in the Lechuguilla Valley.

Brassicaceae—Mustard Family

****Brassica tournefortii***. Sahara mustard, wild turnip. Cool-season annuals, highly variable in size. Flowers pale yellow and apparently selfing (autogamous). Widespread in lowland disturbed and natural sites across the flora area, especially on sandy soils; washes and valley plains, and sometimes extending onto rocky slopes and in canyons even at higher elevations. Sahara mustard is one of the most seriously invasive species in the flora region.

The broad, spreading basal rosette leaves can effectively prevent establishment of other cool-season annuals. "Drought-stressed plants may produce leaves as small as 8 cm and be reproductive, whereas plants on sandy soils with sufficient soil moisture may grow to more than 1 meter across, making this species the largest herbaceous rosette plant in the region" (Felger 2000:186). Native to the Old World, it has spread explosively across the Sonoran Desert (Dimmitt and Van Devender 2009; Felger 2000).

Caulanthus lasiophyllus [*Guillenia lasiophylla. Thelypodium lasiophyllus*]. Cool-season annuals; flowers minute, white to pink-purple; fruits slender, straight, and terete, turning downward as they mature. Widespread; washes, valley plains, and mountains.

Descurainia pinnata. Tansy mustard. Cool-season annuals; leaves highly dissected, the flowers minute, pale yellow. Widespread across the flora area; sandy to rocky soils of washes, plains, bajadas, hills, and mountains. This tansy mustard was an important medicinal plant.

Dimorphocarpa pinnatifida [*Dithyrea wislizeni* in part] Dune spectacle-pod. Cool-season annuals. Flowers white and fragrant. One plant encountered in the canyon above the tinajas, apparently not reproducing due to lack of cross-pollination since they are outcrossers (Andrew Salywon, personal communication 2007).

Desert bladderpod (Physaria tenella). *(Photo by Joan E. Scott)*

Endemic to dunes in nearby northwestern Sonora and southwestern Arizona.

Dithyrea californica. California spectacle-pod. Cool-season annuals; flowers cream white and fragrant. Characteristically on sandy soils. Not known from the core flora area, but recorded east of the Butler Mountains and in the Pinta Sands in Cabeza Prieta.

Draba cuneifolia [*D. cuneifolia* var. *integrifolia. D. cuneifolia* var. *sonorae*] Wedgeleaf draba. Cool-season annuals; plants small; leaves broad and in a basal rosette; flowers small and white. Washes, desert plains, sand flats, canyons, and rocky slopes especially in sheltered niches.

Lepidium lasiocarpum. Sand peppergrass. Cool-season annuals. Flowers small and inconspicuous. Widespread; washes, flats, canyons, and less common on rocky slopes.

Lesquerella, see ***Physaria***

Lyrocarpa coulteri. Lyre-pod. Herbaceous perennials, sometimes flowering in the first season. Flowers yellow to purple-brown, especially fragrant at night. Growing and flowering nonseasonally with sufficient soil moisture. Washes, canyons, gravelly bajadas, and rocky slopes to higher elevations. This is the only perennial member of the mustard family in the flora area.

Physaria tenella [*Lesquerella tenella*] Desert bladderpod. Cool-season annuals; flowers bright yellow; fruits globose, hollow and bladderlike. Mostly at lower elevations; washes, sand flats, valley plains, and mountains.

***Burseraceae*—Torchwood Family**

Bursera microphylla. Elephant tree. Large shrubs or small trees, with thick limbs, and aromatic sap and foliage. Flowers minute, cream-color, with the first summer rains. Rocky slopes to peak elevations and sometimes on upper bajadas. Often freeze damaged. It is a significant medicinal plant and the sap has been used a reddish pigment.

Elephant tree (Bursera microphylla). *(Photo by Bill Broyles)*

Devil's club cholla (Grusonia kunzei) *east of Tinajas Altas. Also called desert club cholla, devil cholla, and a few unprintable names. (Photo by Charles Sheldon. Courtesy Neil B. Carmony)*

Cactaceae—Cactus Family

Carnegiea gigantea [*Cereus giganteus*] Saguaro, sahuaro. Flowering late April and May; fruits ripening in June. Common on upper bajadas and rocky slopes throughout the mountains. The fruits and seeds were major food resources, and the fruit pulp was made into wine. Saguaros are relatively uncommon near the tinajas, undoubtedly due to the desert bighorn that smash the stems with their horns in order to expose and eat the succulent tissue.

Cylindropuntia acanthocarpa var. ***coloradensis*** [*Opuntia acanthocarpa* var. *coloradensis*] Buckhorn cholla. Shrub-sized chollas; flowers yellow-orange, March and April; fruits becoming dry upon ripening in May. Widespread on upper bajadas, canyons, and rocky slopes to higher elevations. The flower buds were prepared as a vegetable.

Cylindropuntia bigelovii [*Opuntia bigelovii*] Teddybear cholla. Shrub-sized chollas. Flowers pale yellow-green, March and April; fruits moderately fleshy, leathery, and yellow. Abundant on desert pavements, sandy plains and flats, bajadas, pediments, and rocky slopes. The plants propagate prolifically from readily rooting fallen joints.

Cylindropuntia echinocarpa [*Opuntia echinocarpa. O. wigginsii*] Silver cholla. Small chollas. Flowers silvery white with green filaments, March and April; fruits becoming dry upon ripening in late May and early June. Upper bajadas at base of the mountains.

Cylindropuntia ramosissima [*Opuntia ramosissima*] Diamond cholla. Slender-stem chollas, sometimes forming spreading colonies. Flowers small, pale yellowish to greenish brown or cream-color, opening mid- to late-afternoon, April and May. Fruits dry at maturity. Especially common on bajadas along the western side of the Lechuguilla Valley.

Echinocactus polycephalus var. ***polycephalus***. Many-headed barrel cactus. Multiple-headed, mound-forming barrel cactus with stout spines. Flowers yellow, late May and June. Scattered mostly on upper bajadas and sandy valley flats.

Echinocereus engelmannii var. ***chrysocentrus***. Strawberry hedgehog cactus. Several to many stems branching mostly from near the base. Flowers large and bright purple-magenta, March and early April; fruits edible and ripening late May and early June. Widespread; upper bajadas and rocky slopes.

Ferocactus cylindraceus [*F. acanthodes*] Mountain barrel cactus. Plants usually with a single stem, becoming taller than wide. Flowers yellow-green. Upper bajadas, slopes and cliffs; larger ones mostly on inaccessible cliffs and rock faces. Barrel cacti are fairly scarce from the vicinity of the waterholes to the peaks, obviously due to bighorn sheep, which dislodge the plants with their horns and eat the fleshy stem.

Grusonia kunzei [*Opuntia kunzei. O. stanlyi* var. *kunzei. Corynopuntia kunzei*] Desert club cholla, devil cholla.

Thick-stemmed club chollas, forming sprawling colonies, and well-armed with rather large and very sharp spines. Flowers pale yellow or cream-color; May and June, the fruits ripening late August through the following May. Scattered on the mid-bajada and upper bajada of the Lechuguilla Valley.

Mammillaria grahamii [*M. microcarpa*] Arizona fishhook cactus. Small cacti with hooked spines. Flowers pink, produced in pulses, generally following rainfall during warmer months. Fruits small, red, and fleshy with black seeds—edible as snack food. Widely scattered; sandy and rocky soils; upper bajadas and rocky slopes at all elevations.

Mammillaria tetrancistra. Corkseed fishhook cactus. Small cacti with hooked spines. Flowers pink, spring and summer; fruits red and fleshy, the basal one third of each seed seated in a corky cup making the fruits unpleasant to eat. This is the most arid-inhabiting of all mammillarias.

Opuntia basilaris var. ***basilaris.*** Beavertail cactus. Dwarf prickly pears with a very thick caudex and short "pads." Areoles bear glochids but no spines. Flowers showy, rose-pink, early April; fruits ripe in late May. Nearly barren desert pavements near the base of the mountain and rocky slopes and canyons nearly to the summit, although generally scarce at higher elevations; especially common in parts of Tinajas Altas Pass.

Peniocereus greggii var. ***transmontanus.*** Desert night-blooming cereus. Stems slender, mostly few-branched, from a single large tuberous root. Flowers large, white, nocturnal and spectacularly fragrant; in early summer. Fruits bright red, juicy and sweet. Tinajas Altas Mountains, apparently scarce, at all elevations. The roots have been utilized medicinally.

Campanulaceae—Bellflower Family

Nemacladus orientalis [*N. glanduliferus* var. *orientalis*] Redtip threadplant. Small, cool-season annuals with wiry stems. Flowers tiny, the corollas white with red-tipped lobes. Known from the flora area by a single record at Coyote Water, although presumably it is more widespread.

Caryophyllaceae—Pink Family

Achyronychia cooperi. Sandmat, frostmat. Cool-season annuals, small and low-growing with conspicuous white stipules. Flowers minute and white. Mostly on sandy soils in valley bottoms, washes, and bajadas.

Loeflingia squarrosa. Cool-season annuals of dwarf, cushionlike plants. Flowers minute and hidden among the leaves. Locally on partially barren pediments of decomposed granite on the east side of Tinajas Altas Mountains.

Chenopodiaceae, see Amaranthaceae

Crassulaceae—Stonecrop Family

Crassula connata [*C. connata* var. *eremica. Tillaea erecta*] Pygmy stonecrop. Winter-spring annuals, the plants diminutive and succulent with beadlike leaves. Flowers white and minute. Sandy-gravelly soils of valley and bajada slopes on west side of Lechuguilla Valley.

Dudleya arizonica [*D. pulverulenta* subsp. *arizonica*] Arizona liveforever. Perennial rosette plants, summer dormant. Leaves thick and succulent. Each flower yellow, orange, and reddish, April and early May. Mountains to peak elevations, mostly on rock slopes and cliffs, often north- and east-facing and wedged in rock crevices.

Crossosomataceae—Crossosoma Family

Crossosoma bigelovii. Ragged rockflower. Unarmed shrubs to 1.5 m tall, evergreen or tardily drought deciduous. Flowers white and fragrant, mid-winter and early spring. Often rooted in bedrock canyons or cliff faces.

Cucurbitaceae—Gourd Family

Brandegea bigelovii. Desert starvine. Cool-season annuals, sometimes seen as early as September; perishing by late spring. Herbaceous vines from a thick root; flowers small, white, and delicately fragrant. Washes and their floodplains, and canyons, often seasonally festooning shrubs and trees in green curtains.

Desert starvine (Brandegea bigelovii). *(Photo by Joan E. Scott)*

Cucurbita digitata. Fingerleaf gourd, coyote gourd. Perennials vines from a large, tuberous root. Growing and flowering during the warmer months. Leaves of mature growth with slender, fingerlike lobes. Flowers large and yellow; fruits of hard-shelled, round gourds. Infrequent along canyons in the Tinajas Altas Mountains.

Limberbush (Jatropha cuneata). *For much of the year it is leafless. (Photo by Bill Broyles)*

Cucurbita palmata. Coyote melon. Similar to *C. digitata* but leaves with broad, mostly triangular lobes. Infrequent on sandy soils of the Lechuguilla Valley.

Cyperaceae—Sedge Family

Cyperus squarrosus [*C. aristatus. Mariscus squarrosus*] Dwarf sedge. Diminutive nonseasonal annuals, tufted and grasslike. Found along Coyote Wash following exceptional rains.

Euphorbiaceae—Spurge Family

Acalypha californica [*A. pringlei*] Copperleaf. Small shrubs. Flowers small, generally reddish, during warmer months with sufficient moisture. Canyons, arroyo bottoms, and protected slopes in the Tinajas Altas mountains.

Ditaxis lanceolata [*Argythamnia lanceolata*] Narrowleaf silverbush. Perennial herbs to subshrubs, herbage silvery pubescent. Flowers inconspicuous, white and green, during the warmer months. Widespread; washes, canyons, bajadas, and rocky slopes to summit elevations. Often heavily browsed, probably by bighorn and other animals.

Ditaxis neomexicana [*Argythamnia neomexicana*] New Mexico silverbush. Nonseasonal annuals or small, short-lived perennials. Flowers green and white, small and inconspicuous. Widespread; washes, valley plains, bajadas, and rocky slopes.

Ditaxis serrata var. ***serrata*** [*Argythamnia serrata*] Yuma silverbush. Nonseasonal annuals or perhaps short-lived herbaceous perennials. Similar to *D. neomexicana* but differing in part by having more robust plants and usually blunter leaf tips; perhaps they are not distinct species. Sandy soils, bajadas and washes at the south end of the Tinajas Altas Mountains.

Euphorbia abramsiana [*Chamaesyce abramsiana*] Desert spurge. Small, prostrate summer annuals and probably also growing in spring; herbage often brownish or reddish. Flowers inconspicuous. Sandy and gravelly soils; washes, sandy flats, and upper bajadas in the Lechuguilla Valley.

Euphorbia eriantha. Beetle spurge. Nonseasonal annuals, mostly encountered in spring, occasionally short-lived perennials. Flowers inconspicuous. Lechuguilla Valley and scattered in canyons and slopes in the mountains.

Euphorbia polycarpa [*Chamaesyce polycarpa*] Common desert spurge. Nonseasonal annuals or small perennials. "Flowers" maroon and white, or drought-stressed plants may fail to develop the white, petal-like appendages. This is the most common and widespread euphorbia in the region; washes, bajada-plains, and rocky slopes.

Euphorbia setiloba [*Chamaesyce setiloba*] Fringed spurge. Nonseasonal annuals, especially with summer-fall rains. Plants glandular-pubescent. "Flower clusters" (cyathia) with toothlike white or pink appendages. The plants are frost-sensitive. Often in broad, sandy-gravelly washes, bajadas, and canyon bottoms probably mostly at low elevations.

Jatropha cuneata. Limberbush. Shrubs, unarmed and with semisucculent stems and thick roots, oozing bloodlike sap when cut or bruised; stems sometimes freeze-damaged in winter. Flowers small and white, with rains from July to October. Widespread on rocky slopes and upper bajadas. This was the most common plant for basket-making by the Hia C'ed O'odham. Hikers brushing by the stems will find bloodlike stains on their clothing after washing them.

Fabaceae (Leguminosae)—Legume Family

Acacia greggii [*Senegalia greggii*] Catclaw acacia. Shrubs or small trees. Flowers pale yellow, in spring and occasional at other seasons; pods ribbonlike. Valley drainageways, and mountain canyons and slopes, especially east- and north-facing, from low to high elevations. Catclaw is a favorite food of desert bighorn.

Acmispon rigidus [*Hosackia rigida. Lotus rigidus*] Desert rockpea. Perennial subshrubs with sparse foliage. Petals bright yellow, becoming red-orange with age. Tinajas Altas Mountains, especially arroyos and canyons among boulders and rocky slopes, often east- and north-facing.

Acmispon strigosus [*Hosackia strigosa. Lotus strigosus* var. *tomentellus*]. Hairy lotus. Winter-spring annuals, mostly low growing or prostrate; leaves often semisucculent. Petals bright yellow with red nectar guidelines on the banner. Washes and canyon bottoms, and perhaps in soil pockets on rocky slopes.

Cercidium, see ***Parkinsonia***

Dalea mollis. Silky dalea. Nonseasonal annuals; flowers small, purple and white. Strangely, there are no modern records of this widespread desert annual from the core Tinajas Altas area, although it occurs in adjacent areas.

Lotus, see ***Acmispon***

Lupinus arizonicus. Arizona lupine. Winter-spring annuals, the plants sometimes semisucculent. Flowers pale lavender-pink. Lowland habitats, especially washes, bajadas, and canyons, and to at least 1,530 feet in the Tinajas Altas Mountains.

Olneya tesota. Ironwood. Trees or large shrubs, the wood extremely hard. Flowers pale lavender; late April and May during years of sufficient rainfall; pods ripening in early summer. Major drainageways, upper bajada-plains, mountain canyons, and infrequent on rocky slopes. Ironwood and both paloverde species were not seen along Coyote Wash (at least from the vicinity of Coyote Water to the Mexican border). Ironwood foliage is a favorite food of desert bighorn and many animals consume the seeds.

The wood has been used for fuel, and some Sonoran Desert people prepared the seeds for food. Ironwood stumps can be seen throughout the region. Bryan (1925:422) noted that at Tinajas Altas "a few palo verde or ironwood trees are still standing and may furnish horse feed, but the traveler should bring his own firewood."

Parkinsonia florida [*Cercidium floridum*] Blue paloverde. Trees often with a well-developed trunk. Flowers bright yellow, in spring and sometimes sporadically in fall. Canyon bottoms and washes, especially the larger ones, along the eastern side of the Tinajas Altas Mountains. Between 2000 and 2006, only two or three blue paloverdes were seen in the wash below (east of) the tinajas, perhaps due to earlier woodcutting—by 2008 these trees were gone, apparently due to drought and firewood collecting by visitors. Although the soft wood is of poor quality for fuel, it was utilized because little else was easily available. The trees were common on the Davis Plain west of the Gila Mountains, although many (or most?) of them perished during extreme drought of 2004 and 2005. The seeds served as food for people.

Parkinsonia microphylla [*Cercidium microphyllum*] Foothill paloverde. Small trees or large shrubs. Flowers pale yellow and white, in April and early May. Upper bajadas on the east the side of Tinajas Altas Mountains and canyons, washes, and rocky slopes from the base to higher elevations on all slope exposures. The seeds were a significant food resource.

Prosopis glandulosa var. ***torreyana*** [*P. juliflora* var. *torreyana*] Western honey mesquite. Trees and shrubs. Leaves appearing in spring and gradually deciduous in winter, with one pair of pinnae. Flowers dull yellow, in April and May and sporadically through early fall; pods mostly ripening in early summer and sporadically through fall. Mostly along the larger drainageways, especially Coyote Wash. Also in the wash below the tinajas and larger canyons and washes running out of the mountains. Scarce in the canyon above the tinajas and not seen at higher elevations. Coyotes feast on fallen pods. Mesquite was the single most useful plant in the Sonoran Desert (Felger 1977, 2007); providing wood for construction, tools, utensils, weapons, and the preferred cooking fuel. The pod mesocarp was a major food resource and some people processed the actual seeds for food. The roots were fashioned into cordage. The whitish gum is edible and was used medicinally. The black pitch oozing onto the bark was also used medicinally and as a dye.

Fouquieriaceae—Ocotillo Family

Fouquieria splendens subsp. ***splendens***. Ocotillo. Long-lived shrubs with wandlike spiny branches. Flowers red-orange, generally in spring. Widespread; sandy soils of upper bajadas and rocky slopes to summit elevations. Ocotillo served as a framework for the traditional Hia C'ed brush house. The flower bases provide sweet nectar for trail snacks.

Geraniaceae—Geranium Family

****Erodium cicutarium***. Filaree, heronsbill, storksbill. Winter-spring annuals; leaves pinnately dissected; flowers small, lavender-pink. In the vicinity of Coyote Water and rare along the canyon above the tinajas. Common in nearby areas, especially in disturbed sites. Native to the Old World and widely naturalized in the New World.

Erodium texanum. False filaree, desert storksbill. Winter-spring annuals; leaf blades ovate, often 3-lobed or 3-parted; flowers small, lavender-pink. Widespread and common; washes, bajadas, plains, and sandy flats.

Hydrophyllaceae, see Boraginaceae

Krameriaceae—Ratany Family

Krameria erecta [*K. parvifolia*] Littleleaf or range ratany. Dwarf woody shrubs with knotty short-shoots. Flowers bright magenta-purple, at various seasons following sufficient rainfall. Scattered on sparsely-vegetated granitic-gravelly dissected bajadas, often with *K. grayi*.

Krameria grayi. White ratany. Sprawling shrubs. Flowers magenta-purple (deep rose), at various seasons following sufficient rainfall, especially in spring. Widespread in lowland areas, especially sandy-gravelly soils of washes, plains, and bajadas, and sometimes on rocky slopes.

Lamiaceae (Labiatae)—Mint Family

Hyptis albida [H. emoryi]. Desert lavender. Multiple-stemmed shrubs. Flowers small, fragrant, and violet-blue; flowering much of the year except during freezing weather and extreme drought. Widespread, especially in washes, also canyon bottoms and on rocky slopes to the mountain summit.

Salazaria mexicana [*Scutellaria mexicana*] Bladder sage. Shrubs, usually with sparse foliage. Flowers white and blue; flowering at various seasons following rains. The calyx enlarges as the fruit develops to become like a paper bag. Upper bajadas, canyons, and rocky slopes, especially north-facing and at higher elevations.

Teucrium glandulosum. Sticky germander. Herbaceous perennials. Flowers attractive, corollas white with lavender markings. Discovered in 2002 in Coyote Wash at Camino del Diablo crossing in clayish soil in dense mesquite brush. Canyon bottoms are usual haunts for this species in nearby northwestern Sonora and mountains in Cabeza Prieta. It is a rather rare species in the United States.

Loasaceae—Stickleaf Family

Mentzelia albicaulis [*M. desertorum*] Whitestem blazingstar. Cool-season annuals. Flowers bright yellow, sometimes with an orange or reddish center. Sandy and gravelly soils of bajadas, arroyos, and washes. *M. desertorum* is a narrowly defined taxon distinguished on technical features.

Mentzelia multiflora [*M. pumila*, in part] Blazing star. Cool-season annuals and perhaps sometimes short-lived perennials. Flowers bright yellow. Lower-elevation canyons and washes in the Tinajas Altas Mountains; apparently not common in the flora area.

Mentzelia puberula. Silver blazingstar. Annuals or short-lived perennials, the plants often bushy; leaves often silvery gray-green. Flowers pale yellow, spring and summer with rains. Canyons and rocky slopes; widely scattered and seldom common.

Petalonyx linearis. Narrowleaf sandpaper plant. Small shrubs or bushes. Flowers white, at least in March and April. Localized and widely scattered, gravelly sandy washes and rocky slopes.

Malvaceae—Mallow Family

Hibiscus denudatus var. ***denudatus***. Rock hibiscus. Perennial subshrubs. Corollas white to pinkish with maroon spots in the center, flowering at various seasons, especially during warmer months. Widespread and common in the Tinajas Altas Mountains; pediments, canyons, and slopes to higher elevations.

Horsfordia alata. Pink velvet-mallow. Tall, spindly shrubs. Flowers pale lavender-white to pink, drying bluish to pale lavender, at various seasons. Small to large washes in upper bajadas, canyons, and rocky slopes to higher elevations.

Horsfordia newberryi. Orange velvet-mallow. Few-branched slender shrubs. Flowers bright yellow-orange, at various seasons. Canyons and rocky slopes in the mountains.

Sphaeralcea ambigua subsp. ***ambigua***. Desert globemallow. Bushy perennials. Flowers orange; various seasons, especially spring. Washes, upper bajadas, canyons, and rocky slopes, and also on well-drained soils of valley plains.

Sphaeralcea coulteri [*S. orcuttii*] Annual globemallow. Cool-season annuals, and sometimes also growing with summer rains; highly variable in size. Flowers orange. Valley floor, bajadas, and lower-elevation canyons. Robust, large plants on sandy soils have been called *S. orcuttii*, which we regard as a synonym (Felger 2000).

Sphaeralcea emoryi. Emory globemallow; *mal do ojo*. Herbaceous perennials; flowers orange to red-orange; spring and summer-fall rainy seasons. Locally in clayish soil in dense mesquite brush at the south end of Coyote Wash at the Camino del Diablo crossing.

Desert four o'clock (Mirabilis laevis). *(Photo by Joan E. Scott)*

Martyniaceae—Devil's Claw Family

Proboscidea altheifolia. Devil's claw. Herbaceous perennials from a deeply buried, large, tuberous root. Flowers showy, yellow with yellow-orange and bronze markings; growing and flowering with hot weather. Sandy to silty soils of washes, floodplains, and valley plains.

Molluginaceae—Carpetweed Family

****Mollugo cerviana.*** Threadstem carpetweed, Indian chickweed. Diminutive summer annuals. Sandy soils along Coyote Wash. Reported as native to the Old World and not native in the New World.

Nolinaceae, see Asparagaceae

Nyctaginaceae—Four-O'clock Family

Abronia villosa. Sand verbena. Cool-season annuals. Flowers fragrant, pinkish purple to pale magenta. Sandy plains and dunes near the Butler Mountains, and not documented elsewhere in the flora area, but widespread on sandy soils in nearby areas.

Allionia incarnata var. ***incarnata*** [*A. incarnata* var. *nuda*] Windmills, trailing four o'clock. Short-lived perennials with trailing stems, and also flowering in the first season. Flowers violet-rose or magenta; flowering nonseasonally in warmer months with sufficient soil moisture. Widespread including floodplains, washes, canyons, flats, and rocky slopes.

****Boerhavia coccinea*** [*B. caribaea*] Scarlet spiderling. Herbaceous perennials, also flowering in the first season; growing with warm weather. Flowers small and bright red-purple. Localized and apparently well-established near the base of the lowermost tinaja; not seen elsewhere in the flora area. This weedy species is probably native in western North America, but does not seem to be native in the Tinajas Altas and nearby regions (Felger 2000).

Boerhavia erecta. Spiderling. Summer annuals. Flowers whitish to pale pink, open in the early morning. Specimens from Coyote Wash are identified as *B. erecta*, but elsewhere in the region, such as washes and canyons, the plants are likely to be *B. intermedia.* The subtle distinctions are discussed by Spellenberg (2003).

Boerhavia spicata [*B. coulteri. B. coulteri* var. *palmeri. B. spicata* var. *palmeri*] Spiderling. Summer annuals; flowering branches conspicuously glandular-sticky, sticking to shoes, socks, and pant legs. Flowers white to pale pink, open in the early morning. Washes, bajadas, flats, and rocky slopes.

Boerhavia wrightii. Large-bract spiderling. Summer annuals. Flowers pinkish white to pink. The plants are often rather robust and the floral bracts, flowers, and fruits are larger than those of the other boerhavias in the region. Sandy to rocky habitats; nearly ubiquitous across the flora area.

Mirabilis laevis var. ***villosa*** [*M. bigelovii*] Desert four o'clock. Herbaceous perennials with glandular-pubescent herbage. Flowers white to pale pink, at various seasons. Widespread but seldom conspicuous; washes of valley plains and bajadas, canyons, and rocky slopes to higher elevations.

Mirabilis tenuiloba. Longlobe four o'clock. Herbaceous perennials, sticky glandular-pubescent throughout. Flowers white. Large canyons in the Tinajas Altas Mountains. The only other known locality in Arizona is Palm Canyon in the Kofa Mountains. The plants are more robust and generally have larger, thicker, and more yellowish leaves and larger "fruits" than those of *M. laevis*.

Onagraceae—Evening Primrose Family

Chylismia arenaria [*Camissonia arenaria*] Desert suncup. Cool season annuals to short-lived perennials. Flowers yellow, becoming pink with age. Rocky slopes and canyons.

Chylismia claviformis subsp. ***peeblesii*** [*Camissonia claviformis* subsp. *peeblesii*] Browneyes. Cool-season annuals. Flowers white, sometimes with pink tinges, and with a red-brown center, the flowers becoming pink with age. Widespread; washes, valley plains and bajadas, canyons, and rocky slopes.

Eremothera chamaenerioides [*Camissonia chamaenerioides*] Willow-herb evening primrose. Cool-season annuals. Flowers very small, at first white, often becoming pink with age. Widespread, especially on rocky slopes from low to high elevations.

Eulobus californicus [*Camissonia californica, Oenothera leptocarpa*] California suncup. Cool season annuals; plants erect and often sparsely branched, the leaves shed by the time of flowering. Flowers yellow, becoming orange with age and often drying pink. Widespread; washes, plains, bajadas, canyons, and rocky slopes.

Oenothera primiveris. Yellow evening primrose. Cool-season annuals. Flowers large and lemon-yellow. Sandy-gravel or silty-clayish soils of washes, sand flats, canyons, and sometimes in soil pockets on rocky slopes. Specimens in the region are generally identifiable as subsp. *bufonis*, distinguished by larger and generally cross-pollinated flowers.

Orobanchaceae—Broom-Rape Family

Orobanche cooperi. Desert broomrape. Cool-season annuals; parasitic on roots of *Ambrosia* shrubs and perhaps *Larrea* (Wiggins 1964). Stems thick and succulent.

Desert broomrape flowers (Orobanche cooperi). *(Photo by Bill Broyles)*

Flowers purple and white, with yellow markings. Locally common on sandy flats and bajadas, sandy-gravelly washes and canyon bottoms. The young plants are edible after thorough roasting.

Papaveraceae—Poppy Family

Eschscholzia minutiflora. Little gold poppy. Cool-season annuals. Flowers small and yellow. Sometimes seasonally common at lower elevations on east side of the mountains.

Plantaginaceae—Plantain Family (includes Scrophulariaceae in part)

Penstemon parryi. Desert penstemon. Usually perennials but at least sometimes functioning as cool-season annuals, flowering during the cooler months. Flowers rose-pink. Known from the flora area by a single plant seen in the wash below the lowermost tinaja.

Penstemon pseudospectabilis var. ***pseudospectabilis***. Mojave beard-tongue. Herbaceous perennials, also flower-

*Woolly plantain (*Plantago ovata*). (Photo by Joan E. Scott)*

ing in the first year. Flowers bright rose-purple; February to May. Canyons and slopes at higher elevations in the mountains, apparently not common.

Plantago ovata var. ***fastigiata*** [*P. insularis* Eastwood. Not *P. insularis* Nyman ex Briquet. *P. fastigiata. P. insularis* var. *fastigiata*] Woolly plantain, Indian wheat. Cool-season annuals; flowers small, straw-colored, and papery. Widespread, especially on sandy soils; washes, valley plains and bajadas, canyons, and soil pockets on rocky slopes. The seeds were significant food and medicinal resources.

Pseudorontium cyathiferum [*Antirrhinum cyathiferum*] Desert snapdragon. Nonseasonal annuals, viscid glandular-pubescent and foul smelling. Flowers small and dark purple. Known from Frontera Canyon, but probably more widespread in the flora area.

Poaceae (Gramineae)—Grass Family

Aristida adscensionis. Sixweeks threeawn. Nonseasonal annuals. Spikelets with three awns or sometimes the two lateral ones reduced or absent on drought-stressed plants. Seasonally common across the flora area, from lowest to highest elevations.

Aristida purpurea var. ***nealleyi.*** Purple threeawn. Tufted perennials; flowering response nonseasonal. Tinajas Altas and Butler Mountains.

Bothriochloa barbinodis [*Andropogon barbinodis*] Cane beardgrass. Robust tufted perennials; growing and reproductive during warmer months. Seen only near the uppermost tinaja.

Bouteloua aristidoides [*B. aristidoides* var. *arizonica*] Sixweeks needle grama. Summer annuals. Widespread from the Lechuguilla Valley to at least 1,700 feet elevation in mountain canyons.

Bouteloua barbata var. ***barbata.*** Sixweeks grama. Summer annuals. Widespread from the Lechuguilla Valley to at least 1,700 feet along mountain canyons.

****Bromus rubens*** [*B. madritensis* subsp. *rubens*] Foxtail brome, red brome. Cool-season annuals. Washes and slopes, but not common. Native to the Old World.

****Cenchrus ciliaris*** [*Pennisetum ciliare*] Buffelgrass; *zacate buffel.* Perennials and flowering in the first season; growing and flowering during warmer months. Abundant along Highway 2 in nearby Sonora, and occasionally northward into washes and canyons at the international border such as in Frontera Canyon. Native to the Old World.

Dasyochloa pulchella [*Erioneuron pulchellum. Tridens pulchellum*] Fluffgrass. Low, tufted perennials. Flowering with warm weather. Mountains, hills, pediments and upper bajadas.

Digitaria californica var. ***californica*** [*Trichachne californica*] California cottontop. Tufted perennials. One plant found at low elevation near the tinajas, but expected at higher elevations.

****Eragrostis cilianensis.*** Stink grass, stinking lovegrass. Summer annuals. Along Coyote Wash. Native to the Old World.

Festuca octoflora [*Vulpia octoflora*] Six weeks fescue, eight-flowered fescue. Small cool-season annuals. Washes, upper bajadas, canyons, and mountains to at least mid-elevations.

Heteropogon contortus [*Andropogon contortus*] Tanglehead. Robust tufted perennials. Reproductive during warmer months. Canyon bottoms in the mountains.

Hilaria rigida [*Pleuraphis rigida*] Big galleta. Large, clumping perennials. Growing and flowering mostly during warmer months. Valley plains and bajadas, occasional on rocky slopes.

Muhlenbergia microsperma. Littleseed muhly. Nonseasonal annuals, mostly seen during cooler seasons. At all elevations; rocky slopes, canyons, bajadas, and washes.

****Pennisetum ciliare***, see ***Cenchrus ciliaris***

****Schismus barbatus*** subsp. ***arabicus*** [*S. arabicus*] Arabian grass. Small, cool-season annuals. Abundant on sandy soils of washes and valley plains, and also on rocky soils on slopes. It was seen in the Tinajas Altas Canyon and on adjacent slopes near the uppermost waterhole, but not at higher elevations. Native to the Old World.

Sporobolus cryptandrus. Sand dropseed. Tufted perennials. Known in the flora area from a single record at Tinajas Altas in 1935.

Stipa speciosa var. ***speciosa*** [*Achnatherum speciosum*] Desert needlegrass. Tufted perennials. Dormant during the coldest and hottest months. Canyons and mountain slopes, mostly north-facing.

Tridens muticus var. ***muticus***. Slim tridens. Small tufted perennials; nonseasonal in growth and flowering. Rocky slopes and shallow soils in canyons, hills, and mountains.

Vulpia octoflora, see ***Festuca octoflora***

Polemoniaceae—Phlox Family

Aliciella latifolia subsp. ***latifolia*** [*Gilia latifolia*] Broadleaf gilia. Cool-season annuals, highly variable in size. Flowers pink. Documented from a mountain near the northern margin of the flora area.

Eriastrum diffusum. Desert woollystar. Cool-season annuals. Flowers small and blue. Washes, bajadas, and low hills.

Gilia stellata. Star gilia. Cool-season annuals. Flowers small, white tinged with violet. Widespread; washes, bajadas, hills, and mountains.

Langloisia setosissima subsp. ***setosissima***. Bristly langloisia. Cool-season annuals. Flowers bright lavender-pink. Sandy or rocky soils; washes, valley plains and bajadas, hills, and mountains.

Linanthus bigelovii. Bigelow's linanthus. Cool-season annuals. Flowers white, nocturnal and powerfully sweet-fragrant, sometimes nauseously sweet. Washes, valley plains and bajadas, canyons, hills, and mountains at least at lower elevations.

Loeseliastrum schottii [*Langloisia schottii*] Schott's calico. Cool-season annuals. Flowers white or pink and showy. Low dunes near the west side of Tinajas Altas Mountains and perhaps elsewhere in the western part of the flora area.

Polygonaceae—Buckwheat Family

Chorizanthe brevicornu subsp. ***brevicornu***. Brittle spineflower, shorthorn spineflower. Cool-season annuals. Flowers minute, white. Sandy, gravelly, and rocky soils; washes, canyons, upper bajadas, pediments, and lower mountain slopes.

Chorizanthe rigida. Devil's spineflower. Cool-season annuals, drying as persistent, spinescent skeletons. Flowers minute, yellow. Bajadas and valley plains, washes, pediments, and lower rocky slopes.

Eriogonum fasciculatum var. ***polifolium***. Flattop buckwheat. Small shrubs. Flowers whitish or pink; February to May. Upper bajadas and washes, canyons, and rocky slopes, especially higher elevations in canyons and east- and north-facing slopes.

Eriogonum inflatum. Desert trumpet, bladder stem. Perennial herbs. Flowers small and yellow, mostly in spring and with the summer rains. Canyons, rocky hills, and mountain slopes.

Eriogonum thomasii. Thomas' buckwheat. Small, delicate, cool-season annuals. Flowers pink and yellow. Washes, valley plains and bajadas, and canyons.

Eriogonum thurberi. Thurber's buckwheat. Small, delicate, cool-season annuals. Flowers white or pink. Plants superficially resembling *E. thomasii*. Sandy soils on west side of the Tinajas Altas Mountains.

Eriogonum trichopes var. ***trichopes***. Little desert trumpet. Cool-season annuals; highly variable in size. Flowers yellow. Sandy, gravelly soils; valley plains, bajadas, and washes.

Eriogonum wrightii var. ***nodosum*** [*E. wrightii* var. *pringlei*] Bastard sage. Small, scruffy shrubs, often nearly leafless. Flowers white or pink, various seasons, but perhaps not during summer. Mountains, especially north-facing slopes and canyons.

Nemacaulis denudata var. ***gracilis***. Woolly heads. Delicate cool-season annuals. Flowers minute, yellow, surrounded by woolly bracts. Dunes and sand flats west of the Tinajas Altas Mountains.

Portulacaceae—Purslane Family

Portulaca halimoides [*P. parvula*] Silkcotton purslane. Succulent summer annuals. Flowers small, reddish and yellow, open in the morning. Sometimes common along Coyote Wash.

Devil's spineflower (Chorizanthe rigida). *(Photo by Joan E. Scott)*

Resedaceae—Mignonette Family

Oligomeris linifolia. Desert cambess. Annuals, recorded October through May. Flowers minute, white and green. Sandy to rocky soils; valley plains, washes, canyons, and bajadas.

Rhamnaceae—Buckthorn Family

Ziziphus obtusifolia var. ***canescens*** [*Condalia lycioides* var. *canescens*] White crucillo, graythorn. Thorny hardwood shrubs. Flowers small, yellow-green, at least in early summer. Not common in washes, canyon bottoms, upper bajadas, and lower slopes on the east side of Tinajas Altas Mountains. Fairly common at the south end of Coyote Wash. Fruits edible and slightly sweet but hardly worth the bother.

Rubiaceae—Madder Family

Galium stellatum var. ***eremicum.*** Starry bedstraw. Small shrubs or subshrubs. Flowers minute, whitish or pale yellow, in spring. Mountains, especially north- and east-facing slopes and canyons.

Rutaceae—Rue or Citrus Family

Thamnosma montana. Turpentine broom. Small aromatic shrubs; stems yellow-green; leaves few and small. Flowers dark indigo-blue and stinky, January to April. Hot, exposed slopes and ridges.

Santalaceae—Sandalwood Family (includes Viscaceae)

Phoradendron californicum. Desert mistletoe. Parasitic on legume trees and shrubs and rarely on *Larrea* and *Asclepias albicans.* Distributed with the hosts. Flowers small, green or yellow and fragrant, at various seasons, often December to February. Fruits ripening late winter to early spring, and are a major food of phainopeplas.

Scrophulariaceae, see Plantaginaceae

Solanaceae—Nightshade or Potato Family

Chamaesaracha coronopus. False nightshade. Herbaceous perennials from deep roots. Flowers cream-yellow at various seasons. Coyote Wash.

Datura discolor. Desert datura. Annuals, nonseasonal, frost sensitive, highly variable in size, and stinky. Flowers white with a purplish throat, nocturnal and fragrant. Washes, valley plains, and canyon bottoms. The flowers are the largest of any plant in the region.

Lycium andersonii var. ***andersonii*** [*L. andersonii* var. *deserticola*] Desert wolfberry. Spinescent hardwood shrubs. Flower lavender, nonseasonal. Fruits orange, fleshy, and edible. Canyons and washes in the mountains.

Lycium fremontii var. ***fremontii.*** Frémont wolfberry. Spinescent hardwood shrubs; flowers lavender, late winter; fruits relatively large, red-orange, fleshy, and edible. Locally along washes.

Lycium macrodon var. ***macrodon.*** Large-leaf desert wolfberry. Spinescent hardwood shrubs. Flowers white and green, in spring. Fruits hard, not edible. Washes, sand flats, and alluvial flats, and sometimes on north-facing slopes at lower elevations.

Lycium parishii var. ***parishii.*** Parish wolfberry. Spinescent hardwood shrubs. Flowers pale lavender, usually appearing after rains. Fruits orange, fleshy, and marginally edible. Washes, flats, canyons, and rocky slopes.

Nicotiana clevelandii. Desert tobacco. Cool-season annuals; flowers white, nocturnal. Mostly along washes in valley plains, upper bajadas, and low hills.

Nicotiana obtusifolia [*N. trigonophylla*] Coyote tobacco, desert tobacco. Perennial herbs. Flowers cream-white, diurnal, almost any time of year depending on soil moisture. Widespread; washes, bajadas, canyons, and rocky slopes. The leaves were smoked as tobacco.

Physalis versicolor [*P. crassifolia* var. *versicolor*] Desert ground cherry. Perennial herbs. Flowers dull yellow, at various seasons. Mountains in canyon washes and slopes.

Themidaceae, see Asparagaceae

Typhaceae—Cattail Family

Typha domingensis. Southern cattail. Robust perennial herbs; winter dormant. One small colony in a mid-elevation tinaja at Tinajas Altas. This is the only wetland plant in the region.

Urticaceae—Nettle Family

Parietaria hespera var. ***hespera.*** Desert pellitory. Delicate cool-season annuals. Flowers minute and green. Coyote Wash and expected in the mountains.

Viscaceae, see Santalaceae

Zygophyllaceae—Caltrop Family

Fagonia californica subsp. ***californica*** [*F. laevis. F. californica* subsp. *laevis*] California otat, California fagonbush. Low-growing perennials, also flowering in the first season, spinescent and essentially glabrous. Flowers lavender-pink, in warm weather except in extreme drought. Widespread; rocky slopes, canyons, and upper bajadas; not seen on open desert flats.

Fagonia pachyacantha. Thickleaf otat. Low-growing perennials, spinescent and glandular pubescent. Flowers lavender-pink, in warm weather except in extreme drought. Rocky, arid slopes, canyons, and upper bajadas; not seen on open desert flats.

Kallstroemia californica. Californica caltrop. Summer annuals. Flowers small, yellow to yellow-orange. Washes and sand flats along Coyote Wash and expected elsewhere.

Larrea divaricata subsp. ***tridentata*** [*L. tridentata*] Creosotebush. Long-lived hardwood shrubs with resinous herbage. Flowers yellow, at various seasons except during extended drought. Abundant across the desert floor, and hills and mountains to about 1,570 feet. Creosotebush has been a major medicinal plant in the Sonoran Desert.

POTENTIAL RECORDS

Amaranthaceae—Amaranth Family

Amaranthus palmeri. Careless weed, pigweed. Summer annuals. The seeds and young plants are edible. Strangely, there are no records for this amaranth in the flora area, but we expect it at least along Coyote Wash. It is common in nearby localities in Yuma County.

Poaceae—Grass Family

Cenchrus palmeri. Giant sandbur. A unique warm-weather annual grass with large burs bearing painfully sharp spines (Felger 2000). Not known for the United States, but widespread on sandy soils in the nearby Gran Desierto nearly to the Arizona border near Frontera Canyon. It should be sought on the Arizona side.

Searching the past. (Photo by Bill Broyles)

References

Oral History Interviews and Personal Communications

Allen, Rufus "Rufe." Line rider for Bureau of Animal Industry; stationed at Tinajas Altas during 1940s campaign against hoof-and-mouth livestock desease.

Atkinson, James. US Fish and Wildlife Service biologist at Cabeza Prieta National Wildlife Refuge.

Begarán, Pancho. Cowboy at Rancho Guadalupe de Victoria in northwest Sonora.

Bell, Fillman. Of Hia C'ed O'odham ancestry; daughter of Marta and Tom Childs Jr.; Ajo native.

Bell, Oscar. Oral historian; son of Charles "Charlie" Bell, early resident of the region; Fillman Bell's husband; Ajo native.

Bibles, D. Dean. Retired executive and state director for Bureau of Land Management and for the US Man in the Biosphere Program.

Bryant, George. Quechan elder from Yuma, Arizona.

Bufkin, Donald H. "Don." Avid historian and cartographer who was also director of Arizona Historical Society.

Burgos, Adolfo. Seri elder who lived at Desemboque, Sonora.

Cachora, Lorey. Quechan elder from Somerton, Arizona.

Coss, Harold "Hal." Retired National Park Service naturalist.

Franks, Glen. Yuma farmer who visited Tinajas Altas many times.

Hartmann, William K. Astronomer, historian, and artist.

Hayden, Julian D. Southwest desert archaeologist who lived in Tucson.

Hays, Richard E. "Hank." Retired US Border Patrol pilot who worked the Yuma area for many years.

Hoy, Wilton "Bill." Retired National Park Service historian.

Irwin, Dr. Ralph. Physician; husband of historian Catherine Irwin.

Jefferson, Preston. Quechan elder who lived at Yuma.

Johnson, Boma. Former Bureau of Land Management archaeologist in Yuma.

Manuel, Francis. O'odham elder who lived at San Pedro village.

McCraw, Joe. Retired chief patrol officer for US Border Patrol, Tacna.

Monson, Gale. Retired US Fish and Wildlife Service manager, including Cabeza Prieta National Wildlife Refuge.

Moser, Mary Beck. Linguist and ethnologist who lived with and studied the Seri of Sonora.

Orosco, Candelaria. Matriarch Hia C'ed O'odham elder who at times lived at Sonoyta, Ajo, and Quitobaquito.

Passos, Claude. Hia C'ed O'odham elder and former resident of Ajo.

Peacock, Doug. Outdoor writer and hiker; resides in Montana.

Peterson, Tom. Retired chief curator, Arizona Historical Society, Tucson.

Pumpelly, Raphael III. Grandson of Raphael Pumpelly and a petroleum engineer.

Ross, Margaret. Hia C'ed O'odham elder from Ajo.

Salywon, Andrew. Research Botanist at Desert Botanical Garden, Phoenix.

Voget, Ken. Retired manager Cabeza Prieta National Wildlife Refuge.

Wadsworth, James. Graduate student in history, University of Arizona, Tucson; assisted Thomas E. Sheridan in research.

Wooley, Walter. Retired miner; resided in Wellton.

Documentaries

The Desert Speaks (KUAT-TV, PBS, University of Arizona, Tucson) offers videos featuring Tinajas Altas in episodes "Desert Water Holes" (program number 607) and "The Devil's Highway" (program number 1108).

Websites

Southwest Environmental Information Network. 2011. http://swbiodiversity.org/seinet/index.php. Accessed March 20.

Stevens, Peter F. 2011. Angiosperm Phylogeny Website. Version 9, June 2008 onward. http://www.mobot.org/MOBOT/research/APweb/. Accessed March 20.

US Environmental Protection Agency website April 2010.

Maps

Barney, James M.

1908 *Map of Yuma County, Arizona.* Arizona State Capitol Archive G4333/Y/1908.

Bryan, Kirk

1922 *Routes to Desert Watering Places in the Papago Country, Arizona.* Maps. U.S. Geological Survey Water-Supply Paper No. 490-D. Washington, D.C.: US Government Printing Office.

1924 *Reconnaissance Geologic Map of the Papago Country, Arizona.* 1: 500,000 scale. US Geological Survey. Washington, D.C.: US Government Printing Office.

General Land Office (GLO)

1879 *Territory of Arizona.* 1: 1,100,000 scale. Compiled by C. Roeser. New York: Julius Bien.

1892 *Territory of Arizona.* Compiled and drawn by A. F. Dinsmore. 1 inch to 13 miles scale. Updated 1894. Phoenix: Arizona State Capitol Archive 4330/1892/US.

Gray, Andrew Belcher

1857 *Map of Proposed Arizona Territory:* from Explorations of A. B. Gray and others, to accompany memoir by Lieut. [Sylvester] Mowry, U.S. Army. 1: 3,300,000 scale. Cincinnati: Midleton, Wallace.

Lumholtz, Carl

1912 *Map of Southwestern Arizona and Northwestern Sonora Comprising the Region Formerly Called Papaguería.* From Original Surveys Made by Carl Lumholtz. Drawn by A. Briesemeister. Insert in *New Trails in Mexico.* New York: Charles Scribner's Sons.

Military Department of New Mexico

1867 Old Territory and Military Department of New Mexico. 1:500,000 scale. Corrected from 1859 edition. Washington, D. C.: US Topographic Bureau.

Stieler, Adolph

1875 Die Vereinigten Staaten Von Nord-Amerika, sheet 4. 1:3,700,000 scale. Gotha: Justus Perthes.

US Bureau of Topographic Engineers

1889 *Map of the New Mexico Territory.*

US Army Corps of Engineers

1942 Yuma, Arizona. 1: 250,000 scale. Number N3200-W11400/60.

US Geological Survey

1990 Tinajas Altas, Ariz.-Son. 7.5' map. 1: 24,000 scale. Number 32114-C1-TF-024.

Walker, Henry P., and Don Bufkin

1986 *Historical Atlas of Arizona.* Norman: University of Oklahoma Press.

Wilson, Eldred D.

1933 *Geologic Map of Southern Yuma County, Arizona.* Phoenix: Arizona Bureau of Mines.

1960 *Geologic Map of Yuma County, Arizona.* Arizona Bureau of Mines, scale 1:375,000. Now available as *Arizona Geological Survey Map* M-3-11. Phoenix: Arizona Bureau of Geology and Mineral Technology.

LITERATURE CITED

Abbey, Edward

1984a This Antique and Empty Land. *Geo* 6:50–59, 114–16.

1984b A Walk in the Desert Hills. *Beyond the Wall.* New York: Holt, Rinehart, and Winston.

Adams, D. K., and A. C. Comrie

1997 The North American Monsoon. *Bulletin of the American Meteorological Society* 78:2197–2213.

Ahlstrom, Richard V. N., editor

2000 *Living in the Western Papaguería: An Archaeological Overview of the Barry M. Goldwater Air Force Range.* SWCA Cultural Resource Report No. 98-186. Phoenix: ARCADIS Geraghty & Miller Environmental Services and Tucson: SWCA Environmental Consultants.

2001 *A Cultural Resources Overview and Assessment for the Cabeza Prieta National Wildlife Refuge.* SWCA Cultural Resource Report No. 01-24. Albuquerque: US Fish and Wildlife Service.

Ahlstrom, Richard V. N., Mark L. Chenault, and David Wrobleski

2000 Chronology and Culture History of the Papaguería. In *Living in the Western Papaguería: An Archaeological Overview of the Barry M. Gold-water Air Force Range,* edited by Richard V.N. Ahlstrom, pp. 67–135. SWCA Cultural Resource Report No. 98-186. Phoenix: ARCADIS Geraghty & Miller Environmental Services and Tucson: SWCA Environmental Consultants.

Ajo Copper News

1917 Ajo's Past, Present, and Future. December 7.

Altschul, Jeffrey H., and Joseph A. Ezzo

1994 The Expression of Ceremonial Space along the Lower Colorado River. In *Recent Research along the Lower Colorado River. Proceedings from a Symposium Presented at the 59th Annual Meeting of the Society for American Archaeology, Anaheim, California, April 1994,* edited by Joseph A. Ezzo, pp. 51–67. Technical Series No. 51. Tucson: Statistical Research, Inc.

Altschul, Jeffrey H., and Bruce A. Jones

1989 *A Cultural Resources Sample Survey of Operation Zones, Barry M. Gold-water Range, Marine Corps Air*

Station, Yuma, Arizona. Technical Series No. 24. Tucson: Statistical Research, Inc.

Altschul, Jeffrey H., and Adrianne G. Rankin

2008 *Fragile Patterns: The Archaeology of the Western Papaguería.* Tucson: Statistical Research, Inc.

Alvarez, Albert, and Kenneth Hale

1970 Toward a Manual of Papago Grammar. *International Journal of Linguistics* 36(2):83–97.

American Ornithologists' Union

1998 *Check-list of North American Birds,* 7th ed. Washington, D.C.: American Ornithologists' Union.

2000 Forty-second Supplement to the American Ornithologists' Union Check-list of North American Birds. *Auk* 117:847–58.

2010 Fifty-first Supplement to the American Ornithologists' Union Check-list of North American Birds. *Auk* 127:726–44.

Andrews, Bob, Bob Righter, and Mike Carter

1992 A Proposed Format for Local Bird Checklists. *Colorado Field Ornithologists Journal* 26(1):12–18. Jamestown, N.D.: Northern Prairie Wildlife Research Center Home Page.

Annerino, John

1996a *Adventuring in Arizona.* San Francisco: Sierra Club Books.

1996b Where Dreams Die. *Arizona Republic* 28 July–22 September, five-part series.

2009 *Dead in Their Tracks,* third edition. Tucson: University of Arizona Press.

Anonymous

1929 Death of John Hohstadt. *Arizona Historical Review* 2(3):6–7 (October).

Archivo Histórico del Estado de Sonora (AHES)

1834 Carpetón 60, Film 45, Microfilm on file at the Arizona Historical Society, Tucson.

Arizona Colorado River Commission

1938 Letter to President Franklin D. Roosevelt, August 4, 1937; letter from Lynn Lockhart, Chairman, Democratic State Central Committee to President Franklin D. Roosevelt, December 30, 1938.

Arizona Game and Fish Department

1997 *Wildlife Water Developments in Arizona: A Technical Review.* Phoenix: Arizona Game and Fish Department.

Austin, Amy T., and Lucía Vivanco

2006 Plant Litter Decomposition in a Semi-arid Ecosystem Controlled by Photodegradation. *Nature* 442(7102/3 August):555–58.

Bailey, L. R., editor

1963 *The A. B. Gray Report: Survey of a Route on the 32nd Parallel for the Texas Western Railroad, 1854, and Including the Reminiscences of Peter R. Brady Who Accompanied the Expedition.* Los Angeles: Westernlore Press.

Bailey, Tom

1959 The Devil's Highway. *True West* November-December:7–8, 42–45.

Baker, T. Lindsay

1980 The Fortune and Fate of the Famous Fortuna Mine. *Arizona Professional Engineer* (July):6–9.

Baldwin, Charlene, and Jack Mount

1991 *Maps of the Pimería.* Tucson: University of Arizona Library.

Barney, James M.

1943 Camino del Diablo [The Devil's Highway]. *Arizona Highways* 19(3):14–19.

Barrios Matrecito, Valdemar

1977 *Por las Rutas del Desierto.* San Luis R.C., Sonora: Patronato de Ediciones Culturales. Reprinted 1980 and 1988, Gobierno del Estado de Sonora.

Bean, Lowell John, and William Marvin Mason

1962 *Diaries and Accounts of the Romero Expeditions in Arizona and California, 1821–1826.* Los Angeles: Ward Ritchie Press.

Beattie, George W.

1928 Diary of a Ferryman and Trader at Fort Yuma, 1855–1857. *Annual Publications of the Historical Society of Southern California,* Part 1, pp. 89–128. Los Angeles: Historical Society of Southern California.

1929 Diary of a Ferryman and Trader at Fort Yuma, 1855–1857. *Annual Publications of the Historical Society of Southern California* Part 11:213–42. Los Angeles: Historical Society of Southern California.

Beck, Daniel D.

2005 *Biology of Gila Monsters and Beaded Lizards.* Berkeley: University of California Press.

Beck, W., D. J. Donahue, A. J. T. Jull, W. S. Broecker, G. Bonani, I. Hajdas, and E. Malotki

1998 Ambiguities in Direct Dating of Rock Surfaces Using Radiocarbon Measurements. *Science* 280 (June 26):2132–35.

Becker, Kenneth M., and Jeffrey H. Altschul

2003 *Historic Context for Prehistoric and Protohistoric Trails and Related Features at Yuma Proving Ground, Arizona.* Technical Report No. 03-13. Tucson: Statistical Research, Inc.

Bell, Fillman, Keith M. Anderson, and Yvonne Stewart

1980 *The Quitobaquito Cemetery and Its History.* Tucson: Western Archeological and Conservation Center, National Park Service, US Department of the Interior.

Bell, Willis H., and Edward F. Castetter

1937 The Utilization of Mesquite and Screwbean by the Aborigines in the American Southwest. *Ethnobiological Studies in the American Southwest.* University of New Mexico Bulletin, Biological Series 4(5):3–63.

Belnap, Jayne

2007 From the Ground Up: Biological Soil Crusts in the Sonoran Desert. In *Dry Borders: Great Natural Reserves of the Sonoran Desert,* edited by Richard S. Felger and Bill Broyles, pp. 343–48. Salt Lake City: University of Utah Press.

Belnap, Jayne, and Otto L. Lange

2001 *Biological Soil Crusts: Structure, Function, and Management.* New York: Springer-Verlag.

Betancourt, Julio L., Thomas R. Van Devender, and Paul S. Martin, editors

1990 *Packrat Middens: The Last 40,000 Years of Biotic Change.* Tucson: University of Arizona Press.

Bolton, Herbert E.

1919a Father Escobar's Relation of the Onate Expedition to California. *Catholic Historical Review* 5:19–41.

1919b *Kino's Historical Memoir of Pimeria Alta,* vols. I–II. Cleveland: Arthur C. Clark.

1930 *Anza's California Expeditions,* vols. I–V. Berkeley: University of California Press. Reprinted 1966, New York: Russell and Russell.

1931 *Outpost of Empire: The Story of the Founding of San Francisco.* New York: Alfred A. Knopf.

1936 *Rim of Christendom: A Biography of Eusebio Francisco Kino, Pacific Coast Pioneer.* New York: MacMillan. Reprinted 1984, Tucson: University of Arizona Press.

1949 *Coronado: Knight of Pueblos and Plains.* Albuquerque: University of New Mexico Press.

Bowden, Charles

1986 *Blue Desert.* Tucson: University of Arizona Press.

1992 *The Sonoran Desert.* New York: H. N. Abrams.

2007 Learning Nothing, Forgetting Nothing: On the Trail of Carl Lumholtz. *Journal of the Southwest* 49(3):356–68.

Bowen, Thomas, Richard Stephen Felger, and R. James Hills

2004 Chollas, Circles, and Seris: Did Seri Indians Plant Cactus at Circle 6? *Desert Plants* 20(2):26–35.

Bowers, Janice Emily

1988 *A Sense of Place: The Life and Work of Forrest Shreve.* Tucson: University of Arizona Press.

2004 *The Best Spring Ever: Why El Niño Makes the Desert Bloom.* Sacramento: California Native Plant Society.

Bradley, W. C., J. T. Hutton, and C. R. Twidale

1978 Role of Salts in Development of Granitic Tafoni, South Australia. *Journal of Geology* 86:647–54.

Brennan, Thomas C., and Andrew T. Holycross

2006 *A Field Guide to Amphibians and Reptiles in Arizona.* Phoenix: Arizona Game and Fish Department.

Briggs, Thomas H., and Karen A. Demsey

2000 Surficial Geology and Geomorphology of the Tinajas Altas Area. In The History of Tinajas Altas, pp. 23–39. *The Only Water for 100 Miles,* vol. 1, edited by Gayle Harrison Hartmann and Mary Charlotte Thurtle. SWCA Cultural Resource Report No. 98-270. Phoenix: ARCADIS Geraghty and Miller and Tucson: SWCA Environmental Consultants. Final draft on file, SWCA Environmental Consultants, Tucson.

Brown, B. T., and R. R. Johnson

1983 The Distribution of Bedrock Depressions (Tinajas) as Sources of Surface Water in Organ Pipe Cactus National Monument, Arizona. *Journal of the Arizona-Nevada Academy of Science* 18:61–68.

Brown, David E.

1998 Water for Wildlife. In *Proceedings of a Symposium on Environmental, Economic, and Legal Issues Related To Rangeland Water Developments, 13–15 November 1997,* pp. 9–16. Tempe: Center for the Study of Law, Science, and Technology, Arizona State University.

Broyles, Bill

1982 Desert Thirst: The Ordeal of Pablo Valencia. *Journal of Arizona History* 23:357–80.

1988 WJ McGee's "Desert Thirst as Disease." *Journal of the Southwest* 30:222–27.

1993 The Devil's Highway. *Arizona Highways* 69(2):4–13.

1995 Desert Wildlife Water Developments: Questioning Use in the Southwest. *Wildlife Society Bulletin* 23:663–75.

1996a *Organ Pipe Cactus National Monument: Where Edges Meet.* Tucson: Southwest Parks and Monuments.

1996b Surface Water Resources for Prehistoric Peoples of Western Papaguería of the North American Southwest. *Journal of Arid Environments* 33:483–95.

1997 Wildlife Water Developments in Southwestern Arizona. *Journal of the Arizona-Nevada Academy of Science* 30(l):30–42.

1998 Reckoning Real Costs and Secondary Benefits of Artificial Game Waters in Southwestern Arizona. In *Proceedings of a Symposium on Environmental, Economic, and Legal Issues Related to Rangeland Water Developments, 13–15 November 1997,* pp. 236–53. Tempe: Center for the Study of Law, Science, and Technology, Arizona State University.

2006 *Sunshot: Peril and Wonder in the Gran Desierto.* Tucson: University of Arizona Press.

Broyles, Bill, and Tricia L. Cutler

1999 Effect of Surface Water on Desert Bighorn Sheep in the Cabeza Prieta National Wildlife Refuge, Southwestern Arizona. *Wildlife Society Bulletin* 27(4):1082–88.

Broyles, Bill, Luke Evans, Richard Stephen Felger, and Gary Paul Nabhan

2007 Our Grand Desert: A Gazetteer. In *Dry Borders: Great Natural Reserves of the Sonoran Desert,* edited by Richard S. Felger and Bill Broyles, pp. 581–679. Salt Lake City: University of Utah Press.

Broyles, Bill, and Mark Haynes

2010 *Desert Duty: On the Line with U.S. Border Patrol.* Austin: University of Texas Press.

Broyles, Bill, Adrianne G. Rankin, and Richard S. Felger

2007 Native Peoples of the Dry Borders Region. In *Dry Borders: Great Natural Reserves of the Sonoran Desert,* edited by Richard S. Felger and Bill Broyles, pp. 128–46. Salt Lake City: University of Utah Press.

Bruder, J. Simon, Diane Fenicle, and Everett Bassett

1988 *Cultural Resources Technical Report for the Goldwater Range Environmental Assessment, Phase I.* Phoenix: Dames & Moore.

Brusca, Richard C.

2007 Invertebrate Biodiversity in the Northern Gulf of California. In *Dry Borders: Great Natural Reserves of the Sonoran Desert,* edited by Richard S. Felger and Bill Broyles, pp. 418–504. Salt Lake City: University of Utah Press.

Bryan, Kirk

1925 *The Papago Country, Arizona: A Geographic, Geologic, and Hydrologic Reconnaissance with a Guide to Desert Watering Places.* US Geological Survey Water-Supply Paper No. 499. Washington, D.C.: US Government Printing Office.

Bufkin, Donald H.

1983 The Making of a Boundary Between the United States and Mexico. *The Cochise Quarterly* 13(1 and 2):3–30.

Bull, William B.

1973 Local Base-level Processes in Arid Fluvial Systems. *Geological Society of America Abstracts with Programs,* vol. 5.

Burrus, Ernest J., SJ

1965 *Kino and the Cartography of Northwestern New Spain.* Tucson: Arizona Pioneers' Historical Society. Reprinted 1991, in *The Jesuit Missions of Northern Mexico,* edited by Charles W. Polzer, Thomas H. Naylor, Thomas E. Sheridan, and Diana Hadley. Spanish Borderlands Sourcebooks, vol. 19, edited by David Hurst Thomas. New York: Garland.

1971 *Kino and Manje: Explorers of Sonora and Arizona.* Rome: Jesuit Historical Institute.

Cabeza Prieta National Wildlife Range

1970 *Master Plan.* Washington, D.C.: US Fish and Wildlife Service.

Cain, James, III, Paul R. Krausman, John R. Morgart, Brian D. Jansen, and Martin P. Pepper

2008 Responses of Desert Bighorn Sheep to Removal of Water Sources. *Wildlife Monographs* 171:1–30.

Campbell, Sean W.

1999 Chemical Weathering Associated with Tafoni at Papago Park, Central Arizona. *Earth Surface Processes and Landforms* 24:271–78.

Castetter, Edward F., and Willis H. Bell

1942 *Pima and Papago Indian Agriculture.* Inter-Americana Studies No. 1. Albuquerque: University of New Mexico Press.

1951 *Yuman Indian Agriculture.* Albuquerque: University of New Mexico Press.

Chadwick, Douglas H.

2006 Songs of the Sonoran. *National Geographic* 210(3):124–49.

Chatwin, Bruce

1987 *The Songlines.* New York: Viking Books.

Chester, Charles C.

2006 *Conservation Across Borders: Biodiversity in an Interdependent World.* Washington D.C.: Island Press.

Childs, Thomas, with Henry F. Dobyns

1954 Sketch of the "Sand Indians." *Kiva* 19(2–4):27–39.

Clotts, Herbert V.

1915 *Nomadic Papago Surveys.* Los Angeles: US Indian Agency.

Colton, Harold S.

1945 The Patayan Problem in the Colorado River Valley. *Southwestern Journal of Anthropology* 1:114-21.

Comrie, Andrew C., and Bill Broyles

2002 Variability and Spatial Modeling of Fine-scale Precipitation Data for the Sonoran Desert of South-west Arizona. *Journal of Arid Environments* 50:573–92.

Corder, L. Gene

1987 *Yuma Border Patrol Sector: Green with Pride.* Yuma: US Border Patrol.

Corman, Troy E., and Cathryn Wise-Gervais, editors

2005 *Arizona Breeding Bird Atlas.* Albuquerque: University of New Mexico Press with Arizona Game and Fish Department.

Coville, Frederick V., and Daniel T. MacDougal

1903 *Desert Botanical Laboratory of the Carnegie Institution.* Publication No. 6. Washington, D.C.: Carnegie Institution.

Cremony, J. C.

1868 *Life Among the Apaches.* San Francisco: A. Roman & Company. Reprinted 1951 and 1954, Tucson: Arizona Silhouettes.

Crosswhite, Frank S.

1981 Desert Plants, Habitat and Agriculture in Relation to the Major Pattern of Cultural Differentiation in the O'odham People of the Sonoran Desert. *Desert Plants* 3(2):47–76.

Davis, Goode P., Jr.

1986 *Man and Wildlife in Arizona: The Exploration Period, 1824–1865,* edited by Neil B. Carmony and David E. Brown. Phoenix: Arizona Game and Fish Department.

Davis, James T.

1961 *Trade Routes and Economic Exchange among the Indians of California.* Reports of the University of California Archaeological Survey No. 54. Berkeley: University of California.

Del Rio, Ignacio

1981 A pretexto de los placeres y el real de la Cieneguilla, Sonora. *Memoria del VI simposio de historia de Sonora.* Hermosillo: Instituto de Investigaciones Históricas.

Devine, David

2004 *Slavery, Scandal, and Steel Rails: The 1854 Gadsden Purchase and the Building of the Second Transcontinental Railroad Across Arizona and New Mexico Twenty-five Years Later.* New York: iUniverse.

Dimmitt, Mark A., and Thomas R. Van Devender

2009 Sahara mustard (*Brassica tournefortii*): A New, Serious Threat to Low Desert Ecosystems in the Southwestern United States and Northwestern Mexico. In *Invasive Plants on the Move: Controlling Them in North America, Weeds Across Borders 2006 Conference, Hermosillo, Sonora, May 25–28, 2006,* edited by T. R. Van Devender, F. J. Espinosa-García, B. L. Harper-Lore, and T. Hubbard, pp. 241–46. Tucson: Arizona-Sonora Desert Museum.

Dobyns, Henry F., Paul H. Ezell, Alden W. Jones, and Greta Ezell

1960 What Were Nixoras? *Southwestern Journal of Anthropology* 16(2):230–58.

Dockal, James A., and Michael S. Smith

2005 Evidence for a Prehistoric Petroglyph Map in Central Arizona. *Kiva* 70:413–20.

Doelle, William H.

1980 *Past Adaptive Patterns in Western Papaguería: An Archaeological Study of Nonriverine Resource Use.* Unpublished Ph.D. dissertation, Department of Anthropology, University of Arizona, Tucson.

Donohue, John A., SJ.

1969 *After Kino: Jesuit Missions in Northwestern New Spain, 1711–1767.* Rome: Jesuit Historical Institute.

Dorn, Ronald I.

2004a Case Hardening. In *Encyclopedia of Geomorphology,* edited by Andrew S. Goudie, pp. 118–19. New York: Routledge.

2004b Rock Coating. In *Encyclopedia of Geomorphology,* edited by Andrew S. Goudie, pp. 870–73. New York: Routledge.

Dunne, Peter Masten

1955 *Jacobo Sedelmayr; Missionary, Frontiersman, Explorer in Arizona and Sonora.* Tucson: Arizona Pioneers' Historical Society.

Dye, Job Francis

1951 *Recollections of a Pioneer, 1830–1852: Rocky Mountains, New Mexico, California.* Los Angeles: Glen Dawson.

Dykinga, Jack

2004 *Jack Dykinga's Arizona.* Englewood, Colo.: Westcliffe.

Eckenburg, Buzz

1984 The Ride of El Camino del Diablo by Motorcycle. Ms. on file, Cabeza Prieta National Wildlife Refuge, US Fish and Wildlife Service, Ajo, Ariz.

Eek, Ann Christine

2007 The Secret of the Cigar Box: Carl Lumholtz and the Photographs from His Sonoran Desert Expedition, 1909–1910. *Journal of the Southwest* 49(3):369–418.

Egerton, Kearney

1974 *Somewhere Out There: Arizona's Lost Mines and Vanished Treasures.* Glendale, Ariz.: Prickly Pear Press.

Eiler, Lorraine Marquez, and David E. Doyel

2008 The Extinct Tribe. In *Fragile Patterns: The Archaeology of the Western Papaguería Archaeology,* edited by Jeffrey H. Altschul and Adrianne G. Rankin, pp. 605–30. Tucson: Statistical Research, Inc.

Eisenhower, John S. D.

1989 *So Far from God: The U.S. War with Mexico, 1846–1848.* New York: Doubleday.

Emory, William H.

1987 *Report on the United States and Mexican Boundary Survey, 1857–1859.* Austin: Texas State Historical Association. Originally published 1857–1859, Washington, D.C.: Cornelius Wendell.

Escandon, Patricia

1985 Economia y sociedad en Sonora: 1767–1821. In *Historia general del Sonora: De la conquista al estado libre y soberano de Sonora,* vol. 2, pp.273–95. Hermosillo: Gobierno del Estado de Sonora.

Etter, Patricia

1998 *The Southern Route, 1849: A History and Annotated Bibliography.* Spokane: Arthur H. Clark.

Evenari, Michael, Leslie Shanan, and Naphtali Tadmor

1982 *The Negev: The Challenge of a Desert.* Cambridge, Mass.: Harvard University Press.

Ezcurra, E., and V. Rodrigues

1986 Rainfall Patterns in the Gran Desierto, Sonora, Mexico. *Journal of Arid Environments* 10:13–28.

Ezcurra, Exequiel, Richard S. Felger, Ann D. Russell, and Miguel Equihua

1988 Freshwater Islands in a Desert Sand Sea: the Hydrology, Flora, and Phytogeography of the Gran Desierto Oases of Northwestern Mexico. *Desert Plants* 9:35–44, 55–63.

Ezell, Paul H.

1954 An Archaeological Survey of Northwestern Papaguería. *Kiva* 19(2–4):1–39.

1955 The Archaeological Delineation of a Cultural Boundary in Papaguería. *American Antiquity* 20(4):367–74.

Felger, Richard Stephen

1977 Mesquite in Indian Cultures of Southwestern North America. In *Mesquite: Its Biology in Two Desert Scrub Ecosystems,* edited by B. B. Simpson, pp. 150–76. Stroudsburg, Penn.: Dowden, Hutchinson, and Ross.

1980 Vegetation and Flora of the Gran Desierto, Sonora, Mexico. *Desert Plants* 2(2):87–114.

2000 *Flora of the Gran Desierto and Río Colorado of Northwestern Mexico.* Tucson: University of Arizona Press.

2007 Living Resources at the Center of the Sonoran Desert: Native American Plant and Animal Utilization. In *Dry Borders: Great Natural Reserves of the Sonoran Desert,* by Richard S. Felger and Bill Broyles, pp. 147–92. Salt Lake City: University of Utah Press.

Felger, Richard S., and Bill Broyles, editors

2007 *Dry Borders: Great Natural Reserves of the Sonoran Desert.* Salt Lake City: University of Utah Press.

Felger, Richard S., Bill Broyles, Michael Wilson, Gary Paul Nabhan, and Dale S. Turner

2007 Six Grand Reserves, One Sonoran Desert. In *Dry Borders: Great Natural Reserves of the Sonoran Desert,* edited by Richard S. Felger and Bill Broyles, pp. 3–26. Salt Lake City: University of Utah Press.

Felger, Richard S., Susan Rutman, Michael F. Wilson, and Kathryn Mauz

2007 Botanical Diversity of Southwestern Arizona and Northwestern Sonora. In *Dry Borders: Great Natural Reserves of the Sonoran Desert,* edited by Richard S. Felger and Bill Broyles, pp. 202–71. Salt Lake City: University of Utah Press.

Felger, Richard S., Dale S. Turner, and Michael F. Wilson

2003 Flora and Vegetation of the Mohawk Dunes, Arizona. *Sida, Contributions to Botany* 20:1153–85.

Felger, Richard S., and Thomas R. Van Devender

In press Flora of Tinajas Altas, Arizona—A Century of Botanical Forays and Forty Thousand Years of Neotoma Chronicles.

Felger, Richard S., Benjamin Theodore Wilder, and Juan Pablo Gallo-Reynoso

2011 Floristic Diversity and Long-term Vegetation Dynamics of Isla San Pedro Nolasco, Gulf of California, Mexico. *Proceedings of the San Diego Society of Natural History* 43:1–43.

Fenneman, Nevin M.

1931 *Physiography of the Western United States.* New York: McGraw-Hill.

Finch, L. Boyd

1996 *Confederate Pathway to the Pacific.* Tucson: Arizona Historical Society.

Flint, Richard

2005 What They Never Told You About the Coronado Expedition. *Kiva* 71(2):203–17.

Flint, Richard, and Shirley Cushing Flint

2005 *Documents of the Coronado Expedition, 1539–1542. "They Were Not Familiar with His Majesty nor Did They Wish to Be His Subjects."* Dallas: Southern Methodist University Press.

Fontana, Bernard L.

1965 An Archaeological Survey of the Cabeza Prieta Game Range, Arizona. Ms. on file, Arizona State Museum, University of Arizona, Tucson.

1974 Man in Desert Lands: The Piman Indians of the Sonoran Desert. In *Desert Biology II*, edited by G. W. Brown Jr., pp. 489–528. New York: Academic Press.

Forbes, Robert H.

1952 *Crabb's Filibustering Expedition into Sonora, 1857.* Tucson: Arizona Silhouettes.

Fox, Kelly M.

1997 Aftosa: The Campaign Against Foot-and-Mouth Disease in Mexico. *Journal of Arizona History* 38(1):23–40.

Franklin, Kevin

1996 Motor Cross. *Tucson Weekly* (March 21–27):42.

Franklin, Kimberly, and Francisco Molina-Freaner

2010 Consequences of Buffelgrass Pasture Development for Primary Productivity, Perennial Plant Richness, and Vegetation Structure in the Drylands of Sonora, Mexico. *Conservation Biology* 24:1664–73.

Gaillard, Capt. David D.

1896 The Perils and Wonders of a True Desert. *Cosmopolitan* 21:592–605.

Gaines, Edmund P., and Guadalupe Sanchez

2009 Current Paleoindian Research in Sonora, Mexico. *Archaeology Southwest* 23(3):4–5.

Gates, Paul W.

1968 *History of Public Land Law Development.* Public Land Law Review Commission. Washington, D.C.: US Government Printing Office.

Gehlbach, Frederick R.

1981 *Mountain Islands and Desert Seas.* College Station: Texas A&M University Press.

Gifford, Edward W.

1931 *Kamia of the Imperial Valley.* Bureau of American Ethnology Bulletin No. 97. Washington, D.C.: Smithsonian Institution.

1933 *The Cocopa.* University of California Publications in American Archaeology and Ethnology 31(5):257–334. Berkeley: University of California Press.

Gil, Patricio Robles, Exequiel Ezcurra, and Eric Mellink, editors

2001 *The Gulf of California: A World Apart.* Mexico City: Pegaso, Casa Lamm, and Agrupación Sierra Madre.

Goebel, Ted, Michael R. Waters, and Dennis H. O'Rourke

2008 The Late Pleistocene Dispersal of Modern Humans in the Americas. *Science* 319(5869):1497–1502.

Gonzales, Dionisio

1846 Guadalupe [de Altar], September 1. Archivo Histórico del Estado de Sonora (AHES), Carpetón 169, Film 46, Microfilm on file at the Arizona Historical Society, Tucson.

Goodwin, Grenville

1971 *Western Apache Raiding and Warfare,* edited by Keith H. Basso. Tucson: University of Arizona Press.

Goudie, A. S.

2004a Tafoni. In *Encyclopedia of Geomorphology,* edited by Andrew S. Goudie, pp. 1034–35. New York: Routledge.

2004b Weathering Pit. In *Encyclopedia of Geomorphology,* edited by Andrew S. Goudie, pp. 1115–16. New York: Routledge.

Grant, Richard

1994 On the Devil's Highway. *Esquire* (September):152–55.

Gregonis, Linda M.

2000 Indigenous Ceramics. In The Tinajas Altas Archaeological Survey, pp. 487–512. *The Only Water for 100 Miles,* vol. 2, edited by Gayle Harrison Hartmann and Mary Charlotte Thurtle. SWCA Cultural Resource Report No. 98-270. Phoenix: ARCADIS Geraghty and Miller and Tucson: SWCA Environmental Consultants. Final draft report on file, SWCA Environmental Consultants, Tucson.

Grey, Zane

1969 *Desert Gold.* New York: Pocket Books.

Griffin, David J.

2007 Birds of Arizona's Southwestern Borderland. In *Dry Borders: Great Natural Reserves of the Sonoran Desert,* edited by Richard S. Felger and Bill Broyles, pp. 280–302. Salt Lake City: University of Utah Press.

Griffith, James S.

2005 Voices from Inside a Black Snake: Religious Monuments of Sonora's Highways. *Journal of the Southwest* 47(2):233–48.

Groschupf, Kathleen D., Bryan T. Brown, and R. Roy Johnson

1988 *An Annotated Checklist of the Birds of Organ Pipe Cactus National Monument, Arizona.* Tucson:

Cooperative National Park Research Studies Unit, University of Arizona.

Guinn, J. M.

1911 The Sonoran Migration. In *Annual Publications of the Historical Society of Southern California, 1909–1910,* pp. 31–36. Los Angeles: J. B. Walters.

Gutmann, James T.

2002 Strombolian and Effusive Activity as Precursors to Phreatomagmatism: Eruptive Sequence at Maars of the Pinacate Volcanic Field, Sonora, Mexico *Journal of Volcanology and Geothermal Research* 113:345–56.

2007 Geologic Studies in the Pinacate Volcanic Field. *Journal of the Southwest* 49(2):189–243.

Hackenberg, Robert A.

1964 Aboriginal Land Use and Occupancy of the Papago Indians. US Department of Justice, Lands Division, Contract No. J-40974, Washington, D.C. Ms. on file, Arizona State Museum Library, University of Arizona, Tucson.

Haley, J. Evetts

1948 *Jeff Milton: A Good Man with a Gun.* Norman: University of Oklahoma Press.

Halloran, Arthur F.

1945 *Narrative Report, Cabeza Prieta Game Range, January-April,* p. 10. Ajo: Cabeza Prieta National Wildlife Refuge.

Hargett, Janet Lee

1967 *Louis John Frederick Jaeger: Entrepreneur at Yuma Crossing.* Unpublished master's thesis, University of Arizona, Tucson.

Hartmann, Gayle Harrison

1989 Does Tinajas Altas Have a Future? *Canyon Echo* (November):1, 2. Newsletter of the Grand Canyon (Arizona) Chapter of the Sierra Club.

Hartmann, Gayle Harrison, and Jane Kolber

2000 Rock Art. In The Tinajas Altas Archaeological Survey, pp. 549–94. *The Only Water for 100 Miles,* vol. 2, edited by Gayle Harrison Hartmann and Mary Charlotte Thurtle. SWCA Cultural Resource Report No. 98-270. Phoenix: ARCADIS Geraghty and Miller, and Tucson: SWCA Environmental Consultants. Final draft report on file, SWCA Environmental Consultants, Tucson.

Hartmann, Gayle Harrison, Jane Kolber, and Mary Charlotte Thurtle

2008 The Rock Art at Tinajas Altas: Cultural and Temporal Affinity. In *Fragile Patterns: Perspectives on Western Papaguería Archaeology,* edited by Jeffrey H. Altschul and Adrianne G. Rankin, pp. 311–28. Tucson: Statistical Research Inc.

Hartmann, Gayle Harrison, and Mary Charlotte Thurtle, editors

2000 *The Only Water for 100 Miles,* vols. 1 and 2. SWCA Cultural Resource Report No. 98-270. Phoenix: Arcadis: Geraghty and Miller, and Tucson: SWCA Environmental Consultants.

Hartmann, Gayle Harrison, and Mary Charlotte Thurtle

2001 The Archaeology of Tinajas Altas, A Desert Water Hole in Southwestern Arizona. *Kiva* 66(4):489–518.

Hartmann, William K.

1989 *Desert Heart: Chronicles of the Sonoran Desert.* Tucson: Fisher Books.

Haury, Emil W.

1975 *The Stratigraphy and Archaeology of Ventana Cave,* second printing. Tucson: University of Arizona Press. Originally published 1950.

Haury, Emil W., and Julian D. Hayden

1975 Preface 1975. In *Stratigraphy and Archaeology of Ventana Cave,* by Emil W. Haury. Tucson: University of Arizona Press.

Hayden, Julian D.

1967 A Summary Prehistory and History of the Sierra Pinacate, Sonora, Mexico. *American Antiquity* 32(3):335–44.

1969 Gyratory Crushers of the Sierra Pinacate. *American Antiquity* 34(2):154–61.

1976 Pre-Altithermal Archaeology in the Sierra Pinacate, Sonora, Mexico. *American Antiquity* 41(3):274–89.

1985 Food Animal Cremations of the Sierra Pinacate, Sonora, Mexico. *Kiva* 50(4):237–49.

1998 *The Sierra Pinacate.* Tucson: Southwest Center and University of Arizona Press.

2007 A Trip to Laguna Prieta. In *Dry Borders: Great Natural Reserves of the Sonoran Desert,* edited by Richard S. Felger and Bill Broyles, pp. 80–84. Salt Lake City: University of Utah Press.

2011 *Field Man: Life as a Desert Archaeologist,* edited by Bill Broyles and Diane E. Boyer. Tucson: University of Arizona Press.

Hayden, Julian, and Friends

1993 On Sheep Cremations and Massacres. In *Counting Sheep: 20 Ways of Seeing Desert Bighorn,* edited by Gary Paul Nabhan, pp. 27–36. Tucson: University of Arizona Press.

Heald, Weldon F.

1965 The Proposed Sonoran Desert National Park. *Arizona Highways* 41(11):2–5, 8, 9.

Hedges, Ken

1975 Notes on the Kumeyaay: A Problem of Identification. *The Journal of California Anthropology* 2(1): 71–83.

Hedges, Ken, and Diane Hamann

1998 Recording the Rock Art at Indian Hill, Anza-Borrego Desert State Park. *American Indian Rock Art* 22:37–56.

Heilen, Michael P.

2004 Julian Hayden's Malpais Model: A Pre-Clovis Claim from the American Southwest. *Kiva* 69(3):305–31.

Henderson, Randall

1940 Watering Places on the Devil's Highway. *Desert Magazine* 3(6):7–10, 34 (April).

Henry, Robert S.

2007 Mammals of the Sonoran Desert Borderland Reserves. In *Dry Borders: Great Natural Reserves of the Sonoran Desert,* edited by Richard S. Felger and Bill Broyles, pp. 272–79. Salt Lake City: University of Utah Press.

Henry, R. S., and L. K. Sowls

1980 *White-tailed Deer of the Organ Pipe Cactus National Monument, Arizona.* Cooperative National Park Resources Studies Unit Technical Report No. 6. Tucson: University of Arizona.

Herring, Patricia

1995 *General José Cosme Urrea: His Life and Times, 1797-1849.* Spokane: Arthur H. Clark.

Hervert, John J., Jill L. Bright, Robert S. Henry, Linden A. Piest, and Mark T. Brown

2005 Home-range and Habitat-use Patterns of Sonoran Pronghorn in Arizona. *Wildlife Society Bulletin* 33(1):8–15.

Hobbs, James

1872 *Wild Life in the Far West: Personal Adventures of a Border Mountain Man.* Hartford: Wiley, Waterman, and Eaton. Reprinted 1977 in *Garland Library of Narratives of North American Indian Captivities,* vol. 87, New York: Garland.

Hodkinson, Ian D., and John K. Jackson

2005 Terrestrial and Aquatic Invertebrates as Bioindicators for Environmental Monitoring, with Particular Reference to Mountain Ecosystems. *Environmental Management* 35(5):649–66.

Hoffmeister, Donald F.

1986 *Mammals of Arizona.* Tucson: University of Arizona Press.

Hogner, Dorothy Childs

1938 *Westward, High, Low, and Dry.* New York: E. P. Dutton.

Holloway, R. G.

2000 Pollen and Flotation Analysis from Ground Stone Artifacts at Sites AZ X:12:2(ASM) and AZ X:12:62(ASM). In The Tinajas Altas Archaeological Survey, pp. 759–70. *The Only Water for 100 Miles,* vol. 2, edited by Gayle Harrison Hartmann and Mary Charlotte Thurtle. SWCA Cultural Resource Report No. 98-270. Phoenix: ARCADIS Geraghty and Miller and Tucson: SWCA Environmental Consultants. Final draft report on file, SWCA Environmental Consultants, Tucson.

Hooper, Mildred, and C. R. Hooper

1977 Ride the Devil's Road. *Outdoor Arizona* May:23–26, 35.

1995 Riding the Devil's Road. *Southwest Passages* 3(2):80–84 (January-February).

Hornaday, William T.

1908 *Camp-Fires on Desert and Lava.* New York: Charles Scribner's Sons. Reprinted 1983, Tucson: University of Arizona Press.

Howell, Alfred B.

1918 Special Reports, 19U. In Field Reports of the U.S. Fish and Wildlife Service, 1860–1961. Field Notes, Record Unit 7176, Box 26, Folders 2–4. Archives and Special Collections, Smithsonian Institution, Washington, D.C.

Hoy, Wilton [Bill]

1990 Sonoyta and Santo Domingo. *Journal of Arizona History* 31:117–40.

1995 Hardscrabble Days at the Ajo Mines: George Kippen's Diary, 1855–1858, *Journal of Arizona History* 36:233–50.

1999 Don Tomas and Tomasito. *Journal of Arizona History* 40(l):l–28.

Huckell, Bruce B.

1982 *The Distribution of Fluted Points in Arizona.* Archaeological Series No. 145. Tucson: Arizona State Museum, University of Arizona.

Hufford, Kenneth

1966 Travelers on the Gila Trail, 1824–1850. *Journal of Arizona History* VII(1):1–8.

Humphrey, Robert R.

1987 *90 Years and 535 Miles: Vegetation Changes Along the Mexican Border.* Albuquerque: University of New Mexico Press.

Huntington, Mary Margaret

1946 El Camino del Diablo. *Westways* 38:9–11.

International Boundary Commission

1898 *Report of the Boundary Commission upon the Survey and Remarking of the Boundary between the United States and Mexico West of the Rio Grande, 1892–1896,* Parts I and II. Senate Document 247, 55th Congress, 2nd session. Washington, D.C: US Government Printing Office.

Irwin, Catherine

ca. 1970 *El Camino del Diablo.* Yuma: Yuma County Historical Society.

Ives, Ronald L.

1936 Melchior Diaz, The Forgotten Explorer. *Hispanic-American Historical Review* 16:86–90.

1960a The Last Journey of Melchior Diaz. *Journal of Geography* 49(2):61–66.

1960b Navigation Methods of Eusebio Francisco Kino, S.J. *Arizona and the West* 2:213–43.

1966a Retracing the Route of the Fages Expedition of 1781 (Part I). *Arizona and the West* 8(1):49–70.

1966b Retracing the Route of the Fages Expedition of 1781 (Part II). *Arizona and the West* 8(2):157–70.

1968 From Pitic to San Gabriel in 1782: The Journey of Don Pedro Pages. *Journal of Arizona, History* 9(4):222–44.

1984 *José Velásquez: Saga of a Borderland Soldier.* Tucson: Southwestern Mission Research Center.

1989 The Grave of Melchior Díaz: A Problem in Historical Sleuthing. In *Land of Lava, Ash, and Sand: The Pinacate Region of Northwestern Mexico,* compiled by James W. Byrkit, edited by Karen Dahood, pp. 155–66. Tucson: Arizona Historical Society. Originally published in *The Kiva* 25(2):31–40.

Jackson, John K., and Stuart Fisher

1986 Secondary Production, Emergence, and Export of Aquatic Insects of a Sonoran Desert stream. *Ecology* 67(3):629–38.

Jackson, John K., and Leopold Füreder

2006 Long-term Studies Of Freshwater Macroinvertebrates: A Review of the Frequency, Duration and Ecological Significance. *Freshwater Biology* 51:591–603.

Jackson, Robert

1994 *Indian Population Decline: The Missions of Northwestern New Spain, 1687–1840.* Albuquerque: University of New Mexico Press.

Jandrey, Gayle

2001 *The Millipede and Other Less Embraceable Friends.* Tucson: Cascabel Press.

Jeffers, Robinson

1928 The Dead Men's Child. In *Cawdor and Other Poems,* pp. 127–29. New York: Horace Liveright.

Johnson, Boma

1986 *Earth Figures of the Lower Colorado and Gila River Deserts: A Functional Analysis.* Arizona Archaeologist No. 20. Phoenix: Arizona Archaeological Society.

Jones, Susan C., and William L. Nutting

1989 Foraging Ecology of Subterranean Termites in the Sonoran Desert. In *Special Biotic Relationships in the Arid Southwest,* edited by Justin O. Schmidt, pp. 79–106. Albuquerque: University of New Mexico Press.

Karns, Harry J., editor and translator

1954 *Luz de Tierra Incognita: Unknown Arizona and Sonora, 1693–1721, by Captain Juan Mateo Manje.* Tucson: Arizona Silhouettes.

Kearney, Thomas H., and Robert H. Peebles

1960 *Arizona Flora.* Berkeley: University of California Press.

Kelly, William H.

1977 *Cocopa Ethnography.* Anthropological Papers of the University of Arizona No. 29. Tucson: University of Arizona Press.

Keith, Stanton B.

1978 *Index of Mining Properties in Yuma County, Arizona.* Bulletin No. 192. Tucson: Geological Survey Branch, Arizona Bureau of Geology and Mineral Technology.

Knipe, Theodore

1956 *Javelina in Arizona.* Phoenix: Arizona Game and Fish Department.

Krausman, Paul R., Andrew V. Sandoval, and Richard C. Etchberger

1999 Natural History of Desert Bighorn Sheep. In *Mountain Sheep of North America,* edited by Raul Valdez and Paul R. Krausman, pp. 137–91. Tucson: University of Arizona Press.

Kresan, Peter L.

2007 A Geologic Tour of the Lower Colorado Region of Arizona and Sonora. In *Dry Borders: Great Natural Reserves of the Sonoran Desert,* edited by Richard S. Felger and Bill Broyles, pp. 31–45. Salt Lake City: University of Utah Press.

Kroeber, Clifton B., and Bernard L. Fontana

1986 *Massacre on the Gila: An Account of the Last Major Battle Between American Indians, with Reflections on the Origin of War.* Tucson: University of Arizona Press.

Kubly, Dennis M.

1990 Limnological Features of Desert Mountain Rock Pools. In *Wildlife Water Development,* edited by G. K. Tsukamoto and S. J. Stiver, pp. 103–20. Proceedings of 1988 Symposium, Las Vegas, Nevada, sponsored by the US Bureau of Land Management, Nevada Department of Wildlife, and Nevada Wildlife Symposium.

1992 Aquatic Invertebrates in Desert Mountain Rock Pools: The White Tank Mountains, Maricopa County, Arizona. *Journal of the Arizona-Nevada Academy of Science* 26:55–67.

L'Amour, Louis

1957 *Last Stand at Papago Wells.* New York: Fawcett.

1966 *Kid Rodelo.* New York: Bantam Books.

Latta, Frank F.

1980 *Joaquin Murrieta and His Horse Gangs.* Santa Cruz, Calif.: Bear State Books.

Lawrence, James

2000 *Celebration of the Land: The Last Sanctuaries.* Mexico City: Agrupación Sierra Madre and CEMEX.

Ledesma-Vásquez, Jorge, and Ana Luisa Carreño

2010 Origin, Age, and Geological Evolution of the Gulf of California. In *The Gulf of California: Biodiversity and Conservation,* edited by Richard C. Brusca, pp. 7–23. Tucson: University of Arizona Press.

Lee, Raymond M.

1989 *The Desert Bighorn Sheep in Arizona.* Phoenix: Arizona Game and Fish Department.

Lenon, Robert

1987 The Routes of the Explorers in Pimeria Alta. Paper presented at the Arizona Historical Convention, Flagstaff, May 1987.

Linton, D. L.

1955 The Problem of Tors. *Geographical Journal* 121(4):470–86.

Lockwood, Frank

1968 *Pioneer Portraits.* Tucson: University of Arizona Press.

Logan, Kenneth A., and Linda L. Sweanor

2001 *Desert Puma: Evolutionary Ecology and Conservation of an Enduring Carnivore.* Washington D.C.: Island Press.

Lowe, Charles H., Cecil R. Schwalbe, and Terry B. Johnson

1986 *The Venomous Reptiles of Arizona.* Phoenix: Arizona Game and Fish Department.

Lucchitta, Ivo, Kristin McDougall, Donald G. Metzger, Paul Morgan, Gerald R. Smith, and Barry Chernoff

2001 The Bouse Formation and Post-Miocene Uplift of the Colorado Plateau. In *Colorado River Origin and Evolution,* edited by Richard A. Young and Earle E. Spamer, pp. 173–78. Grand Canyon: Grand Canyon Association.

Lumholtz, Carl

1889 *Among Cannibals: Account of Four Years' Travels in Australia, and of Camp Life with the Aborigines of Queensland.* London: Murray. Reprinted 1980, Canberra: Australian National University Press.

1902 *Unknown Mexico: A Record of Five Years Exploration among the Tribes of the Western Sierra Madre; in the Tierra Caliente of Tepic and Jalisco; and among the Tarascos of Michoacan.* New York: Charles Scribner's Sons.

1912 *New Trails in Mexico: Travels among the Papago, Pima and Cocopa Indians.* New York: Charles Scribner's Sons. Reprinted 1971, Glorieta, N.M.: Rio Grande Press; also reprinted 1990 Tucson: University of Arizona Press.

Lyon, Jerry, editor

2000 *Procurement, Processing, and Passing Through: An Archaeological Survey of 14,390 Acres in the Central Sauceda Valley (East TAC) on the Barry M. Goldwater Air Force Range in Southwestern Arizona.* SWCA Cultural Resource Report No. 98–267. Phoenix: ARCADIS Geraghty & Miller Environmental Services and Tucson: SWCA Environmental Consultants.

MacArthur, R. H., and E. O. Wilson

1967 *The Theory of Island Biogeography.* Princeton: Princeton University Press.

MacKay, William P.

1991 The Role of Ants and Termites in Desert Communities. In *The Ecology of Desert Communities,* edited by Gary A. Polis, pp. 113–50. Tucson: University of Arizona Press.

MacKay, William P., John C. Zak, and Walter G. Whitford

1989 The Natural History and Role of Subterranean Termites in the Northern Chihuahuan Desert. In *Special Biotic Relationships in the Arid Southwest,* edited by Justin O. Schmidt, pp. 53–77. Albuquerque: University of New Mexico Press.

Mallery, T. D.

1936 Rainfall Records for the Sonoran Desert. *Ecology* 17:110–21, 212–15.

Mares, Michael A.

2002 *A Desert Calling: Life in a Forbidding Landscape.* Cambridge, Mass.: Harvard University Press.

Martínez, Lorenzo

1838 Letter to José Urrea, Guadalupe de Altar, August 11. Archivo Histórico del Estado de Sonora (AHES), Carpetón 90, Film 45, Microfilm on file at the Arizona Historical Society, Tucson.

Mason, William

1963 The Sonora Trail. *Terra* 2(1):10–13 (Summer).

Mathiot, Madeleine

1973 *A Dictionary of Papago Usage.* Bloomington: University of Indiana Linguistic Monographs.

Matson, Daniel S., and Bernard L. Fontana, editors

1996 *Before Rebellion: Letters and Reports of Jacobo Sedelmayr, S.J.* Tucson: Arizona Historical Society.

McCarthy, Helen, Robert A. Hicks, and Clinton M. Blount

1985 *Functional Analysis of Bedrock Mortars: Western Mono Food Processing in the Southern Nevada.* Fair Oaks, Calif.: Theodoratus Cultural Research.

McCarty, Kieran, O.F.M., editor and translator

1975 The Colorado Massacre of 1781. *Journal of Arizona History* 16:221–25.

McClintock, James H

1916 *Arizona: Prehistoric, Aboriginal, Pioneer, Modern-The Nation's Youngest Commonwealth within a Land of Ancient Culture,* vol. 1. Chicago: S. J. Clarke.

McDonald, James E.

1959 "It Rained Everywhere But Here!"—The Thunderstorm-Encirclement Illusion. *Weatherwise* 12(4):158–60, 174.

McGee, Emma R.

1915 *Life of WJ McGee.* Farley, Iowa: private printing.

McGee, W. J.

1901 The Old Yuma Trail. *National Geographic* 12(3,4):103–7, 129–43.

1906a Climatology of Tinajas Altas, Arizona: A Preliminary Report. *Science* 23:721–30.

1906b Desert Thirst as Disease. *Interstate Medical Journal* 13:1–23. Reprinted 1988, *Journal of the Southwest* 30:228–53.

1912 Symptomatic Development of Cancer. *Science* 36:348–50 (September 13).

McGinnies, William G.

1981 *Discovering the Desert.* Tucson: University of Arizona Press.

McGuire, Randall H.

1982 Problems in Culture History. In *Hohokam and Patayan; Prehistory of Southwestern Arizona,* edited by Randall H. McGuire and Michael B. Schiffer, pp. 153–222. New York: Academic Press.

McGuire, Randall H., and Michael B. Schiffer, editors

1982 *Hohokam and Patayan: Prehistory of Southwestern Arizona.* New York: Academic Press.

McIvor, D. E., J. A. Bissonette, and G. S. Drew

1994 *A Critical Review of the Status of the Yuma Mountain Lion.* Logan: Utah Cooperative Fish & Wildlife Research Unit, Utah State University.

Mearns, Edgar A.

1907 *Mammals of the Mexican Boundary of the United States: A Descriptive Catalogue of the Species of Mammals Occurring in That Region; with a General Summary of the Natural History, and a List of Trees.* US National Museum Bulletin No. 56. Washington, D.C.: Smithsonian Institution. Reprinted 1974, New York: Arno Press.

Meier, Marilyn

1990 El Camino del Diablo. *Lapidary Journal* August:91–96.

Merlin, Pinau

1999 *A Field Guide to Desert Holes.* Tucson: Arizona-Sonora Desert Museum Press.

Michler, Lt. Nathaniel

1857–1859 Report. In *Report on the United States and Mexico Boundary Survey,* by William H. Emory, pp. 101–25. Washington, D.C.: Cornelius Wendell. Reprinted 1987, Austin: Texas State Historical Association.

Miller, George

1938 Letter from Boy Scout Executive to William Sawtelle, Arizona State Game Department, October 26, 1938.

Miller, Glenn A., and J. H. Jett

1979 *Status of Mineral Resource Information for the Luke Air Force Range, Arizona.* Phoenix: Arizona Department of Mineral Resources.

Mitchell, John D.

1952 Lost Ledge of Tinajas Altas. *Mining World* 14:45.

Mittermeier, Russell A., Cyril F. Kormos, Christina G. Mittermeier, Patricio Robles Gil, Trevor Sandwith, and Charles Besançon

2005 *Transboundary Conservation: A New Vision for Protected Areas.* Mexico City: CEMEX, Agrupación Sierra Madre, and Conservation International.

Monson, Gale, and Lowell Sumner

1980 *The Desert Bighorn: Its Life History, Ecology, and Management.* Tucson: University of Arizona Press.

Monson, Gale, and Allan R. Phillips

1981 *Annotated Checklist of the Birds of Arizona.* Tucson: University of Arizona Press.

Moraga, Raphael

1838 Diary, April 24–May 1; Guadalupe [de Altar], May 2. Archivo Histórico del Estado de Sonora (AHES), Carpetón 92, Film 45, Microfilm on file at the Arizona Historical Society, Tucson.

Murbarger, Nell

1960 Mexico's Desert Highway. *Desert Magazine* 23(1):27–29 (January).

Nabhan, Gary Paul

1980 *Ammobroma sonorae,* an Endangered Parasitic Plant in Extremely Arid North America. *Desert Plants* 2(3):188–96.

1982 *The Desert Smells Like Rain: A Naturalist in Papago Indian Country.* San Francisco: North Point Press.

1985 *Gathering the Desert.* Tucson: University of Arizona Press.

1994 *Desert Legends: Re-storying the Sonoran Borderlands.* New York: Henry Holt.

Nabhan, Gary Paul, Wendy Hodgson, and Frances Fellows

1989 A Meager Living on Lava and Sand? Hia C'ed O'odham Food Resources and Habitat Diversity in Oral and Documentary Histories. *Journal of the Southwest* 31(4):508–32.

Nabhan, Gary Paul, and Jacquie Gage Kahn

2000 Native American Uses of Tinajas Altas: An Ethnobotanical Perspective. In The History of Tinajas Altas, pp. 55–56. *The Only Water for 100 Miles,* vol. 1, edited by Gayle Harrison Hartmann and Mary Charlotte Thurtle. SWCA Cultural Resource Report No. 98-270. Phoenix: ARCADIS Geraghty and Miller and Tucson: SWCA Environmental Consultants. Final draft on file, SWCA Environmental Consultants, Tucson.

Nichol, A. A.

1939 O'Neill's Grave in O'Neill's Pass in the O'Neill Mountains. *Random Papers* [of the Southwest National Monuments, Casa Grande, Arizona] (July):65–67.

Nichols, Tad

2007 Afield with Desert Scientists. In *Dry Borders: Great Natural Reserves of the Sonoran Desert,* edited by Richard S. Felger and Bill Broyles, pp. 85–91. Salt Lake City: University of Utah Press. Reprinted from "Dry Borders," edited by Richard S. Felger and Bill Broyles. 1997. *Journal of the Southwest* 39(3–4):353–70.

North, Diane M. T.

1980 *Samuel Peter Heintzelman and the Sonora Exploring & Mining Company.* Tucson: University of Arizona Press.

Nye, William F.

1861 *Sonora: Its Extent, Population, Natural Productions, Indian Tribes, Mines, Mineral Lands, Etc., Etc.*, translated from the Spanish of Francisco Velasco. San Francisco: H. H. Bancroft.

Office of Information Services

1957? *Brief History - Vincent AFB, Arizona.* Yuma: Ariz.: Vincent Air Force Base.

Officer, James E.

1981 La Nacionalidad de Joaquin Murrieta. *Memoria de VT Simposio de Historia de Sonora.* Hermosillo: Institute de Investigaciones Historicas.

1987 *Hispanic Arizona, 1536–1856.* Tucson: University of Arizona Press.

Olszewski, Deborah I., Glenn P. Darrington, and Sharon K. Bauer

1995 *From the Aguila Mountains to the Crater Range: Cultural Resources Sample Survey of the North Tactical Range, Barry M. Goldwater Air Force Range, Southwestern Arizona.* Intermountain Cultural Resource Services Research Paper No. 25. Phoenix: Dames & Moore.

Ortiz, Alfonso, editor

1983 *Southwest. Handbook of North American Indians,* vol. 10, William G. Sturtevant, general editor. Washington, D.C.: Smithsonian Institution.

Palmer, J., and R. A. Neilson

1962 The Origin of Granite Tors on Dartmoor, Devonshire. *Proceedings of the Yorkshire Geological Society* 33(3):315–39.

Patzman, Stephen N.

1963 Louis John Frederick Jaeger: Entrepreneur of the Colorado River. *Arizoniana* (*The Journal of Arizona History*) IV(1):31–36.

Peacock, Doug

1986 A Lonely Desert Rendezvous. *American West* 23:62–71.

2005 *Walking It Off: A Veteran's Chronicle of War and Wilderness.* Spokane: Eastern Washington University Press.

Pendleton, Lorann

1984 Archaeological Investigations in the Picacho Basin. Report on file, Southeast Information Center, Imperial Valley College Museum, El Centro.

Penfield, Thomas

1973 *A Guide to Treasure in Arizona.* Deming, New Mexico: True Treasure Publications.

Phillips, Allan, Joe Marshall, and Gale Monson

1963 *The Birds of Arizona.* Tucson: University of Arizona Press.

Phillips, Steven J., and Patricia Wentworth Comus

2000 *A Natural History of the Sonoran Desert.* Tucson: Arizona-Sonora Desert Museum and Berkeley: University of California Press.

Polzer, Charles W., SJ

1982 *Kino Guide II: His Missions—His Monuments.* Tucson: Southwestern Mission Research Center.

Polzer, Charles W., SJ., and Thomas E. Sheridan, editors

1997 *The Presidio and Militia on the Northern Frontier of New Spain.* The Californias and Sinaloa-Sonora,1700–1765, vol. 2, part 1. Tucson: University of Arizona Press.

Priestley, Herbert, translator

1913 The Colorado River Campaign, 1781–1782: Diary of Pedro Fages. *Academy of Pacific Coast History Publications* 3(May):135–233.

Pumpelly, Raphael

1870 *Across America and Asia.* New York: Leypoldt and Holt.

1918 *My Reminiscences,* vol. 1. New York: Henry Holt. Reprinted 1965 as *Pumpelly's Arizona,* Tucson: Palo Verde Press.

Radding, Cynthia

1997 *Wandering Peoples: Colonialism, Ethnic Spaces, and Ecological Frontiers in Northwestern Mexico, 1700–1850.* Durham: Duke University Press.

Rappole, John H.

2000 *Birds of the Southwest: Arizona, New Mexico, Southern California, and Southern Nevada.* College Station: Texas A&M University Press.

Rea, Amadeo M.

1997 *At the Desert's Green Edge: An Ethnobotany of the Gila River Pima.* Tucson: University of Arizona Press.

1998 *Folk Mammalogy of the Northern Pimans.* Tucson: University of Arizona Press.

Redondo, Santiago

ca. 1838 Incomplete document. Archivo Histórico del Estado de Sonora (AHES), Carpetón 92, Film 45, Microfilm on file at the Arizona Historical Society, Tucson.

Reid, Jefferson, and Stephanie Whittlesey

1997 *The Archaeology of Ancient Arizona.* Tucson: University of Arizona Press.

Roberts, Heidi, Richard V. N. Ahlstrom, and David Tucker, editors

2000 *East TAC: An Archaeological Survey of 7,792 Acres on the Barry M. Goldwater Range in Southwestern Arizona.* SWCA Cultural Resource Report No. 99-7. Phoenix: ARCADIS Geraghty & Miller Environmental Services and Tucson: SWCA Environmental Consultants.

Rodack, Madeline

1968 Ghosts; Do Not Disturb! *Desert Magazine* 31(4):23–27 (April).

Rogers, Malcolm J.

1929 A-56 file. Museum of Man, San Diego.

1966 *Ancient Hunters of the Far West.* San Diego: Union-Tribune.

Rojas, Manuel

1986 *Joaquin Murrieta, El Patrio: El "Far West" Del Mexico Cercenado.* Mexicali: Gobierno del Estado de Baja California.

Roosevelt, Kermit

1920 *The Happy Hunting Grounds.* New York: Charles Scribner's Sons. Partially reprinted, 1993, in *Counting Sheep: 20 Ways of Seeing Desert Bighorn,* edited by Gary Paul Nabhan, pp. 75–89. Tucson: University of Arizona Press.

Rosen, Philip C.

2007 Reptiles and Amphibians of Arid Southwestern Arizona and Northwestern Sonora. In *Dry Borders: Great Natural Reserves of the Sonoran Desert,* edited by Richard S. Felger and Bill Broyles, pp. 310–37. Salt Lake City: University of Utah Press.

Rosenstock, S. S., W. B. Ballard, and J. C. deVos Jr.

1999 Viewpoint: Benefits and Impacts of Wildlife Water Developments. *Journal of Range Management* 52:302–11.

Rosenstock, Steven S., Chantal S. O'Brien, Robert B. Waddell, and Michael J. Rabe

2004 *Studies of Wildlife Water Developments in Southwestern Arizona: Wildlife Use, Water Quality, Wildlife Diseases, Wildlife Mortalities, and Influences on Native Pollinators.* Technical Guidance Bulletin No. 8. Federal Aid in Wildlife Restoration Project W-78–R. Phoenix: Arizona Game and Fish Department.

Rosenstock, Steven S., Vernon C. Bleich, Michael J. Rabe, and Carlos Reggiardo

2005 Water Quality at Wildlife Water Sources in the Sonoran Desert, United States. *Rangeland Ecology and Management* 58(6):623–27.

Rosentreter, Roger, Mathew Bowker, and Jayne Belnap

2008 *A Field Guide to Biological Soil Crusts of Western U.S. Drylands: Common Lichens and Bryophytes.* Denver: US Government Printing Office.

Rosenus, Alan

1995 *General M. G. Vallejo and the Advent of the Americans.* Albuquerque: University of New Mexico Press.

Russell, Steven M., and Gale Monson

1998 *The Birds of Sonora.* Tucson: University of Arizona Press.

Sanchez, M. Guadalupe

2001 A Synopsis of Paleo-Indian Archaeology in Mexico. *Kiva* 67(2):119–36.

Sanchez, Guadalupe, Edmund P. Gaines, and Vance T. Holliday

2009 El Fin del Mundo. *Archaeology Southwest* 23(3):6-7.

Santiago, Mark

1998 *Massacre at the Yuma Crossing: Spanish Relations with the Quechans—1779–1782.* Tucson: University of Arizona Press.

Sastre, Governor Mateo, and Viceroy Bucareli y Ursúa

1772–1773 Correspondence. February 12, 1772–December 28, 1773. AGN, PI, ff. 309–505. DRSW 41–1972. Microfilm copy available at Office of Ethnohistorical Research (OER), Arizona State Museum (ASM), University of Arizona.

Sauer, Carl. O.

1937 Communication (in reply to one by Ronald L. Ives regarding Melchior Díaz). *The Hispanic American Historical Review* 16:146–49.

Saxton, Dean, and Louise Saxton

1969 *Pima-Papago Dictionary.* Tucson: University of Arizona Press.

Schaefer, Jerry

1994 The Stuff of Creation: Recent Approaches to Ceramics Analysis in the Colorado Desert. *Recent Research along the Lower Colorado River,* edited by Joseph A. Ezzo, pp. 81–94. Technical Series No. 51. Tucson: Statistical Research, Inc.

Schneider, Joan S.

1993a Antelope Hill: A Cultural Resources Inventory and Inquiry into Prehistoric Mill Implement Quarrying and Production Behaviors Along the Lower Gila River, Yuma County, Arizona. In *Glyphs and Quarries of the Lower Colorado River Valley,* edited by Joseph A. Ezzo and Jeffrey H. Altschul, pp. ii–79 (part 1). Technical Series No. 44. Tucson: Statistical Research, Inc.

1993b *Milling-Implement Quarrying and Production in Eastern California and Western Arizona: A Behavioral Perspective.* Unpublished Ph.D. dissertation, University of California, Riverside.

Schroeder, Albert H.

1952 *A Brief Survey of the Lower Colorado River from Davis Dam to the International Border.* Santa Fe: National Park Service, US Department of the Interior.

1979 Prehistory: Hakataya. In *Southwest,* edited by Alfonso Ortiz, pp. 100–107. Handbook of North American Indians, vol. 9, William G. Sturtevant, general editor. Washington, D.C.: Smithsonian Institution.

Sedelmayr, Jacobo

1996 *Before Rebellion; Letters & Reports of Jacobo Sedelmayr, S.J.* Translated by Daniel S. Matson, with an introduction and annotations by Bernard L. Fontana. Tucson: Arizona Historical Society.

Segerstrom, K., and H. Henriquez

1964 *Cavities, or "Tafoni," in Rock Faces of the Atacama Desert, Chile.* Professional Paper No. 501-C, C121–C125. US Geological Survey, US Department of the Interior.

Sellers, William D., and Richard H. Hill

1974 *Arizona Climate, 1931–1972.* Tucson: University of Arizona Press.

Shackley, M. Steven

1993 Lithics and Obsidian Analysis. In *Technical Report.* Hunter-Gatherer Settlement, Subsistence, and Symbolism at White Tanks, Yuma Proving Ground, Arizona, vol. 1, edited by Jerry Schaefer, pp. 109–34. San Diego: Brian F. Mooney Associates.

2005 *Obsidian: Geology and Archaeology in the North American Southwest.* Tucson: University of Arizona Press.

Shafiqullah, M., P. E. Damon, D. J. Lynch, S. J. Reynolds, W. A. Rehrig, and R. H. Raymond

1980 K-Ar Geochronology and Geologic History of Southwestern Arizona and Adjacent Areas. In *Studies in Western Arizona,* edited by J. P. Jenney and C. Stone, pp. 201–43. Arizona Geological Society Digest No. 12. Tucson: Arizona Geological Society.

Shaw, Charles E., and Sheldon Campbell

1974 *Snakes of the American West.* New York: Alfred A. Knopf.

Sheldon, Charles

1993 *The Wilderness of the Southwest: Charles Sheldon's Quest for Desert Bighorn Sheep and Adventures with the Havasupai and Seri Indians,* edited by Neil Carmony and David E. Brown. Salt Lake City: University of Utah Press.

Sheridan, Thomas E.

1992 The Limits of Power. *Antiquity* 66(250):153–71.

1995 *Arizona: A History.* Tucson: University of Arizona Press.

1996 The O'odham (Pimas and Papagos): The World Would Burn Without Rain. In *Paths of Life: American Indians of the Southwest and Northern Mexico,* edited by Thomas E. Sheridan and Nancy J. Parezo, pp. 115–40. Tucson: University of Arizona Press.

1999 *Empire of Sand: The Seri Indians and the Struggle for Spanish Sonora, 1645-1803.* Tucson: University of Arizona Press.

Shreve, Forrest

1934 Rainfall, Runoff and Soil Moisture under Desert Conditions. *Annals of the Association of American Geographers* 24:131–56.

1951 *Vegetation of the Sonoran Desert.* Publication No. 591. Washington, D.C.: Carnegie Institution of Washington. Reprinted as vol. 1, Forrest Shreve and I. L. Wiggins, *Vegetation and Flora of the Sonoran Desert.* Stanford: Stanford University Press.

Shreve, Forrest, and Ira L. Wiggins

1964 *Vegetation and Flora of the Sonoran Desert,* vols. 1–2. Stanford: Stanford University Press.

Simmons, Norman M.

1966. Flora of the Cabeza Prieta Game Range. *Arizona Academy of Science* 4:93–104.

Slaughter, Mark C., David Tucker, and Annick Lascaux, editors

2000 *Trade Corridors and Ethnic Boundaries: An Archaeological Survey of 12,089 Acres along the San Cristobal and Growler Washes on the Barry M. Goldwater Range in Southwestern Arizona.* SWCA Cultural Resource Report No. 98–181. Phoenix: ARCADIS Geraghty & Miller Environmental Services and Tucson: SWCA Environmental Consultants.

Smith, Olga Wright

1956 *Gold on the Desert.* Albuquerque: University of New Mexico Press.

Sowls, Lyle K.

1997 *Javelinas and Other Peccaries: Their Biology, Management, and Use.* College Station: Texas A&M University Press.

Spencer, Jon E., and Philip A. Pearthree

2001 Headward Erosion Versus Closed-basin Spillover as Alternative Causes of Neogene Capture of the Ancestral Colorado River by the Gulf of California. In *Colorado River Origin and Evolution,* edited by Richard A. Young and Earle E. Spamer, pp. 215–19. Grand Canyon: Grand Canyon Association.

Spier, Leslie

1933 *Yuman Tribes of the Gila River.* Chicago: University of Chicago Press. Reprinted 1970, New York: Cooper Square; 1978, New York: Dover.

Spellenberg, R.

2003 *Boerhavia.* In *Flora of North America,* vol. 4, edited by Flora North America Editorial Committee, pp. 17–28. New York: Oxford University Press.

Stern, Peter, and Robert Jackson

1988 Vagabundaje and Settlement in Colonial Northern Sonora. *The Americas* 44(4):461–81.

Stewart, Bob, Priscilla Bodkin, and Hank Bodkin

2001 *Butterflies of Arizona: A Photographic Guide.* Arcata, Calif.: West Coast Lady Press.

Strand, Jennifer G.

2000 The Faunal Remains from the Tinajas Altas Archaeological Survey. In The Tinajas Altas Archaeological Survey, pp. 753–58. *The Only Water for 100 Miles,* vol. 2, edited by Gayle Harrison Hartmann and Mary Charlotte Thurtle. SWCA Cultural Resource Report No. 98-270. Phoenix: ARCADIS Geraghty and Miller and Tucson: SWCA Environmental Consultants. Final draft report on file, SWCA Environmental Consultants, Tucson.

Sykes, Godfrey

1927 The Camino del Diablo: With Notes on a Journey in 1925. *Geographical Review* 17:62–74.

1931 Rainfall Investigations in Arizona and Sonora by Means of Long-period Rain Gauges. *Geographical Review* 21:229–33.

1951 Summer Journey on the Devil's Road. *Desert Magazine* 14(6): 5–6 (April).

Taylor, David

2010 *Working the Line.* Santa Fe, New Mexico: Radius Books.

Thomas, Bob

1980 Death Toll Builds Every Year on 'Devil's Highway.' *Arizona Republic* July 10:B1,3.

1981 Highway of Death. *Arizona Republic* September 17:F9–10.

1987 Devil's Highway Holds Puzzles of Ancient Past [dateline Tinajas Altas]. *Arizona Republic* (August 9):Sl,6–7.

1989 Tinajas Altas: Life Hid in the Stones. *Arizona Republic* 18 June:T5.

1992 Devil's Highway: Hell on Earth. *Arizona Republic* 16 August: T2.

Thomas, Michael F.

2004 Deep Weathering. In *Encyclopedia of Geomorphology,* edited by Andrew S. Goudie, pp. 228–33. New York: Routledge.

Thomas, Robert K

1953 Papago Land Use West of the Papago Indian Reservation, South of the Gila River and the Problem of Sand Papago Identity. Ms., Arizona State Museum, Tucson.

1991 Papago Land Use West of the Papago Indian Reservation, South of the Gila River and the Problem of Sand Papago Identity. In *Ethnology of Northwest Mexico: A Sourcebook,* edited by Randall H. McGuire, pp. 357–99. New York: Garland.

Thompson-Olais, Laura

1994 *Sonoran Pronghorn Recovery Plan Revision.* Albuquerque: US Fish and Wildlife Service.

Townsend, Ben

1969 Graves of Gold. *True Treasure* January-February: 17–19.

Tucker, W. C., Jr.

1980 Tectonic Geomorphology of the Luke Air Force Range, Arizona. In "Studies in Western Arizona," edited by J. P. Jenney and C. Stone. *Arizona Geological Society Digest* 12:63–87.

Tucker, David B., editor

2000a *Footsteps on the Bajada: An Archaeological Survey of 15,813 Acres on the North Tactical Range of the Barry M. Goldwater Air Force Range in Southwestern Arizona.* SWCA Cultural Resource Report No. 99-140. Phoenix: ARCADIS Geraghty & Miller Environmental Services and Tucson: SWCA Environmental Consultants.

2000b *South TAC II: An Archaeological Survey of 7,083 Acres in the Northeastern Growler Valley on the Barry M. Goldwater Air Force Range in Southwestern Arizona.* SWCA Cultural Resource Report No.99-203. Phoenix: ARCADIS Geraghty & Miller, Environmental Services and Tucson: SWCA, Environmental Consultants.

Tunnicliff, Brock, Stanley K. Brickler, Margot W. Garcia, Jack Utter, Robert C. Johnson, Harold T. Coss, Gayle Harrison Hartmann, David F. Jickling, Mary E. Wilkosz, Jeffrey M. Sanders, Steven D. Moore, and Jack D. Doyle

1986 *Natural Resources Management Plan for Luke Air Force Range.* Tucson: School of Renewable Natural Resources, University of Arizona. Submitted to USAF-Tactical Air Command, Luke Air Force Base, Arizona.

Turnage, William V., and T. D. Mallery

1941 *An Analysis of Rainfall in the Sonoran Desert and Adjacent Territory.* Publication No. 529. Washington, D.C.: Carnegie Institution.

Turner, Raymond M., and David E. Brown

1982 Sonoran Desertscrub. In Biotic Communities of the American Southwest–United States and Mexico, edited by David E. Brown, *Desert Plants* 4(1–4):81–221.

Turner, Raymond M., Janice E. Bowers, and Tony L. Burgess

1995 *Sonoran Desert Plants: An Ecological Atlas.* Tucson: University of Arizona Press.

Udall, Morris K.

1967 A National Park for the Sonoran Desert. *Sierra Club Bulletin* 52(9):105–9 (October).

Udall, Stewart, with Jack Loefler

2007 The Sonoran Desert National Park. In *Dry Borders: Great Natural Reserves of the Sonoran Desert,* edited by Richard S. Felger and Bill Broyles, pp. 533–36. Salt Lake City: University of Utah Press.

Unitt, Phillip

1984 *The Birds of San Diego County.* Memoir No. 13. San Diego: San Diego Society of Natural History.

US Bureau of Land Management

1940–1942 Survey notes. In microfiche file for T13S, R17W GSRM AZ, vols. 3558 and 4276. Public Room, Bureau of Land Management, Phoenix.

US Department of the Interior

1997 Letter of Intent between the Department of the Interior of the United States and the Secretariat of Environment, Natural Resources, and Fisheries of the United Mexican States for Joint Work in Natural Protected Areas on the U.S.-Mexico Border, May 5.

US Public Law No. 99-606

1986 99th Cong., 6 Nov. 1986. Stat. 3457. *The Military Lands Withdrawal Act.*

US Public Law 106-65

1999 106th Cong. *The Military Lands Withdrawal Act of 1999.*

Urrea, Antonio

1851 Letter to the Governor of Sonora, Guadalupe [de Altar], September 19.

Van Devender, Thomas R.

1990 Late Quaternary Vegetation and Climate of the Sonoran Desert, United States and Mexico. In *Packrat Middens: The Last 40,000 Years of Biotic Change,* edited by Julio L. Betancourt, Thomas R. Van Devender, and Paul S. Martin, pp. 134–65. Tucson: University of Arizona Press.

2007 Ice Ages in the Sonoran Desert: Pinyon Pines and Joshua Trees in the Dry Border Region. In *Dry Borders: Great Natural Reserves of the Sonoran Desert,* edited by Richard S. Felger and Bill Broyles, pp. 58–68. Salt Lake City: University of Utah Press.

Van Valkenburgh, Richard

1945 Tom Childs of Ten-Mile Wash. *Desert Magazine* 9(2):3–6 (December).

1946 Notebook of Interviews with Tom Childs, December 29–31,1945. Ms. on file, Amerind Foundation, Dragoon, Arizona.

Vivian, R. Gwinn

1965 An Archaeological Survey of the Lower Gila River, Arizona. *Kiva* 30:95–146.

Voss, Stuart

1982 *On the Periphery of Nineteenth-Century Mexico: Sonora and Sinaloa, 1810-1877.* Tucson: University of Arizona Press.

Wallace, Henry D., and James P. Holmlund

1986 *Petroglyphs of the Picacho Mountains, South Central Arizona.* Anthropological Papers No. 6. Tucson: Institute for American Research.

Watchman, Alan

2000 A Review of the History of Dating Rock Varnishes. *Earth-Science Reviews* 49:261–77.

Waters, Michael

1982a Appendix H: Ceramic Data from Lowland Patayan Sites. In *Hohokam and Patayan: Prehistory of Southwestern Arizona,* edited by Randall McGuire and Michael B. Schiffer, pp. 571–88. New York: Academic Press.

1982b The Lowland Patayan Tradition, In *Hohokam and Patayan: Prehistory of Southwestern Arizona,* edited by Randall McGuire and Michael B. Schiffer, pp. 275–97. New York: Academic Press.

Waters, Michael R., and Thomas W. Stafford Jr.

2007 Redefining the Age of Clovis: Implications for the Peopling of the Americas. *Science* 315(5815):1122–26.

Weber, David J.

1992 *The Spanish Empire in North America.* New Haven: Yale University Press,.

Weight, Harold O.

1949 Gems on the Devil's Highway. *Desert Magazine,* 12(11):9–14 (September).

1952 Las Tinajas Altas. *Pacific Discovery,* 5:4–11.

1955 Waybill to the Lost Jabonero. *Desert Magazine,* 18(1):10–14 (January).

1968 Tin Lizzies and Iron Men. *Westways,* 32–35 (November).

Werner, Floyd, and Carl Olson

1994 *Insects of the Southwest.* Tucson: Fisher Books.

West, Robert C.

1993 *Sonora: Its Geographical Personality.* Austin: University of Texas Press.

Wheelock, Walt

1988a California Indian Trade Routes. *The Wrangler* 2(l):l, 4–5. San Diego: Westerners San Diego Corral.

1988b Indian Trade Routes. *CMSA Newsletter* 5(l):3–4 (November). San Diego: California Mission Studies Association.

1989 Foreword. In *Mohave Desert OHV Trails,* by Florine Lawler, pp. 7–14. Glendale, Calif.: La Siesta Press.

White, Chris

1974 Lower Colorado River Area Aboriginal Warfare and Alliance Dynamics. In *Antap California Indian Political and Economic Organization,* edited by Lowell John Bean and Thomas F. King, pp. 111–35. Ramona, Calif.: Ballena Press.

Whittlesey, Derwent

1951 Kirk Bryan, 1888–1950. *Annals of the Association of American Geology* 41(March):88.

Wiggins, Ira L.

1964 *Flora of the Sonoran Desert.* In *Vegetation and Flora of the Sonoran Desert,* F. Shreve and I. L. Wiggins, pp. 189–1740. Palo Alto: Stanford University Press.

Wilder, Benjamin Theodore, and Richard Stephen Felger

2010 Dwarf Giants, Guano, and Isolation: Vegetation and Floristic Diversity of Isla San Pedro Mártir, Gulf of California, Mexico. *Proceedings of the San Diego Society of Natural History* 42:1–24.

Wilder, Benjamin T., Richard S. Felger, H. Romero-Morales, and A. Quijada-Mascareñas

2007 New Plant Discoveries for the Sonoran Islands, Gulf of California, Mexico. *Journal of the Botanical Research Institute of Texas* 1:1203–27.

Williams, Anita Alvarez

1974 *The Cocopah People.* Phoenix: Indian Tribal Series.

1975 *Travelers Among the Cucupa.* Los Angeles: Dawson's Book Shop.

1983 Cocopa. In *Southwest,* edited by Alfonso Ortiz, pp. 99–112. Handbook of North American Indians, vol. 10, William G. Sturtevant, general editor. Washington, D.C.: Smithsonian Institution.

2004 *Primeros Pobladores de la Baja California.* Mexicali, BC: CONACULTA-INAH.

Williams, Jeanne

1976 *Lady Bought With Rifles.* New York: Coward, McCann & Geoghegen.

Williams, Kent

2004 *Border Gold.* Bloomington, Ind.: 1st Books Library.

Willson, Roscoe

1948 Another Lost Mine of State is Hidden in Desolate Area: Arizona Days. *Arizona Republic* (February 29).

1970 Father Kino Was Here: Arizona Days. *Arizona Republic* (March 15).

Wilson, Eldred D.

1931 New Mountains in the Yuma Desert, Arizona. *Geographical Review* 21:221–28.

1933 *Geology and Mineral Deposits of Southern Yuma County, Arizona.* Tucson: Arizona Bureau of Mines Bulletin No. 134.

Winship, George P., editor

1990 *The Journey of Coronado 1540–1542.* Golden, Colo.: Fulcrum.

Winters, Wayne

1963 *Campfires Along the Treasure Trail.* Tucson: Tombstone Nugget.

Witzeman, Janet L., Salome R. Demaree, and Eleanor L. Radke

1997 *Birds of Phoenix and Maricopa County, Arizona,* second edition. Phoenix: Maricopa Audubon Society.

Wolcott, Marjorie Tisdale

1929 *Pioneer Notes from the Diaries of Judge Benjamin Hayes, 1849–1875.* Los Angeles: private printing.

Wright, Doris Marion

1940–1941 The Making of Cosmopolitan California: An Analysis of Immigration, 1848–1870. *California Historical Society Quarterly* 19 (December) and 20 (March).

Young, Richard A., and Earle E. Spamer, editors

2001 *Colorado River Origin and Evolution.* Grand Canyon: Grand Canyon Association.

Zepeda, Ofelia

1985 *The Sand Papago Oral History Project.* Tucson: Western Archeological and Conservation Center, National Park Service, US Department of the Interior.

About the Authors

Bill Broyles is a research associate at the University of Arizona's Southwest Center. He taught in the public high school classroom for 31 years and now studies the Sonoran Desert and its people. His bookwork includes *Sunshot* and *Our Sonoran Desert*, co-authoring *Desert Duty*, and co-editing *Dry Borders* and *Field Man*. Nearly 40 years ago his imagination was captured by Tinajas Altas and its enduring story.

Gayle Harrison Hartmann first visited Tinajas Altas in 1971 with her husband Bill and has been fascinated by it ever since. In 1998 she had the good fortune to spend three months camping there while working on an archaeological survey. Over the last 35 years she has worked in a variety of archaeological capacities throughout southern Arizona including a stint as editor of *Kiva*, the quarterly journal of the Arizona Archaeological and Historical Society. Currently she is a research associate at the University of Arizona's Arizona State Museum.

Gary Paul Nabhan is an Arab-American writer, lecturer, food and farming advocate, rural lifeways folklorist, and conservationist whose work has long been rooted in the US-Mexico borderlands region he affectionately calls "the stinkin' hot desert." He recently accepted a tenured professorship as a Research Social Scientist based at the Southwest Center of the University of Arizona—his alma mater. His many books include *The Desert Smells Like Rain*, *Counting Sheep*, *Gathering the Desert*, *Desert Legends*, and *Cultures of Habitat*. When not gardening, caring for heritage breeds of sheep and turkeys, or hiking with his dogs, he is active in the Order of Ecumenical Franciscans, the Orion Society, and the local foods movement.

Thomas E. Sheridan holds a joint appointment as professor of Anthropology at the Southwest Center and the Department of Anthropology at the University of Arizona. He has authored or co-edited 12 other books, including *Arizona: A History*, *Los Tucsonenses: The Mexican Community of Tucson*, and *Landscapes of Fraud: Mission Tumacacori, the Baca Float, and the Betrayal of the O'odham*, which recently won the Past Presidents' Gold Award from the Association of Borderlands Studies. He directed the Mexican Heritage Project at the Arizona Historical Society from 1982 to 1984, and was Director of the Office of Ethnohistorical Research at the Arizona State Museum from 1997 to 2003.

Mary Charlotte Thurtle was the field director for the Tinajas Altas Archaeological Survey in 1998. She spent 13 years as a professional archaeologist researching the cultures of the Southwest authoring numerous reports and publications. Recently she turned her attention to documentary filmmaking and is the Executive Director of Pan Left Productions.

About the Appendix Contributors

Richard S. Felger. University of Arizona, Botany Department Herbarium; and San Diego Natural History Museum

David J. Griffin. Griffin Biological Services

Robert S. Henry. Arizona Game and Fish Department

Jacquie Gage Kahn. Environmental educator

Philip C. Rosen. University of Arizona, School of Natural Resources and the Environment

Thomas R. Van Devender. Sky Island Alliance

Index

Page numbers in italic indicate illustrations.

Abronia villosa, 231
acacia, catclaw (*Acacia greggii*), 11, 59, 146, 188, 200, 229
Acalypha californica, 228
Acanthaceae, 198, 219
Accipiter spp., 209
Achyronychia cooperi, 227
Acmispon spp., 229
Actitis macularia, 210
Aeronautes saxatalis, 210
agave, desert (*Agave deserti*), 64, 198, 220
Agelaius phoeniceus, 214
agriculture, 45, 52, 69, 159
Agua Dulce Spring, 28, 31, 58, *92*, 120, 130
Agua Escondida, 31, 34, 35, 41, 74, 77, 83, 85, 86. *See also* Cabeza Prieta Tanks; Spook Tank
Aguaje (Agua) de la Luna, 28, 31, 35, 41, 74, 77, 85. *See also* Heart Tank
Aguaje Empinado. *See* El Aguaje Empinado
Agua Salada, 28, 31, 78
Aguila Mountains, 45, 121, 159
Aguirre, Pedro, 135
Ainsa, Agustín, 122
Ajo, 48, 112, 115; mines at, 117–19
Akimel O'odham, 3, 50, *51*, 52, 55, 66, 91, 99
Alamo Muerto, 103
Alarcón, Hernando de, 71
algae, 185, 188
Aliciella latifolia, 234
Allen, Nadyne & Rufus "Rufe," 157, 159
Allionia incarnata, 231
Alta California. *See* California
Altar, 99, 101, *102*, 105, 156
Altar district, 101, 103, 108
amaranth (*Amaranthus* spp.), 219, 237
Ambrosia spp., 198, 218, *220*–21
Ammospermophilus harrisii, 202, 206
amphibians, 185, 203, 215
Amphispiza bilineata, 213
Amsinckia spp., 223
Anacardiaceae, 219
Anas spp., 208
Andrew, James W., 170
Ankrim, Captain (William?), 105
Annerino, John, 109, 169
Antelope Hill, 35, 48, 53, 96, 118, 136
Anthus spp., 213
Antrozous pallidus, 205
Anza, Juan Bautista de, 3, 66; California expeditions, 80–91
Apaches, 28, 48–49, 107, 110–11, 112, 118–19
Apocynaceae, 219
Aquila chrysaetos, 209
Aquiles Cerdán (Serdán), 159
archaeology: Archaic & Ceramic periods, 45–46, *47*; Paleoindian, 43–45; at Tinajas Altas, 9, 25, 59–63, 146–47, 169
Archaic period, 44, 45, 61
Archilochus alexandri, 210
Ardea spp., 208
Arenas, Luis, 101
Areneños. *See* Hia C'ed O'odham
Aristida spp., 233
Arizona Archaeological Society, 169
Arizona Colonization Company, 122
Arizona elegans, 216
Arizona Game & Fish Department, *25*–26, 35, *37*, 193
Arizona Game Protective Association, 170
Arizona Historical Society, 148, *175*
Arizona State Natural Areas, 173
arrowleaf, 223
Asclepias spp., 219–20
Asio spp., 210
Asparagaceae, 198, 220
Aspidocelis tigris, 203, 215
Asteraceae, 198, 218, 220–23
Astrolepis cochisensis, 218
Athene cunicularia, 210
Atriplex polycarpa, 198, 219
Auriparus flaviceps, 212
automobile: travel by, 145–46, 159–62, 163

Baccharis spp., 221
backswimmers, 185–87
badgers, 182, 202, 206
Bahiopsis parishii, 221
BAI. *See* US Bureau of Animal Industry
Bailey, Tom, 106
Baileya pleniradiata, 221
Baja California, 14, 73–74, 78, 80
Baker Peaks, 96
Baker Tanks, 40, 58, *75*, 78, 82, 96, 97, 131; Kino at, 35, *36*, 41, 77; travel routes, 28, 115, 116, 117
bandits, 28, 112, 113, 124

Barlow, John W., 169
Barney, James M., 161
barrel cactus, 59, 183, *184*, 199, 226
Basin-&-Range, 10, 14–15
Bassariscus astutus, 206
bats, 188, 205
beardgrass, cane, 233
beard-tongue, 232–33
beargrass. *See* tree-beargrass
beavertail cactus, 199
Bebbia juncea, 198, 221
bedrock grinding features, *55*, 59, *62–63*, 69, 146–47, 148, *175*, plate; pollen analysis of, 64–65
bedstraw, starry, 236
beetles, 183; aquatic, 185, *186–87*
Bell, Fillman & Mary, *55*
Bell Gang, 124
Benton, Thomas Hart, 169
Berry family, 154
Betty Lee Mine, 154
Bibles, Dean, 174
bighorn sheep, desert, 11, 82, 144, *178*, 179, *180–81*, 182, 201, 206, plate; cremation of, 59, 68; food plants, 219, 220, 226; hunting of, 66, 69, 124, 126, 131, 162
biological soil crust, 184
biological surveys, 142–43, 144–45
biosphere reserves, 174
birds, 182, 188. *See also by type*
blackbirds, 214
Black Tanks, 28, 115, 117
bladderpod, desert, *225*
blazingstar, 230
bluebirds, 212
boa, rosy, 216
bobcat, 144–45, 179, 202, 206
Boerhavia spp., 200, 218, 231–32
Bolton, Herbert, 31, *35*, 71, 83, 85
Bombycilla cedrorum, 213
Boraginaceae, 218, 223–24
border, 170; fence/wall, 127, 167, *171*; survey of, 128, 131–33, 135–36, *137*, *138*, 169
Border Gold (Williams), 165
Borrego Canyon, 15, 154–55
Borrego Tank, *25*
Botello, Narciso, 101
Bothriochloa barbinodis, 233
Bouchard, Hippolite, 95
boundary markers, 130, *131*, *132–33*
Bouse Formation, 14
Bouteloua spp., 233
Bowden, Charles, 169
Brady, Peter Rainsford, 128
Brandegea bigelovii, *227*
Brassicaceae, 199, 218, 224–25
brickell-bush, spiny-leaf, 221
brightwhite, 223
brittlebush, 59, 65, 198, 221
brome, red/foxtail, 233
broomrape, desert, 200, 232
Brown, David, 105
browneyes, 232
Bryan, Kirk, 19, *21*, 24, 124, 141, 160, *161*; maps by, 136, 139
Bryant, George, 58
Bubo virginianus, 210
Bubulcus ibis, 208
buckwheat, 218, 234
Buenoa sp., 185–86
buffelgrass, 218, 233
Bufo spp., 185, 203, *215*
Bunker, N. S., 122
buntings, 214
burrobush, 221
bursage, 11, 146, 183, 220
Bursera microphylla, 185, 199, *225*
Buteo spp., *209*

Caballero, Felix, 96–97, *98*
Caballo en Pelo (Cebello Pola), 154
Cabeza Prieta Mountains, 28, 95, 140
Cabeza Prieta National Wildlife Refuge, 5, 26, 43, *44*, 54, 77, 169, 170, 172, 173, 174, 193
Cabeza Prieta Pass, 83
Cabeza Prieta Peak, 10, 14, 34, 39, 89, 140
Cabeza Prieta Tanks, 28, *35*, 40, 68, *80*, 92, 97, 106, 112, 115, 119, 120, 122; Anza at, 35, 82, 83, 86, 87, 90, 91; Kino at, 3, 31, 34, 36, 74; Sedelmayr at, 77, 78
Caborca, 48, 58, 71, *102*, 122, *124*, 156
Caborca-Yuma Trail, 115
Cactaceae, 183, 199, 218, 226–27
Calamospiza melanocorys, 214
calico, Schott's, 234
Calidris spp., 210
California, 3, 14, 78, 103, 112; Anza &, 80–90; Gold Rush, 5, 9, 104–8
California, Gulf of, 9, 28, 52, *53*, 71, 73
Callipepla gambelii, 182, 208
Callisaurus draconoides, 203, 216
caltrop, California, 237
Calycoseris wrightii, 221
Calypte spp., 210
cambess, desert, 236
Camino del Diablo, El, 3, *7*, 28, 71, 101, 103, *104*, 124, 141, *150*, 194; Anza's travels on, 81–91; Apaches on, 49, 110–11; auto travel on, 145–46, 159–63; gold-seekers on, 105–12; horse gangs on, 112–13; Kino on, 31, 36, 73–77; Kippen on, 117–18; livestock on,

155–57; maps of, *4*, 139–40; Native American use of, 53, 66; routes across, 97, 99, 115–16, 119; Sedelmayr on, 77–78; water on, 9, 40–41
Campanulaceae, 227
Camp Calhoun, 106
Camp Grip, 127
campsites, 64, 131, 156; BAI, 148, *157*; O'odham, 68, 69; prospectors', 154–55; rock circles as, 60–61
Campylorhynchus brunneicapillus, 212
Canis latrans, 11, 179, 182, 201, 205
Caravajales, Juan, 156
Cardinalis sinuatus, 214
careless weed, 237
Carnegiea gigantea, 11, 183, *190*, 199, 226
Carnegie Desert Laboratory, 126, 145
carpetweed, threadstem, 231
Carpodacus spp., 214
Carr, Carolyn Ann, 43
Carreon, Francisco, 103
Carrizal. *See* El Carrizal
carrizo, 31
cartography, 127–28, 136–40
Caryophyllaceae, 227
Casares, Eustacio, 162
Castro, José, 104
Cathartes aura, 209
Catharus spp., 212
Catherpes mexicanus, 212
cattail: southern, 201, 236
cattle drives, 29, 156–57
cattle inspection stations, 157–59
Caulanthus lasiophyllus, 224
cedar, pygmy-, 198, *222*
Celaya, Alberto, 155, 156
Celaya, Diego, 103
cemetery: at Tinajas Altas, *133, 134*, 141, 144, 147–49
Cenchrus palmeri, 237
ceramics, *45–46, 48*, 52, 60, 66, 68
Cercidium. See Parkinsonia spp.
cereus: night-blooming, 199, 227
Chaenactis spp., 221
Chaetodipus spp. 206
Chaetura vauxi, 210
Chamaesaracha coronopus, 236
Chappel, Murray, *27, 114*
Charadrius vociferus, 209
chat, yellow-breasted, 213
Cheilanthes parryi, 218
Chenopodium murale, 198, 219
chickweed: Indian, 231
chicory: desert, 223
Childs family, Tom, 54, 116–*17*
Chilomeniscus cinctus, 216
chinchweed: desert, 222
Chionactis occipitalis, 216
cholla, 59, 183, 199, 218, 226
Chondestes grammacus, 213
Chordeiles acutipennis, 210
Chorizanthe spp., 234
chuckwalla: common, 66, *182*, 203, 216
chuckwalla delight, 198, 221
chuparosa, 183, 198, 219
Chylismia spp., 232
Cieneguilla, 3, 99–100, 101, *102*, 112
Cipriano Hills, 156
Cipriano Pass, 14, 28, 78, 86, 97, 156
Circus cyaneus, 209
climate, 11, 45, 185
Clovis points, 43, *44*
coachwhip, 216
Coccyzus americanus, 210
Cocomaricopa, José, 95
Cocomaricopas. *See* Pee-Posh
Cocopah, 48, 52, 53, 68, 71, 96, 97, 99; animal & plant use, 198–204
Cojuat, 96
Colaptes spp., 211
Coleonyx variegatus, 215
Collaer, Nick, 127
Colorado River, 9, 14, 48, 53, 66, 71, 73, 119, 156; gold on, 108–9, 152
Columba fasciata, 210
Colville, Frederick W., 145
comb-bur, 224
compass plant, 222
composites (Compositae), 220–23
Contopus spp., 211
Cooke, Philip St. George, 104
copepods, 187, 188
copperleaf, 228
Copper Mountains, 28, 35, 95, 96, 154
coralsnake: Sonoran, 216
Cordova, Alejandro, 110
Coronado expedition, 46, 71–73
Coronel, Antonio Franco, 101
Corvus corax, 182, 212
Costello, Miguel, 110
cottontail: desert, 11, 66, 182, 202, 206
cottontop: Arizona, 233
Couts, Cave Johnson, 108
cowbird: brown-headed, 214
coyotes, 11, 179, 182, 201, 205
Coyote Wash, *19*, 97, 219, 222, 230, 231, 233; Anza at, 83, 87
Coyote Water, *19, 20*, 28, 83, 86, 160
Coz-Villa, Rosalia, 116
Crabb, Henry Alexander, 121–22
Crassula connata, 227

Cremony, John, 48
creosotebush, 11, 90, 183, 201, 237
Crescent group, 154
crinklemat: Palmer's, 224
Croix, Teodoro de, 91
Crossosoma bigelovii, 227
Crotalus spp., 182, 204, 216
Crotaphytus nebrius, 216
crucillo: white, 201, 236
crustaceans, 184–85, 187, 188
cryptantha (*Cryptantha*) spp., 218
cuckoo: yellow-billed, 210
Cucurbita spp., 199, 227–28
curlew: long-billed, 210
Customs Patrol, 126
cyanobacteria, 185
Cylindropuntia spp., 59, 183, 199, 218, *226*
Cyperus squarrosus, 228

dalea: silky (*Dalea mollis*), 229
damselflies, 185, 187
Dart Tank, *25*
Dasyochloa pulchella, 233
datura: desert (*Datura discolor*), 236
Dendroica spp., 213
Descurainia pinnata, 199, 224
desert broom, 221
desert gold, 221
Desert Magazine, 163
desert-marigold: woolly, 221
Desert Speaks, The (TV show), 165
desertstar: Mohave, 222
desert varnish: dating, 60
desert velvet, 223
devil's claw, 231
Devil's Highway. *See* Camino del Diablo, El
Devil's lettuce, 223
Díaz, Juan, 85–86
Díaz, Melchior: travels of, 71–73
Digitaria californica, 233
Dimorphocarpa pinnatifida, 224
Dipodomys spp., 90, 179, 182, 202, 206
Dipsosaurus dorsalis, 66, 182, 203, 216
Ditaxis spp., 228
Dithyrea californica, 225
Dolores group, 154
Dome, 152
Dominguez, Cipriano, 162
doves, 182, 210, *211*
dowitcher: long-billed, 210
draba: wedgeleaf (*Draba cuneifolia*), 225
dragonflies, 185, 186, 187
Dripping Springs, 28–29, 35, *37*, 75
dropseed, 234
drug traffickers, 5
Dudleya arizonica, 227
Duncan, Dan, 165
Dunne, Peter Masten, 77, 78
Dye, Job Francis, 105
Dykinga, Jack, 165

eagle: golden, 209
earthquakes, 15
Echinocactus polycephalus, *184*, 226
Echinocereus engelmannii, 199, 226
egrets, 208
Egretta thula, 208
El Aguaje de la Purificación, 82, 83, 86. *See also* Cabeza Prieta Tanks
El Aguaje Empinado, 82, 83, 86, 90. *See also* Heart Tank
Elanus leucurus, 209
Elapidae, 216
El Camino del Diablo. *See* Camino del Diablo, El
El Capitán, 29, 154, 160
El Carrizal, *32*; Spanish use of, 31, 35, 40, 74, 77, 81, 89, 92, 95
elephant tree, 183, 185, 199, *225*
El Sahuaro, 159
Emory, William H., 104, 128, 169
Empidonax spp., 211
Encelia spp., 59, 65, 198, 221
ephedra: boundary (*Ephedra aspera*), 65, 199, 219
Eptesicus fuscus, 205
Eragrostis cilianensis, 233
Eremophila alpestris, 212
Eremothera chamaenerioides, 232
Erigeron lobatus, 221
Eriogonum spp., 218, 234
Erodium spp., 230
Eschscholzia minutiflora, 232
Eublepharidae, 215
Eucrypta micrantha, 224
Eulobus californicus, 232
Eumops perotis, 205
Euphagus cyanocephalus, 214
Euphorbiaceae, 199, 218, 228–29
Euphorbia spp., 218, 228
evening primroses, 232
Executive Order 6910, 170
Ezcura, Exequiel, 174
Ezell, Paul, 43, 60, 68

Fabaceae, 200, 218, 229
Fages, Pedro: campaigns of, 91–95
fagonbush: California, 236
Fagonia spp., 236–37
falcons, 182, 209
Fallis, Earl, 127

Felger, Richard, 143, *146*
Fer, Nicolas de, 136
ferns: brake/cloak, 218
Ferocactus cylindraceus, 199, 226
fescue: six weeks/eight-flowered, 234
fiddleneck, 223
filaree, 230
Filibuster Camp, 122
finches, 214
fir: desert, 198, *222*
fishhook cactus, 199, 227
fleabane, 221
Fleagle, Jake, 106
flickers, 211
floods, 19
Flores, José María, 104
fluffgrass, 233
fluffweed, 222
flycatchers, 211
Fly Field, 170
Font, Pedro, 87, 89, 90–91, 136
Foreign Miners Tax Law, 108
Forster, Juan, 101
Fortuna District, 154
Fortuna Mine, 124, 146, 154, 160
Fort Yuma, 119
49ers, 9; Anglo, 105–7; Sonoran, 107–8
Foster, Ol' Charlie, 154
Fouquieria splendens, 183, 200, 229–30
four o'clocks, *231*, 232
foxes, 179; gray, 205; kit, *192*, 201, 206
Franciscans, 3, 78
Franks, Glen, 163
Franks, Judy, 163
Free, Charlie, 106, 107
freighting, 119–21
frostmat, 227

Gadsden, James, 169
Gadsden Purchase, 115, 127–28, 169
Gaillard, David Du Bose, 141; border survey, 131–32, 133, 135–36
Galium stellatum, 236
galleta: big, 90, 234
Gallinago gallinago, 210
Gambelia wislizenii, 216
Gándara, Manuel María, 121, 122
Garcés, Francisco, 48, 80, 86, 87, 91
Garcia, Victoriano, 54
Gastelum, Francisco, 103
gecko: western banded, 182, 203, 215
Gehlbach, Frederick, 175
General Land Office, 136, 137, 169
Geococcyx californianus, 182, 210
geology, 10–11, 146, 152, 154; of granite, 15–18; landscape, 14–15; maps, 136, 139; water &, 18–19
Geothlypis trichas, 213
Geraea canescens, 65, 183, 221, *222*
germander: sticky, 230
Gila City, 108, 152
Gila monster, 204, 215
Gila Mountains, 14, 15, 53, 58, 75, 131
Gila River, 14, 48; travel routes, 28, 36, 53, 74, 75, 77–78, 90, 91, 97, 99, 108, 115, 116, 117, 119, 159
Gila Tanks, 117, 119
Gila Trail, 41, 87, 107
Gilg, Adam, 31
gilia, 234
Gilia stellata, 234
Glanton, John, 106
globemallow, 200, 230
gnatcatchers, 212
goldeneye: Parish's, 221
goldenweed: spiny, 223
goldfinches, 182, 214
Goldman, Edward Alphonso, 145
gold strikes, 103, 152, 155; California, 5, 9, 104–7; Sonorans at, 107–12; Spanish, 3, 99–101
Goldwater Range, Barry M., 5, 10, 44, 151, 169, 170, 174, 193
Gonzales, Dionisio, 104
Gonzalez, Faustino, 103
Gonzalez, Manuel, 74, 75
goosefoot: netleaf, 198, 219
gopher: Botta's pocket, 206
Gopherus agassizii, 66, 182, 203, 215
gourd: coyote/fingerleaf, 227
grackle: great-tailed, 214
grama, 233
Gran Desierto, El, 26, 28, 220
granite, 14; tafoni in, 15, 17, 18
Grant, Richard, 169
grasses, 218, 233–234
Grave of Eight, *152, 153*
graves, *152–53*; at Tinajas Altas, 109, 110, 126, *133, 134*, 141, 144, *147–49*, 151, 160
Gray, Andrew Belcher: maps by, 136, *138*; survey by, 127–28
graythorn, 201, 236
grinding stones, *55, 62–63*, 45, 48
grosbeaks, 214
ground cherry: desert, 236
groundsel: Mojave, 223
ground squirrels, 66, 182, 202, 206
Grusonia kunzei, 226
Guadalupe Victoria Ranch, 157
Guiraca caerulea, 214
gumhead, 221

Gunnery Range granite, 14, 15
Gymnosperma glutinosum, 221

Halchidhoma (Jalchedune), 3, 91
Hargrave, Lyndon, 45–46
Harlan, Annita, 143
harrier: northern, 209
HAVE HOST, 167
hawks, *209*
Hayden, Julian, *44*, 60, 63
Hayes, Benjamin, 108
Heald, Weldon, 173
Heart Tank, *34*, 40, 85, 89, 90, 92, 97, 112, 115, 157; Spanish use of, 3, 28, 31, 74, 77, 78, 82, 86
hedgehog cactus: strawberry, 199, 226
heliotrope: desert, 224
Heloderma suspectum, 204, 215
Henderson, Randall, 148, 163
heron: great blue, 208
heronsbill, 230
Hesperocallis undulata, 220
Heteropogon contortus, 234
Hia C'ed O'odham, 28, 50, 52, 62, 99, 140; & Apaches, 49, 112; & Cabeza Prieta Tanks, 34, 83; plant & animal use, 59, 65–66, 198–204, 229, 230; & Tinajas Altas, 9, 53–55, 58, 67, 68, 69, 103
hibiscus: rock (*Hibiscus denudatus*), 230
hierba del venado, 198, 223
High Tanks. *See* Tinajas Altas
Hilaria rigida, 90, 234
Hirundo rustica, 212
Hobbs Drag, 167
Hobbs, James, 105–6
Hoffmeister, Donald F., 144–45
Hogner, Dorothy, 163
Hohokam culture, 46, *48*, 60
Hohstadt, John, 156
Holzner, Frank Xavier, 141, 142
honeysweet, 219
hoof-&-mouth disease, 157
Hornaday, William T., 115, 162
horned lizards, 203, 216
horse gangs, 112–13
Horse Trail, 112
horseweed: Coulter's, 222
Horsfordia spp., 230
Howell, Alfred Brazier, 144
Hoy, Bill, 117
Hubbard, Cal, 106–7, 113
hummingbird-bush: desert, 198, 219
hummingbirds, 210, 219
Humphrey, Robert, 132–33
Hunter, Sherod, 116
hunting, 43, 45, 66; bighorn sheep, 69, 124, 126, 131, 162
Huntington, Mary, 163
Hypsiglena torquata (*chlorophaea*), 216
Hyptis albida, 230

ibis: white-faced, 209
ICBM Silo Superhardened Technology (ISST), 167
Icteria virens, 213
Icterus spp., 214
iguana: desert, 66, 182, 203, 216
inscriptions: historical, 150–51, 160
insects, 126, 183, 184, 185, 186; life cycles, 187–88
inspection station: BAI, 157–59
International Boundary Survey, 141–42
invasive species, 218, 224, 233
invertebrates, 183; & tinajas, 184–89
ironwood, 59, 146, *176–77*, 183, 200, 229
Irwin, Catherine, 148, 154
Ives, Ronald, 71, 83

Jabonero, El, 155
jackrabbit: black-tailed, 11, 66, 182, 202, 206
Jaeger, Louis J. F.: trips by, *119*–21
Jalchedune (Halchidomas), 3, 91
Jatropha cuneata, 185, 199, *228*, 229
javelina, 11, 179, 202, 206
Jeffers, Robinson, 165
Jefferson, Preston, 58
Jesuits, 77, 78
Johnson, Lyndon, 173
Joshua trees, 146
Joya Vieja mine, 154
junco: dark-eyed (*Junco hyemalis*), 214; California, 146; juniper, 11
Justicia californica, 198, 219

Kallstroemia californica, 237
Kamia, 48
kangaroo rats, 90, 179, 182, 202, 206
Kearney, Thomas H., 143
Kearny, Stephen Watts, 104
kestrel: American, 209
killdeer, 209
kingbird: western, 211
kinglets, 212
kingsnake: common, 204, 216
Kino, Eusebio Francisco, 46, 48; maps by, *70*, 136, 139; travels of, 3, 30–31, 34–36, *38*–41, 67, 73–77
Kinsey, Benjamin J., 136
Kippen, George, 112, 117–19
kissing bugs, 126
kite: white-tailed, 209
Klett, Mark, 163, 165

Krameria spp., 59, 200, 230
Kumeyaay, 48, 53, 61, 68

La Basura, 101
Lactuca serriola, 222
Laennecia coulteri, 222
Laguna Prieta, 29, 71, 78, *81*, 96, 97, 156, 157, 160
Lagunas, 152
Laguna Salada, 87, 91
La Joyita, 159
Lamiaceae (Labiatae), 230
Lampropeltis getula, 204, 216
langloisia: bristly (*Langloisia setosissima*), 234
Lanius ludovicianus, 211
La Posa District, 154
Laramide Orogeny, 14
lark: horned, 212
Larrea divaricata, 237
Las Cuevitas, 110
Las Palomas, 103
Las Playas, 14, 40, 95, 118, *121*, 156; travel routes, 28, 29, 31, 115
La Tinaja: Anza's, 85; Kino's, 35, 41, 74, 75, 77. *See also* Baker Tanks
La Tinaja Alta (Fages's), 92, 95
lavender: desert, 183, 230
Lechuguilla Mountains, 29, 156
Lechuguilla Tank, 156
Lechuguilla Valley, *12–13*, 15, 19, 43, 53, 71, 83, 95, 97, 122, 154, 167; plants found in, 220, 227, 228
Lepidium lasiocarpum, 199, 225
Leptotyphlops humilis, 216
Lepus californicus, 11, 66, 182, 202, 206
lettuce: prickly, 222
Levy, Mikul, 68
lianthus: Bigelow's (*Linanthus bigelovii*), 234
Libellula sp., 186
Lichanura trivirgata, 216
lily: Ajo/desert-, 183, 220
limberbush, 185, 199, *228*, 229
Limnodromus scolopaceus, 210
liveforever: Arizona, 227
livestock, 90, 112; BAI &, 157–59; on Camino del Diablo, 105, 155–57; & travel routes, 28, 29
lizards, 182, 203, 215–16
Llano de Tuzal, 90
Loasaceae, 230
Loeflingia squarrosa, 227
Loeseliastrum schottii, 234
Logfia spp., 222
Lopez, Francisco, 107
Los Alquives, 85. *See also* Dripping Springs
lotus: hairy, 229
lovegrass: stinking, 233
Lower Colorado Buff Ware, *46*, 60, 66, 68
Lower Colorado River Valley subdivision, 11, 218
Lower Drag, 167
Luke Air Force Gunnery Range, 151
Lumholtz, Carl, 22, 90, 91, 140, *144*, 154, 155, 169; on Hia C'ed O'odham, 68, 69; maps by, 136, *139*; *New Trails in Mexico*, 143; on Tinajas Altas graves, 141, 147
Lunt, Scott, *18*
lupine: Arizona, 229
Lycium spp., 65, 201, 218, 236
lycophytes, 218–19
lyre-pod (*Lyrocarpa coulteri*), 225

MacDougal, Daniel T., 145
Macrotus californicus, 202, 205
mallards, 208
Mallery, T. Dwight, *145*, 146
Malpais complex, 44
Malvaceae, 200, 230
mammals, 182, 205–6; human use of, 59, 201–2; scientific collection of, 142, 144–45. *See also by type*
Mammillaria spp., 199, 227
Manje, Juan Mateo, 31, 34, 46, 59, 73
Maricopa. *See* Pee-Posh
Marine Corps Air Station Yuma, 151, 170
Martin, Josefa Jacoba, 150–51
Martínez, Lorenzo, 103
Martyniaceae, 231
Masticophis flagellum, 216
mayflies, 187, 188
McCraw, Joe, 151
McGee, William John (WJ) "Don," 5, 68, 140–*41*, 146–47, 162
McKinney, Capt., 122, 124
meadowlark: western, 214
Mearns, Edgar Alexander, 141–*43*, 218
medicinal plants, 219, *225*, 229
Melanerpes uropygialis, 182, 210
melon: coyote, 228
Melospiza lincolnii, 214
Mendoza, Francisco, 124, 126
Mentzelia spp., 230
merlin, 209
Merrill, Augustine, 54
Merrill, William, 54
Mesa of the Dead, *134*, 141, 147, 219
mesquite: western honey, 11, 59, 63, 183, 188, 200, 229
Mexicali, 71, 156
Mexican Revolution, 124, 126
Mexican War, 104–5
Mexico 2 highway, 162
mice, 66, 182, 206
Michler, Nathaniel, 31, 128, 130, 135

Micrathene whitneyi, 210
microorganisms: in tinajas, 22, 23
Micruroides euryxanthus, 216
midges, 185, 187
Mid-Tertiary Orogeny, 14
migration: O'odham seasonal, 58–59, 66; to gold fields, 107–12
Military Lands Withdrawal Act of 1986 (Public Law No. 99-606), 170
Military Lands Withdrawal Act of 1999 (Public Law 106-65), 170
milkweeds, 219–20
Miller, Glenn, 154
Milton, Jefferson Davis, 126–*27*
Mimus polyglottos, 212
Mina del Desierto, 154
mines, 154, 155
Mirabilis spp., *231*, 232
Mission Camp, 41
missions, 3, 77, 101
mistletoe: desert, 201, 236
mockingbird: northern, 212
Mohave, 3, 47–48, 52, 99
Mohawk Dunes, 15
Mohawk Mountains, 14, 96
Mohawk Station, 41
Mollugo cerviana, 231
Molothrus ater, 214
Monoptilon bellioides, 222
Monterey (Calif.), 3, 78, 95
moonshiners, 159
Moraga, Rafael, 103
Morales, Francisco, 106
Mormon tea, 65, 199, 219
mortars, *55*; bedrock, 59, *62–63*, 146–47, 148, *175*, plate; pollen in, 64–65
Morton, Rogers, 173
mosquitoes, 185, 187, 188
motorcyclists, 148
mountain lions, 11, 179, 202, 206
mountain sheep. *See* bighorn sheep
muhly: littleseed (*Muhlenbergia microsperma*), 234
mules, 103, 105–6
Murrieta, Joaquín, 112, 113
Mustang Trail, 71, 112
mustards, 199, 218, 224
Myadestes townsendi, 212
Myiarchus spp., 211
Myotis spp., 202, 205

nama (*Nama* spp.), 224
National Park Service, 173, 174
Native Americans, 140, 197; animal & plant use, 11, 142–43, 182, 198–204, 219, 220, *225*, 226, 227, 229, 230; interrelationships, 52–53, 99; travel routes, 26, 28; at Tinajas Altas, 9, 64–66. *See also various groups; tribes*
needlegrass: desert, 234
Nemacaulis spp., 227, 234
Neotoma spp., 66, 145, 182, 202, 206
neststraw: desert, 223
New Trails in Mexico (Lumholtz), 143
Nichols, Tad, *145*, 146, 147, 163, 167
Nicotiana spp., 201, 236
nighthawk: lesser, 210
nightshade: false, 236
Níjoras, 101
Nolinaceae, 198
Notholaeana californica, 218
Notonecta sp., 185
Number 14 sites, 148–49
Numenius americanus, 210
Nyctinomops spp., 205

ocotillo, 183, 200, 229–30
Oenothera primiveris, 232
Ol' Charlie's Camp, 154, *155*
Old Smokey group, 154
Old Yuma Trail, 140–41, 157
Oligomeris linifolia, 236
Olneya tesota, 59, 146, *176–77*, 183, 200, 229
Onagraceae, 232
Oñate, Juan de, 48, 73
O'Neill, Dave, 150; grave of, *149*
O'Neill Hills, 83, 89
Onychomys torridus, 206
O'odham, 3, 50–52, 60, 68, 103, 131; on Anza's trip, 85, 86; plant use, 65–66; raids by, 112, 116; seasonal travel, 58–59. *See also* Akimel O'odham; Hia C'ed O'odham; Tohono O'odham
O'ovak, 55, 58, 68. *See also* Tinajas Altas
Oporornis tolmiei, 213
Opuntia basilaris, 199, 227
Oreoscoptes montanus, 213
Oreothlypis spp., 213
Organ Pipe Cactus National Monument, 5, 44, 145, 170, 172, 174
orioles, 214
Orobanche cooperi, 200, 232
Orosco, Candelaria, 49, 55
Ortega, Bartolo, 113, 157
Ortega, Cipriano, 113, 154, 155, 156, 157
Ortega, Francisco, 54
osprey, 209
ostracods, 185, 187
otat, 236–37
Otus kennicottii, *207*, 210

Ovis canadensis. See bighorn sheep
owls, *207*, 210

packrat middens, 11, 146, 218
packrats, 66, 145, 182, 202, 206
Paipai, 48, 53
Palafoxia arida, 222
paleoecology, 11, 14, 146
Paleoindians, 43–45
Palma, Captain, 91
Palo Ensebado, 100
paloverde, 11, 14, 146, 200, 229
Pandion haliaetus, 209
Papago. *See* Tohono O'odham
Papago Country, Arizona, The (Bryan), 19, *21, 135*, 136, 139, 160
Papago Tank, 156
Papago Well, 157, 159
Papaveraceae, 232
Parabuteo unicinctus, 209
Parastrellus hesperus, 205
Parietaria hespera, 236
Parker, Charlie, 122
Parkinsonia spp., 11, 14, 146, 200, 229
Passerculus sandwichensis, 214
Passerina amoena, 214
Patayan, 45–46, 60, 68
Peacock, Doug, 169
peccary: collared (*Pecari tajacu*), 11, 179, 202, 206
Pectis papposa, 222
Pectocarya spp., 224
Peebles, Robert H., 143
Pee-Posh, 3, 48, 52, 53, 58, 91, 99, 97
pellitory: desert, 236
Peniocereus greggii, 199, 227
penstemon (*Penstemon* spp.), 232–33
peppergrass: sand, 199, 225
Perityle emoryi, 222
Perognathus spp., 145, 202, 206
Peromyscus spp., 202, 206
Pesqueira, Ignacio, 121
Petalonyx linearis, 230
Petrochelidon pyrrhonota, 212
petroglyphs. *See* rock art
Peucephyllum schottii, 198, *222*
phacelia (*Phacelia* spp.), 224
phainopepla (*Phainopepla nitens*), 213
Phalaenoptilus nuttallii, 210
Pheucticus melanocephalus, 214
phoebes, 211
Phoradendron californicum, 201, 236
Phrynosoma spp., 203, 216
Phyllorhynchus spp., 204, 216
Physalis versicolor, 236
Physaria tenella, 225
Picoides scalaris, 210
pictographs. *See* rock art
pilgrimages, 116
pilot training, 151
Pima. *See* Akimel O'odham
Pinacate volcanic region, 14, 26, 66, 77, 126
pincushion, 221
piñon: single-leaf, 11, 146
Pinta Playa, 14
Pipilo spp., 213
pipistrelle, 205
pipits, 213
Piranga spp., 214
Pitiquito, 95
Pituophis catenifer, 216
Plantaginaceae, 232–33
plantain: wooly (*Plantago ovata*), 65, 200, 233
plants: Native American use of, 59, 64–66, 142–43, 219, 220, 225, 226, 227, 229, 230. *See also by genera; type*
plate tectonics, 14
Plegadis chihi, 209
Pleistocene: vegetation, 11, 146
Pleurocoronis pluriseta, 223
Poaceae, 218, 233–34
pocket mice, 145, 202, 206
Polemoniaceae, 234
Polioptila spp., 212
Polygonaceae, 218, 234
Pooecetes gramineus, 213
poorwill: common, 210
poppy: little gold, 232
Porophyllum gracile, 198, 223
Portilla, Pablo de la, 103
Portulaca halimoides, 234
Poston, Charles Debrille, 116
pottery. *See* ceramics
Pozo Blando, 86, 87
pozos, 9, *26*
Pozos de en Medio, 28, 83, 86, 87, 89, 90. *See also* Coyote Water
Pozos (Pozas) de San Miguel, 28, 91–92, *93*, 95
Prenanthella exigua, 223
presidios, 3, 99
prickly pear: dwarf, 199, 227
primrose, 183
Proboscidea altheifolia, 231
Proebstel, Ike and Win, 162
projectile points, 43–*44*, 45, 61
pronghorn, 66, 179, 182, 201, 206
Prosopis glandulosa, 11, 59, 63, 183, 188, 200, 229
prospectors, 103, 151–52; campsites, 154–*55*; Spanish-era, 100–101
Psathyrotes ramosissima, 223

Pseudorontium cyathiferum, 233
Pteridaceae, 218
Public Land Order No. 5493, 173
Puerto Blanco, 87, *89*, 90
Puerto Blanco Mountains, 159
Puffer, Charles, 161
Pumpelly, Raphael, 22, 54, 113, 115–*16*, 124, 159–60
Purísima Concepción mission, La, 91
purplemat, 224
purslane: silkcotton, 234
pyrrhuloxia, 214

quail: Gambel's, 182, 208
Quechan, 3, 47, 48, *49*, 52, 53, 58, 71, 78, 99, 106; animal & plant use, 198–204; Spanish &, 91, 93, 95; travel routes, 96, 97
Quechan trail, 96, 97
Quiscalus mexicanus, 214
Quitobaquito, 48, 93, 110, 155; Hia C'ed O'odham at, 52, 68, 69
Quitovac, 58, 69, 71, *102*, 103

Rafinesquia neomexicana, 223
ragweed: canyon, 59, 65
raiding: Apache, 28, 48–49, 118–19; O'odham, 112, 116
rainfall, 11, *12–13*, 146, 185, 188
Rancho La Berruga (El Bámori), 112
ratany, 59, 200, 230
rattlesnakes, 182, 204, 216
raven: common, 182, 212
Raven Butte Tank, *114*
rebellions: O'odham, 103; Quechan, 91
recreation, 162–63, *166*
Redondo, José María, 155, 156
Redondo, Santiago, 103
Regulus calendula, 212
repeat-photography analysis, 132–33
reptiles, 59. *See also by species; type*
Resedaceae, 236
revolutionaries, 124, 126
Rhamnaceae, 201, 236
Rhinocheilus lecontei, 216
Rhus kearneyi, 219
Ringler, Mike, *128*
ringtail, 206
Riparia riparia, 212
roadrunner: greater, 182, 210
roads, 160–62; bulldozed, *167*, 169
roasting pits, 64
Roberson, David F., 151
robin: American, 212
rock art, 9, 61–*62*, 68, 69, 136, 147. *See also* inscriptions
rock circles, 46, 60–*61*, 68
rock-daisy: desert, 222
rockflower: ragged, 227
rockpea: desert, 229
rodents, 11, 145, 179, 182, 206. *See also by species; type*
Rogers, Malcolm, 43, *44*, 60, 147
Rojas, Manuel, 71, 112
Romero, José, 53, 95–97, *98*
Roosevelt, Franklin, 170
Roosevelt, Kermit, 126, 141, 155, 162
Roosevelt, Theodore, 170
rotifers, 185, 186, 187, 188
Rubiaceae, 236
Rutaceae, 236

S.A. Tank, 83
sage: bladder, 230; Mojave, 146
saguaro, 11, 183, *190*, 199, 226
Salazar, Francisco, 49, 110–12
Salazaria mexicana, 230
Salinas de Gortari, Carlos, 174
Salpinctes obsoletus, 212
saltbush: desert, 198, 219
salt expeditions: Tohono O'odham, 53
Salvadora hexalepis, 216
San Antonio strike, 103
sandbur: giant, 237
sandfood, *53*
sand-lily: blue, 220
sandmat, 227
Sandoval, Sr., 156
Sand Papago. *See* Hia C'ed O'odham
sandpaper plant: narrowleaf, 230
sandpipers, 209–10
San Francisco de Asís, 100
San Ildefonso de Cieneguilla. *See* Cieneguilla
San Marcelo de Sonoyta. *See* Sonoyta
Sanora Mendez, Ramón, 156–57
San Perfecto, 103
Santalaceae, 201, 236
Santa Rosa de Buenavista, 100
Santo Domingo hacienda, 156
sapsucker: red-naped, 210
Sastre, Mateo, 112
Sauceda Valley, 45
Sauromalus ater, 66, *182*, 203, 216
Sayornis spp., 211
Scaphiopus couchii, 203, 215
Sceloporus magister, 216
Schismus barbatus, 218, 234
Schott, Arthur, 130, 131, 140
Sedelmayr, Jacob, 77–78, *79*
sedge: dwarf, 228
Selaginella eremophila, 219
Selasphorus rufus, 210
Senecio mohavensis, 223

Seri, 49, 52, 71
Sheldon, Charles, 124, *125*, 126, 155, 160, 162
Shoot, Piny, 106, 107
shoveler: northern, 208
Shreve, Forrest, *145*
shrike: loggerhead, 211
shrimp, 185, 187, 188, *189*
Shuchard, Charles: drawings, 128, *129*
Sialia spp., 212
Sierra del Rosario, 126
Sierra Pinta, 28, *89*
silverbush, 228
silver strikes, 103
siskin: pine, 214
skunks, 182; spotted, 206
sleeping circles, 46
Smith, Thomas, 105
Smith Mine, 155
Smoke Tree Wash, 40
smuggling, 124
snails: pulmonate, 187
snakes, 182, 204, 216
snapdragon: desert, 233
snipe: Wilson's, 210
Solanaceae, 201, 218, 236
solitaire: Townsend's, 212
Sonchus spp., 223
Sonora, 124; gold in, 99–101, 103
Sonoran Desert National Park & Preserve, 5, 170, 172–73, 174
Sonora semiannulata, 216
Sonora Trail, 101, *102*, 116
Sonoyta, 93, 110, 156; travel through, 27–28, 31, 36, 58, 71, 73, 74, 77, 78, 81, 89, 115
Sonoyta River, 9, 14, 28, *32*, 89
Southern Overland Trail, 105
Southern Route, 5, 105
sowthistle, 223
spadefoot: Couch's, 203, 215
Spanish needles, 222
sparrows, 213–14
spectacle-pod, 224, 225
Spermophilus: *harrisii saxicolus*, 145; *tereticaudus*, 202, 206
Sphaeralcea spp., 200, 230
Sphyrapicus nuchalis, 210
spiderling, 200, 218, 231–32
spiders: crab, 183
spikemoss, 219
Spilogale putorius, 206
spineflower, 234, *235*
Spinus spp., 182, 214
Spizella spp., 213
Spook Tank, 28, 83, *85*, 86, *100*
Sporobolus cryptandrus, 234
spurge, 228
starling: European, 213
starvine: desert, *227*
Steinmetz, George, 165
Stelgidopteryx serripennis, 212
Stellula calliope, 210
Stephanomeria pauciflora, 223
Stieler, Adolf, 118
Stipa speciosa, 234
stonecrop: pygmy, 227
storksbill (*Erodium* spp.), 230
Sturnella neglecta, 214
Sturnus vulgaris, 213
Stylocline micropoides, 223
sumac: desert/Kearney, 183, 219
suncups, 232
sunflower: desert, 65, 183, 221, *222*
Suni, Francisco "Tulpo," 54
surveyors, surveys: border, 127–36
Surveyors Canyon, 66
Surveyors Tank, 132, 160
Suvuk, 69
swallows, 212
sweetbush, 198, 221
swifts, 210
Sykes, Godfrey, 5, 145, 146
Sylvilagus audubonii, 11, 66, 182, 202, 206

Tachycineta spp., 212
tackstem, 221
Tacna, 117, 127
Tadarida brasiliensis, 205
tafoni, 15, *16–17*, 18
tanagers, 214
tanglehead, 234
Tarabal, Sebastián, 80
tarantulas, 183
Taylor, David, 133
teal, 208
termites, 183
Teucrium glandulosum, 230
Texas Hill, 96
Thamnosma montana, 236
Thomas, Bob, 167
Thomomys bottae, 206
thrashers, 182, 213
threadplant: redtip, 227
threadsnake, 216
threeawn, 233
threefold: California, 223
thrushes, 212
Thryomanes bewickii, 212
Tidestromia lanuginosa, 219
Tinaja Corazón, 112. *See also* Heart Tank

Tinaja de Eusebio, 87. *See also* Tinajas Altas
Tinaja del Tule, 85, 110, 111. *See also* Tule Tank
Tinaja Escondida, 112. *See also* Cabeza Prieta Tanks
tinajas, 9; accessibility of, 24–25; geology of, 18–19; invertebrates in, 184–89; water in, 21–24
Tinajas Altas, *4*, 5, *8*, 10, 19, 31, 49, 78, 95, 97, 99, 107, 116, 122, *125*, *129*, 131, 162, *164–65*, 167, *168*, 191, plates; Anza &, 86, 87; archaeology of, 43, 45, 46, 59–*63*, 146–47, 169; BAI camp at, 157–59; biological surveys of, 142–43, 144–45; boundary survey, 130, 133; catchments at, 9, *21–22*, *23*, 24–25; 49ers at, 106, 110, 111; freighters at, 119, 120, 121; graves at, *133*, *134*, 141, 144, *147*–49, 151, 160; Kippen at, 117–18; livestock at, 112, 155–56; maps of, *135*, 136–40; Mexican revolutionaries at, 124, 126; Native Americans at, 53–55, 58, 64–66, 68–69; recreation at, 163, *166*; travel routes &, 28, 29, 58
Tinajas Altas Archaeological Survey, 144, 150
Tinajas Altas Area of Critical Environmental Concern, 174
Tinajas Altas Mountains, 9, 25, 43, 54, *56–57*, 58, 66, *67*, 96, 97, 131, 154, 173; geology, 10–11, 14, 15, 152; plants in, 227, 228, 229, 230, 232
Tinajas Altas Pass, 14, 28, 43, 58, 64, 66, 78, 83, 86, 97, 124
Tinajas Cojuat, 96
Tinajas de la Candelaria, 87, 90–91. *See also* Cabeza Prieta Tanks
Tinajas de los Papagos, *67*
Tinajas del Tule, 68, 110, 111
Tiquilia palmeri, 224
toads, 185, 203, *215*
tobacco, 201, 236
Tobar, Juan José, 103
Tohono O'odham, *42*, 50, *51*, 52, 53, *58*, 66, *68*, *69*, 82, 99, 112; animal & plant use, 198–204
Tomaso, 54
tortoise: desert, 66, 182, 203, 215
tourism, 5, 160–62, 163, 169, 193
towhees, 213
Toxostoma spp., 182, 213
trail systems, 3, 95; Native American, 58–59, *65*, 66, 69, 97
travel routes, 3, 115; Díaz's, 71–73; Kino's, 30–31, 34–36, 37–41; & waterholes, 26–30, 97, 99. *See also* Camino del Diablo, El
treasure hunters, 147, 148, 149–50, 154, 155
tree-beargrass: desert, 146, 183, 198, *217*, 220
Trichoptilium incisum, 223
tridens: slim (*Tridens muticus*), 234
Trimorphodon biscutatus, 216
Trincheras culture, 46, *48*, 60
Tringa spp., 209
Triteleiopsis palmeri, 220
Trixis californica, 223
Troglodytes aedon, 212
Tule Desert, 28, 83, 95, 110, 116, 119
Tule Tank, *2*, *6*, 22, 28, 41, 85, 87, 97, 121, 122, *123*, 155, 156; 49ers at, 106, 110, 111
Tule Well, 97, 108, *109*, 142, 156, 157
Turdus migratorius, 212
turpentine broom, 236
Typha domingensis, 201, 236
Tyrannus verticalis, 211
Tyto alba, 210

Udall, Morris K., 170, 172–73
Udall, Stewart, 170, 174
Uma notata, 216
Urocyon cinereoargenteus, 179, 205
Urosaurus graciosus, 216
Urrea, Antonio de, 108, 112
Urrea, José, 103
Urticaceae, 236
US Air Force, 10, 43, 151, 167
US Army Corps of Engineers, 131–32, 136
US Border Patrol, 5, 127, 151, 167, 193
US Bureau of Animal Industry (BAI), 9; inspection station, 148, *157*–59
US Bureau of Land Management, 169, 174
US Department of Defense, 174
US Fish & Wildlife Service, 26, 167
US Geological Survey, 136, *161*
US Marine Corps, 10, 151, 167, 170, 193
Uta stansburiana, 203, 216

Valencia, Pablo, 141, 154
Valenzuela, Teodoro, 112–13
Valor, Palmer, 28
Van Devender, Rebecca, 143
Van Devender, Thomas, 143, *146*
Velasco, Francisco, 101, 108
velvet ants, 183
velvet-mallow, 230
verbena: sand, 183, 231
verdin, 212
Vincent Air Force Base, 170
vireos (*Vireo* spp.), 212
Viscaceae, 201
volcanism, 14
Volcano Lake, 71
Vopoki Ridge, 15, 83
Vulpes velox, 179, *192*, 201, 206
vulture: turkey, 209

wagon trains, 106–7, 119–21
warblers, 213

warfare, 97; O'odham-Mexican, 103; Quechan-Spanish, 91, 93; Yuman-O'odham, 3; among Yuman-speaking groups, 47, 96
wasps: tarantula, 183
water: geology of, 10–11, 18–19; quality of, 22–24; seasonal migration &, 58–59; at Tinajas Altas, 21–22
water boatmen, 185, 186
water bugs, 186
water catchments: constructed, *25–26*
waterholes, 33; distances between, *39*, 40(table); travel routes &, 9, 26–30, 97, 160; water quality at, 22–23. *See also* tinajas
water striders, 185, 187
waxwing: cedar, 213
Wellton, 127, 160, 163
Wellton District, 154
Wellton Hills, 28
Western Apache: raids by, 48–49, 112
wheat: Indian, 65, 200, 233
Wheelock, Walt, 71
whiptail: tiger, 203, 215
White Tanks, 45; organisms in, 184–85
Wiggins, Ira L., *145*
wildflowers, 183, 230, 234, 236; Aster Family, 220–23; Borage Family, 223–24; Four-O'clock Family, 231–32; Mustard Family, 224–25; Plantain Family, 232–33
Williams, Kent, 165
Wilson, Eldred D., 139, 146, 152, 154
Wilsonia pusilla, 213
wind caves. *See* tafoni
windmills, 231
Winters, Wayne, 148
wirelettuce: brownplume, 223
wolfberry, 65, 201, 218, 236
woodpeckers, 182, 210
woodrats, 66, 145, 182, 202, 206
woolly heads, 234
woollystar: desert, 234
World War II, 5, 170
wrens, 182, 212

Xanthisma spinulosum, 223
Xanthocephalus xanthocephalus, 214
Xantusia vigilis, 182, 203, 215
Xenodontidae, 216

Yaquis, 100, 101
yellow head, 223
yellowlegs, 209
yellowthroat: common, 213
Yodaville, 151
Yoeme. *See* Yaquis
Yuma Army Air Base, 5, 170
Yuma Chamber of Commerce, 160–61
Yuma Crossing, 3, 5, 91, 97
Yumas. *See* Quechan
Yuman speakers, 3, 68, 74, 99; ancestral, 45–46; & Garcés, 80, 91; historical period, 46–48. *See also various groups*
Yuma Trail, 116

Zenaida spp., 210, *211*
Zizyphus obtusifolia, 201, 236
Zonothrichia leucophrys, 214
Zygophyllaceae, 201(table), 236–37

The Southwest Center Series

Joseph C. Wilder, Editor

Ignaz Pfefferkorn, *Sonora: A Description of the Province*

Carl Lumholtz, *New Trails in Mexico*

Buford Pickens, *The Missions of Northern Sonora: A 1935 Field Documentation*

Gary Paul Nabhan, editor, *Counting Sheep: Twenty Ways of Seeing Desert Bighorn*

Eileen Oktavec, *Answered Prayers: Miracles and Milagros along the Border*

Curtis M. Hinsley and David R. Wilcox, editors, *Frank Hamilton Cushing and the Hemenway Southwestern Archaeological Expedition, 1886–1889,* vol. 1: *The Southwest in the American Imagination: The Writings of Sylvester Baxter, 1881--1899*

Lawrence J. Taylor and Maeve Hickey, *The Road to Mexico*

Donna J. Guy and Thomas E. Sheridan, editors, *Contested Ground: Comparative Frontiers on the Northern and Southern Edges of the Spanish Empire*

Julian D. Hayden, *The Sierra Pinacate*

Paul S. Martin, David Yetman, Mark Fishbein, Phil Jenkins, Thomas R. Van Devender, and Rebecca K. Wilson, editors, *Gentry's Rio Mayo Plants: The Tropical Deciduous Forest and Environs of Northwest Mexico*

W J McGee, *Trails to Tiburón: The 1894 and 1895 Field Diaries of W J McGee,* transcribed by Hazel McFeely Fontana, annotated and with an introduction by Bernard L. Fontana

Richard Stephen Felger, *Flora of the Gran Desierto and Río Colorado of Northwestern Mexico*

Donald Bahr, editor, *O'odham Creation and Related Events: As Told to Ruth Benedict in 1927 in Prose, Oratory, and Song by the Pimas William Blackwater, Thomas Vanyiko, Clara Ahiel, William Stevens, Oliver Wellington, and Kisto*

Dan L. Fischer, *Early Southwest Ornithologists, 1528--1900*

Thomas Bowen, editor, *Backcountry Pilot: Flying Adventures with Ike Russell*

Federico José María Ronstadt, *Borderman: Memoirs of Federico José María Ronstadt,* edited by Edward F. Ronstadt

Curtis M. Hinsley and David R. Wilcox, editors, *Frank Hamilton Cushing and the Hemenway Southwestern Archaeological Expedition, 1886–1889,* volume 2: *The Lost Itinerary of Frank Hamilton Cushing*

Neil Goodwin, *Like a Brother: Grenville Goodwin's Apache Years, 1928--1939*

Katherine G. Morrissey and Kirsten Jensen, editors, *Picturing Arizona: The Photographic Record of the 1930s*

Bill Broyles and Michael Berman, *Sunshot: Peril and Wonder in the Gran Desierto*

David W. Lazaroff, Philip C. Rosen, and Charles H. Lowe Jr., *Amphibians, Reptiles, and Their Habitats at Sabino Canyon*

David Yetman, *The Organ Pipe Cactus*

Gloria Fraser Giffords, *Sanctuaries of Earth, Stone, and Light: The Churches of Northern New Spain, 1530–1821*

David Yetman, *The Great Cacti: Ethnobotany and Biogeography*

John Messina, *Álamos, Sonora: Architecture and Urbanism in the Dry Tropics*

Laura L. Cummings, *Pachucas and Pachucos in Tucson: Situated Border Lives*

Bernard L. Fontana and Edward McCain, *A Gift of Angels: The Art of Mission San Xavier del Bac*

David A. Yetman, *The Ópatas: In Search of a Sonoran People*

Julian D. Hayden, *Field Man: Life as a Desert Archaeologist,* edited by Bill Broyles and Diane E. Boyer

Bill Broyles, Gayle Harrison Hartmann, Thomas E. Sheridan, Gary Paul Nabhan, and Mary Charlotte Thurtle, *Last Water on the Devil's Highway: A Cultural and Natural History of Tinajas Altas*